Dahlem Workshop Reports
Life Sciences Research Report 32
Exploitation of Marine Communities

The goal of this Dahlem Workshop is:
to evaluate the ability of fishery science and management to deal with changes in the marine ecosystems

Life Sciences Research Reports
Editor: Silke Bernhard

Held and published on behalf of the
Stifterverband für die Deutsche Wissenschaft

Sponsored by:
Senat der Stadt Berlin
Stifterverband für die Deutsche Wissenschaft

# Exploitation of Marine Communities

## R. M. May, Editor

Report of the Dahlem Workshop on
Exploitation of Marine Communities
Berlin 1984, April 1-6

Rapporteurs:
J.R. Beddington · R.J.H. Beverton · P.A. Larkin
G. Sugihara

Program Advisory Committee:
R.M. May, Chairperson · J.R. Beddington
J.A. Gulland · J.H. Steele · F.R.M. Thurow

Springer-Verlag
Berlin Heidelberg New York Tokyo 1984

Copy Editors: K. Geue, J. Lupp
Text Preparation: M. Böttcher, J. Lambertz, M. Lax, D. Lewis
Photographs: E. P. Thonke



With 4 photographs, 42 figures, and 14 tables

ISBN 3-540-15028-5 Springer-Verlag Berlin Heidelberg New York Tokyo
ISBN 0-387-15028-5 Springer-Verlag New York Heidelberg Berlin Tokyo

CIP-Kurztitelaufnahme der Deutschen Bibliothek

Exploitation of marine communities:
report of the Dahlem Workshop on Exploitation of Marine Communities
Berlin 1984, April 1 - 6 /
R.M. May, ed. Rapporteurs: J.R. Beddington ...
(Held and publ. on behalf of the Stifterverb. für d. Dt. Wiss.
Sponsored by: Senat d. Stadt Berlin; Stifterverb. für d. Dt. Wiss.).-
Berlin; Heidelberg; New York; Tokyo: Springer, 1984.
(Life sciences research report; 32)
(Dahlem workshop reports)

NE; May, Robert M. (Hrsg.): Beddington, J.R.
(Mitverf.): Workshop on Exploitation of Marine Communities
<1984, Berlin, West>; GT



Printing: Color-Druck, G. Baucke, Berlin
Bookbinding: Lüderitz & Bauer, Berlin
2131/3020-5 4 3 2 1 0

# Table of Contents

**The Dahlem Konferenzen**
S. Bernhard ix

**Introduction**
R.M. May 1

**Dynamics of Single Species**
Group Report
*R.J.H. Beverton, Rapporteur*
*J.G. Cooke, J.B. Csirke, R.W. Doyle, G. Hempel, S.J. Holt, A.D. MacCall, D.J. Policansky, J. Roughgarden, J.G. Shepherd, M.P. Sissenwine, P.H. Wiebe* 13

**Why Do Fish Populations Vary?**
*M.P. Sissenwine* 59

**The Availability and Information Content of Fisheries Data**
*J.G. Shepherd* 95

**Dynamics and Evolution of Marine Populations with Pelagic Larval Dispersal**
*J. Roughgarden, S. Gaines, and Y. Iwasa* 111

**Ecosystems Dynamics**
Group Report
*G. Sugihara, Rapporteur*
*S. Garcia, J.A. Gulland, J.H. Lawton, H. Maske, R.T. Paine, T. Platt, E. Rachor, B.J. Rothschild, E.A. Ursin, B.F.K. Zeitzschel* 131

**Observed Patterns in Multispecies Fisheries**
*J.A. Gulland and S. Garcia* 155

**Some Approaches to Modeling Multispecies Systems**
*R.T. Paine* 191

**The Response of Multispecies Systems to Perturbations**
*J.R. Beddington* 209

**Management under Uncertainty**
Group Report
*J.R. Beddington, Rapporteur*
*W.E. Arntz, R.S. Bailey, G.D. Brewer, M.H. Glantz, A.J.Y. Laurec, R.M. May, W.P. Nellen, V.S. Smetacek, F.R.M. Thurow, J.-P. Troadec, C.J. Walters* 227

**Kinds of Variability and Uncertainty Affecting Fisheries**
*J.H. Steele* 245

**Managing Fisheries under Biological Uncertainty**
*C.J. Walters* 263

**The Wider Dimensions of Management Uncertainty in World Fisheries**
*G.D. Brewer* 275

**Strategies for Multispecies Management**
Group Report
*P.A. Larkin, Rapporteur*
*C.W. Clark, N. Daan, S. Dutt, V. Hongskul, S.A. Levin, G.G. Newman, D.M. Pauly, G. Radach, H.K. Rosenthal* 287

**Strategies for Multispecies Management: Objectives and Constraints**
*C.W. Clark* 303

**Management Techniques for Multispecies Fisheries**
*G.G. Newman* 313

**Epilogue**
*J.A. Gulland* 335

**Geographical Glossary** 339

**Glossary of Technical Terms**
*J.G. Cooke* 341

**List of Participants with Fields of Research** 349

**Subject Index** 355

**Author Index** 367

# The Dahlem Konferenzen

**Founders**

Recognizing the need for more effective communication between scientists, especially in the natural sciences, the Stifterverband für die Deutsche Wissenschaft*, in cooperation with the Deutsche Forschungsgemeinschaft**, founded Dahlem Konferenzen in 1974. The project is financed by the founders and the Senate of the City of Berlin.

**Name**

Dahlem Konferenzen was named after the district of Berlin called "Dahlem", which has a long-standing tradition and reputation in the arts and sciences.

**Aim**

The task of Dahlem Konferenzen is to promote international, interdisciplinary exchange of scientific information and ideas, to stimulate international cooperation in research, and to develop and test new models conducive to more effective communication between scientists.

**Dahlem Workshop Model**

Dahlem Konferenzen organizes four workshops per year, each with a limited number of participants. Since no type of scientific meeting proved effective enough, Dahlem Konferenzen had to create its own concept. This concept has been tested and varied over the years, and has evolved into its present form which is known as the *Dahlem Workshop Model*. This model provides the framework for the utmost possible interdisciplinary communication and cooperation between scientists in a given time period.

---

**The Donors Association for the Promotion of Sciences and Humanities*

***German Science Foundation*

The main work of the Dahlem Workshops is done in four interdisciplinary discussion groups. Lectures are not given. Instead, selected participants write background papers providing a review of the field rather than a report on individual work. These are circulated to all participants before the meeting to provide a basis for discussion. During the workshop, the members of the four groups prepare reports reflecting their discussions and providing suggestions for future research needs.

**Topics**
The topics are chosen from the fields of the Life Sciences and the Physical, Chemical, and Earth Sciences. They are of contemporary international interest, interdisciplinary in nature, and problem-oriented. Once a year, topic suggestions are submitted to a scientific board for approval.

**Participants**
For each workshop participants are selected exclusively by special Program Advisory Committees. Selection is based on international scientific reputation alone, although a balance between European and American scientists is attempted. Exception is made for younger German scientists.

**Publication**
The results of the workshops are the Dahlem Workshop Reports, reviewed by selected participants and carefully edited by the editor of each volume. The reports are multidisciplinary surveys by the most internationally distinguished scientists and are based on discussions of new data, experiments, advanced new concepts, techniques, and models. Each report also reviews areas of priority interest and indicates directions for future research on a given topic.

The Dahlem Workshop Reports are published in two series:
1) Life Sciences Research Reports (LS), and
2) Physical, Chemical, and Earth Sciences Research Reports (PC).

**Director**
Silke Bernhard, M.D.

**Address**
Dahlem Konferenzen
Wallotstrasse 19
1000 Berlin (West) 33

Exploitation of Marine Communities, ed. R.M. May, pp. 1-10. Dahlem Konferenzen 1984. Berlin, Heidelberg, New York, Tokyo: Springer-Verlag.

# Introduction

R.M. May
Biology Dept., Princeton University
Princeton, NJ 08544, USA

Humans have harvested the sea for a long time. Some of the earliest traces left by our species are the shellfish middens of coastal dwellers, and some of the earliest and most delicate tools made by humans are bone fishhooks and other artifacts associated with fishing. Such exploitation of the fruits of the sea was not a trivial accomplishment; it appears, for example, that Australian aboriginals who were isolated on islands in the Bass Straight by the rising seas following the last ice age underwent a cultural degeneration on their way to eventual extinction, and one of the first skills they appear to have lost is fishing. Today, with the tools provided by modern technology, there is no danger of our losing the ability to catch the fish. Instead we have the problem that aggressive overexploitation may lead - and in some cases arguably has already led - to our losing the fish themselves.

Recognition of the dangers of excessive harvesting of particular fish stocks has led to the formation of national and international organizations whose aim is to regulate fishing in specific regions (the North Sea, the North Atlantic, the Southern Ocean, and so on) or of particular species (the International Whaling Commission, the original Pacific Tuna Commission). The outcome today is an international but relatively small cadre of fisheries scientists (most of whom are scientific civil servants), whose work is a fascinating - if sometimes uneasy - blend of basic science and direct application. Much of this activity consists of short-term recommendations about harvest levels for particular stocks, and these biologically-based recommendations necessarily roil together with all kinds of economic, social, and political complications en route to the final setting of catch quotas or levels of fishing effort. A lot of the scientific research in this general area is published as *samizdat* for specific meetings, or at best in specialized journals with relatively limited circulation. This publication pattern stems naturally from the social dynamics of the management process with which the research is associated, but it is nevertheless a pity: these papers often contain data and theory that would be of great interest to the wider community of population biologists and ecologists who often are unaware of the work because of its relative inaccessibility.

Conversely, much of the pursuit of fundamental understanding of biological populations and ecosystems by people in universities, museums, nature conservancies, and the like, deals with questions that are likely to be relevant to fisheries management, particularly long-term management. The papers of *American Naturalist, Evolution, Ecology, Journal of Animal Ecology* (and even of the more abstract journals such as *Mathematical Biosciences* or *Journal of Theoretical Biology*) abound with empirical and theoretical studies of such questions as: what is the genetic and dynamic response of a population to natural or artificial disturbance; what governs the observed patterns of relative abundance of species in ecosystems; how do we aggregate variables in a complex food web, to arrive at some simple and manageable description of the essentials of the system; how does environmental variability and unpredictability affect the persistence of species or the composition of communities; what is the relation between an ecosystem's complexity and its stability (and what do we mean by stability)? This is only a partial sampling from a long list of fundamental questions, some of which are now on center stage in fisheries management, and others of which possibly should be.

Of course, this dichotomy between fisheries scientists and academic ecologists that I have just sketched is a gross caricature. On the one hand, fisheries models (such as those of Beverton and Holt) have found their way into most introductory ecology texts, and, on the other hand, most fisheries people are well acquainted with the literature on basic ecology and have indeed made important contributions to it. It nevertheless seemed a good idea to bring together an assortment of people - some knowledgeable about fisheries (from biology through to politics and marketing) and some unversed in fishery lore (but including population geneticists, community ecologists, and resource economists) - to think about problems in the exploitation of marine fisheries, in an atmosphere free from any of the overtones that often constrain professional fisheries meetings. The present volume is the result.

This volume aims to be useful as an appraisal of the state of the art by a mixed collection of insiders and outsiders. Most interestingly, I think, it aims (especially in the four group reports) to identify some of the major areas of unresolved controversy and some of the major questions yet unanswered. I see the book as essentially a tentative statement - often by several dissonant voices - about directions in which we may be heading; the book is emphatically not a canonical utterance on how to do things. It is intended to stimulate, not to codify.

Following the usual Dahlem Workshop format, the discussions were organized under four themes. Although crisp demarcation is not possible, the first two themes broadly deal with biological aspects of the dynamics of single populations and the dynamics of systems with many species. The later two themes take up questions of management under uncertainty and multispecies management. In all this, the word "fish" is interpreted broadly to include such taxonomically varied beasts as whales, shrimp, crabs, shellfish, and squid, along with fish in a strictly zoological sense.

## DYNAMICS OF SINGLE POPULATIONS

As background to this section, Sissenwine surveys current efforts to understand the factors affecting the birth, death, and general life history within fish populations. Shepherd similarly reviews the kinds of data we have, and the kinds of data we do not have but need, for making short- and long-term predictions about harvested fish stocks. Coming at some of these questions from a different angle, Roughgarden et al. discuss new kinds of models that are appropriate to populations with dispersed larvae; some of these ideas are applied, by way of explicit illustration, to experiments on the life history of barnacles.

Against this background, the group report (Beverton et al.) explores several issues that arise in the harvesting of single populations. Some of these biological problems have implications for evolutionary biology and ecology far outside the practical confines of maximizing profits or minimizing the risks of overfishing.

Although little considered in conventional fisheries management, fishing can be a major selective force acting on the genetic composition of fish stocks. It is, for example, possible that fishing a previously unexploited stock may lead to faster growth and earlier breeding amongst members of the surviving population. Predictions of yield, profits, and the population's capacity to withstand exploitation hinge on a sound understanding of these changes in demography under exploitation. The changes can happen quickly and have generally been attributed to an increase in the per capita food supply of individuals in the exploited population. As stocks are fished down, food supplies per individual increase, leading to faster growth and earlier maturity. However, an explanation based on such density-dependent effects in the population dynamics is not the only possibility; the response could also be genetic. If, as seems likely, growth rates and age of maturity are genetically determined and there is heritable variation for these traits within the population, then fishing represents a massive and novel selection on the population. Individuals that grow faster and/or mature earlier may be at an advantage in the fished population because, other things being equal, they are more likely to reproduce before being caught. In most populations we might expect to see both population dynamic and genetic responses to fishing, working either in concert or in opposition to one another to influence the demography of the exploited population. The consequences for managers wishing to predict the long-term dynamics of fish stocks are obvious. But over and above these practical considerations, it seems likely that evolutionary biologists interested in life-history strategies may be able to draw inspiration and information from the way fish life histories can evolve in response to fishing (2).

Density-dependent and genetic phenomena, moreover, do not exhaust the catalogue of mechanisms that could be responsible for observed changes in vital rates. Another possibility is that changes in environmental variables may more or less accidentally coincide with changes in fishing intensity. It may even be that apparent changes may be artifacts of the way the data are gathered. In short, apparent changes in the dynamical characteristics of fish populations under exploitation may come from density-dependent responses, from genetic responses, from environmental changes, or they may be artifacts of data collection. We clearly need better ways of discriminating among these possibilities.

## DYNAMICS OF MULTISPECIES SYSTEMS

Here Gulland and Garcia provide a synoptic account of the patterns that have been observed in multispecies fisheries under exploitation. Paine gives the general perspective of a community ecologist on different approaches to studying such multispecies food webs, and Beddington catalogues the kinds of responses to disturbance that multispecies systems may show.

The task of constructing useful models for complex multispecies communities is another area where practical problems of fisheries management intersect with basic questions in theoretical ecology (4, 6). Multispecies problems are particularly conspicuous in tropical waters: off the Senegal-Mauritanian coast, for example, as many as 174 species are involved in the demersal fishery, and a single trawl haul can contain 20-50 species. Official statistics for this area tend to group catches into commercial categories, but still 48 species are identified in the records for Senegal. It goes without saying that classical single-species fishery models are of little use under the circumstances. But detailed simulation models involving all the commercial species and their food supplies are not a solution either; even if they were, the sensitivity of very complex models to the errors inherent in estimating the plethora of parameters necessary for their construction makes them unreliable.

Simplified models are less sensitive to parameter variation. The problem, however, is how to go about simplifying, and how to aggregate the variables (5). Grouping species into manageable sets could be done in several ways: taxonomically, for example; or by habitat; or by size categories; or by commercial value. At the moment, there are no reliable theoretical rules to guide such endeavors, and the best solution in any particular case rests on the circumstances and the objectives of the modeller. One promising approach may be to use some combination of body size and habitat. Even the very complex West African fishery can be grouped into a limited number of species assemblages on the basis of habitat, specifically water depth and bottom type. Within habitat divisions, grouping individuals by body size (irrespective of taxonomy) may not seem the most obvious thing to do, but since body size has rather predictable effects on an animal's vital rates, the approach has much to commend it (7, 10). In the long term, successful fisheries management predictions using aggregative models of complex, multispecies fisheries could have profound implications for ecologists struggling to make sense of, and to model, equally complex communities in other habitats.

Ultimately, of course, even classical "single species" fisheries (such as cod or herring in the North Sea, or anchovies off Peru) are members of complex communities harboring, and providing food and shelter for, predators and competitors. A broader "ecosystem" perspective applied even to a simple single-species fishery may suggest ways of using indicator species - seabirds or penguins, for example - to monitor the health of fish stocks before things start to go seriously wrong. The conferees (Sugihara et al.) disagreed about the feasibility of such an approach, but I think the idea has merit. It certainly seems that reduced breeding success, or failure of seabird colonies to recover from a natural disaster that in the past has presented no problems, could be clear indications that all is not well with the birds' food supply.

**MANAGEMENT UNDER UNCERTAINTY**

Under this theme, Steele outlines the kinds of environmental uncertainty and variability that can affect fisheries; he develops the interesting idea that such noise may be "colored," with the shape of the noise spectrum depending on the spatial and/or temporal scale on which it is considered. (Steele elsewhere conjectures that marine and terrestrial noise spectra may characteristically have different colors, "red" in the sea versus "white" on land (11).) Walters summarizes methods that can be used to manage fisheries under uncertainty, and Brewer widens the discussion to explore some of the social and political dimensions of management under uncertainty.

The essential theme here is that unpredictable environmental events can lead to significant fluctuations in fish stocks, and thus to uncertainty in the catch from year to year, or over longer periods. In particular, it can be that recruitment relations are affected differently by some kinds of environmental fluctuations when the stock density is relatively low (as a result of heavy exploitation) than when it is relatively high (in its pristine state); this, in turn, may result in the yield becoming significantly more variable as fishing pressures increase. Such effects can lead to situations where it is better to fish less intensely, diminishing the fishing effort below that estimated to give the maximum average yield, in order substantially to reduce the severity of fluctuations in the yield (1, 8, 9). The trading of risk against maximum possible return in this way is more familiar in the management of university endowments and other economic situations.

In this general context, the report of the group (Beddington et al.) sought to identify some tentative patterns in the degree to which different kinds of fish stocks exhibit such variability. One suggestion was that species high in the food web tend to live longer, which in turn tends to make their population densities relatively less variable. Another suggestion was that tropical species tend to live longer than ecologically similar temperate ones, again tending to make for relatively steadier dynamics. There appears to be some evidence in support of this idea for fish, although for terrestrial insects Wolda and others have concluded that tropical populations do not fluctuate significantly less than temperate ones; it may be that there are differences between terrestrial and marine habitats, or it may simply be that comparisons between tropical and temperate fish communities yet await studies as careful as that of Wolda (3, 12). A more specific suggestion, tied to the details of life histories, is that species such as plaice or sole that make their living on the essentially two-dimensional world of the seafloor are thereby in a less chancy environment, and that this is why their stocks appear to fluctuate less than do many pelagic fish such as herring (see Table 1 in the Beverton et al. report on the Dynamics of Single Species).

The group report on Management Under Uncertainty also emphasizes that fluctuations in the processing and marketing of fish are every bit as worrisome as fluctuations in the biological stock or in the catch. For fisheries that are relatively predictable, it may be desirable to bring industrial capacities for treatment, packaging, and marketing fish products close to the minimum required to exploit the resource at some set average level, allowing for fluctuations but not subject to gross overexpansion in good times and collapse in bad ones. It is easier to say this than to do it, but, in the absence of such efforts at control, all factors - from biological harvesting to marketing - can create lags and inertia that end up greatly amplifying such fluctuations as are inherent in the system. Conversely, for fisheries that are intrinsically highly variable (either in yield or in species composition), as some pelagic fish seem to be, it may be best to aim for flexibility in harvesting, processing, and marketing, given that heavy exploitation in the good years will apparently have little repercussion on future biological events. Tidy though such a dichotomy is in principle, we really need to understand multispecies effects better before we can implement the ideas with any confidence.

Real uncertainties in fish catches and marketing are further complicated by psychological aspects of uncertainty. Thus, for example, the macho acceptance of risks by some kinds of fishermen can confound policies based on conservative rationality.

## MULTISPECIES MANAGEMENT

This central theme is ineluctably interwoven with all the preceding three. As background for the discussion, Clark reviews some of the strategic objectives of, and tactical constraints on, the management of multispecies fisheries. In a complementary paper, Newman surveys the policy instruments - quotas, explicit limits on effort, taxes, licenses, and the like - that may be available to managers.

The group report (Larkin et al.) suggests that the kind of multispecies management is likely to depend on the character of the ecosystem. Partly because temperate fish communities are relatively less complex, and partly because they are better known, we have some idea of the interactions among the constituent species and of how the food web is put together. Temperate zone fishermen often target a specific stock for a specific market. The more troublesome problems of multispecies management thus tend to be social and political (do we harvest cod for direct human consumption, or "industrial" fish to feed to chickens?), rather than intrinsically biological. Similarly, the system of krill and the baleen whales and other creatures that eat them in the Southern Ocean is relatively simple, and we may hope to describe its dynamics at least to a crude approximation (4). In contrast, tropical communities are typically more rich in species, and the catch is usually an aggregation of species. In deliberately oversimplified terms, management of temperate fisheries relies essentially on population dynamics or at worst community dynamics; tropical fisheries must be treated as ecosystems.

A potentially powerful idea is "management by experiment" in which deliberately different policies are employed in different regions. For instance, quotas may be set very high in some places and very low in others, so that learning and harvesting march together. The intellectual attractions of such techniques of exploratory and adaptive management are great, but in practice the methods are likely to be resisted by fishermen (who will understandably tend to accept overexploitation and resist underexploitation). There are some success stories of experimental management, such as the removal of large carnivorous fish off parts of the coast of Spain to create a shrimp fishery, but as yet such stories are few.

## CONCLUSION

My impression of the general tenor of this volume is that our understanding of the population biology of most fish stocks is good enough to offer reasonably reliable advice to managers. Clearly some aspects are much more fully comprehended than others: not surprisingly, we have a better understanding and a more satisfactory base of data for short-term predictions than for long-term ones; well-defined single species stocks are better understood than are situations where several distinct species or even aggregates of species are harvested simultaneously; long established temperate zone fisheries are better understood than tropical ones. Much basic research remains to be done. But our grasp of the essentials of many fisheries is quite good.

It is in the translation of the biological understanding into effective management action that things keep going wrong. Gulland catches the spirit of many of the discussions in his concluding essay when he writes of the failure of many bankers and politicians to grasp that fish stocks do not automatically and invariably renew themselves simply because they are called "renewable resources" in elementary economics texts. The need for scientific advice to be carried convincingly to policymakers is emphasized in several of the book's papers and group reports, but no golden road to this Nirvana is offered (other than persistent and sympathetic effort, acknowledging the problem to be more complicated than just "their" failure to listen).

For me, this contrast between the essential adequacy of our scientific understanding of fisheries and the often distressing inadequacy of our machinery for translating this understanding into effective action was crystallized epigrammatically by Daniel Pauly. To pursue excellent research that is then disregarded is, he said, like "using a large, modern hospital only for diagnosis, and not for treatment." The moral, of course, is not to close the hospital, but to run it better.

**REFERENCES**


(1) Beddington, J.R., and May, R.M. 1977. Harvesting natural populations in a randomly fluctuating environment. Science 197: 463-465.

(2) Istock, C.A. 1984. Boundaries to life history variation and evolution. In A New Ecology: Novel Approaches to Interactive Systems, eds. P.W. Price, C.N. Slobodchikoff, and W.S. Gaud, pp. 143-168. New York: Wiley.

(3) May, R.M. 1979. Fluctuations in abundance of tropical insects. Nature 278: 505-507.

(4) May, R.M.; Beddington, J.R.; Clark, C.W.; Holt, S.J.; and Laws, R.M. 1979. Management of multispecies fisheries. Science 205: 267-277.

(5) O'Neill, R.V., and Rust, B. 1979. Aggregation error in ecological models. Ecol. Model. 7: 91-105.

(6) Paine, R.T. 1980. Food webs: linkage, interaction strength and community infrastructure. J. Anim. Ecol. 49: 667-685.

(7) Peters, R.H. 1983. The Ecological Implications of Body Size. Cambridge: Cambridge University Press.

(8) Shepherd, J.G., and Horwood, J.W. 1979. The sensitivity of exploited populations to environmental "noise", and the implications for management. J. Cons. Int. Explor. Mer 38: 318-323.

(9) Sissenwine, M.P. 1977. The effect of random fluctuations on a hypothetical fishery. Int. Comm. Northw. Atlant. Fish. Sel. Paper 2: 137-144.

(10) Southwood, T.R.E. 1981. Bionomic strategies and population parameters. In Theoretical Ecology: Principles and Applications, ed. R.M. May, pp. 30-52. Oxford: Blackwell.

(11) Steele, J.H., and Henderson, E.W. 1984. Modeling long-term fluctuations in fish stocks. Science 224: 985-987.

(12) Wolda, H. 1978. Fluctuations in abundance of tropical insects. Am. Natur. 112: 1017-1045.

Standing, left to right:
John Shepherd, Roger Doyle, David Policansky, Justin Cooke, Mike Sissenwine, Alec MacCall, Peter Wiebe

Seated, left to right:
Gotthilf Hempel, Ray Beverton, Sidney Holt, Jonathan Roughgarden, Jorge Csirke

Exploitation of Marine Communities, ed. R.M. May, pp. 13-58. Dahlem Konferenzen 1984. Berlin, Heidelberg, New York, Tokyo: Springer-Verlag.

# Dynamics of Single Species
## Group Report

R.J.H. Beverton, Rapporteur
J.G. Cooke
J.B. Csirke
R.W. Doyle
G. Hempel
S.J. Holt
A.D. MacCall
D.J. Policansky
J. Roughgarden
J.G. Shepherd
M.P. Sissenwine
P.H. Wiebe

**INTRODUCTION**

The group resolved to interpret its terms of reference by considering the following question: "How well can contemporary fisheries science detect, measure, interpret, and predict significant changes in the dynamics of single-species populations in the context of the broader span of marine ecosystems and having regard to both scientific and practical requirements?"

The "context of marine ecosystems" is a reminder that two other discussion groups were dealing with ecosystems dynamics and multispecies management, respectively (see Sugihara et al. and Larkin et al., both this volume). Specifically, on multispecies questions we decided to go only as far as establishing where the science of single-species dynamics and resulting assessments are most likely to be defective because, in reality, no species is an island.

We return to this point in the final section of our report. Here it is sufficient to note that study of the variability and self-regulative capacity of a single-species population does, of course, require recognition of the influence of physical and biological factors of its environment. In the single-species approach these biological influences, such as food supply and predation, are treated implicitly by general theory or by empirical evidence of density-dependent processes, operating through one or more of the appropriate population parameters (i.e., recruitment, growth, mortality, etc.). A multispecies approach, on the other hand, deals explicitly with the dynamics of particular interacting species in the same or different trophic levels and perhaps of the dynamics of the larger system, insofar as this is feasible.

Our deliberations also excluded substantive discussion of the practical requirements of fishery management, since this was the remit of the group on "MANAGEMENT UNDER UNCERTAINTY" (Beddington et al., this volume). We concluded, nevertheless, that it was necessary to take into account the main kinds of questions asked and information needed by the decision-makers, particularly as regards precision and time scales; it would otherwise have been difficult to evaluate the adequacy for practical purposes of present scientific knowledge about fish resources.

Of the various themes running through our deliberations, two deserve special mention. One is the problem of distinguishing the role of natural and man-made factors in generating the observed changes in exploited populations. This arises both in the interpretation of historical data and in the framing of advice for future management. It can be regarded as a particular case of the general signal vs. noise problem, but the analogy should not be pressed too far. Usually we found it best to ask the direct question: how would we expect the particular feature of the dynamics of the population in question to be affected by a given form and intensity of exploitation?

The second theme relates to time scales. The dimension of time is, of course, inherent in the concept of population dynamics, but our study repeatedly brought out the great importance of long time-series of data. Valuable though studies of particular processes (e.g., recruitment mechanisms) are for interpretation, that kind of information alone cannot provide a quantitative picture of the dynamics of the population as an entity. No less significant, though in a different way, is the time dimension of the decision-making process in fisheries management and its implications to the nature and precision of the scientific knowledge required. This question is developed by Shepherd (this volume) and was taken up at several points in our deliberations.

## IDENTITY OF SINGLE-SPECIES POPULATIONS

### Definition of Population

It is obviously necessary for both scientific and management purposes to be clear as to what constitutes the population unit, in the biological sense, compared with the fished stock. The two need not be identical and often are not; but in that case it is as well to be aware of how they differ. The biological concept of population is based on the integrity and continuity of the reproductive unit. There may be genetic separation from other such units of the same taxonomic species, but this is not obligatory. In contrast, the definition of stock is essentially operational, being the fished part of one or more biological populations.

A great deal of literature exists on methods of identifying biological populations of fish of the same species, ranging from morphometric or physiological characteristics of the whole fish (e.g., shape or fecundity) through structural features (i.e., vertebral or fin ray counts) to biochemical and serological techniques (24, 59). The practical objective is usually to find diagnostic population characteristics which would enable an individual fish in a mixed sample (e.g., a catch) to be assigned unambiguously to its population of origin; but few of the conventional techniques meet this criterion and, indeed, the aim may be illusory unless there is true and sustained genetic separation between the populations in question.

If genetic separation exists, it may well not be manifest in terms of easily measured characteristics. Even the most recent electrophoretic and mitochondrial DNA-sequencing techniques have no logical priority in this respect over the more traditional methods based on meristics or morphometrics. Stocks which have similar frequencies of electrophoretic markers may, in fact, be isolated genetically and demographically, while other stocks may differ genetically despite high rates of interchange between them. In any event, effective genetic interchange can be achieved by relatively slow mixing rates of individuals from populations (as defined above) whose average characteristics (including location) may be different enough to be recognized easily.

Not infrequently, as knowledge about the structure of the large-fish populations has increased, what was originally thought to be a single homogeneous population has proved to consist of two or more reproductive units whose spawning is to a greater or lesser extent separated in time or space (or both), with a remarkably high degree of "return" to the spawning grounds. The fact that regular sampling to distinguish fish originating from one or another of such subpopulations is usually difficult and costly raises the question: does it matter?

For short-term assessments the answer is probably no. In the longer term, the answer will depend on the relative importance of three factors: a) the consistency of the pattern of fishing relative to the distribution of the constituent subpopulations; b) the extent to which the dynamics of the various subpopulations are different; and c) the capacity of the subpopulations to maintain their identity over long periods.

The North Sea fisheries for plaice (Pleuronectes platessa) and herring (Clupea harengus) provide two contrasting examples among many variants that can be found. The stocks (as defined above) consist in each case of fish originating from three main reproductive centers, roughly 100-150 miles apart. In plaice, spawning is nearly synchronous (January-March) but there is a high degree of return of mature fish to their previous spawning ground (27). The spawning populations appear to have maintained their identity and relative importance for as long as records have been kept (since about the beginning of the century). The fishery exploits a mixture of fish from all the spawning groups, mainly outside the spawning areas and seasons. There is no reason to believe that in this case significant bias is caused by treating the fished stock as an entity.

In contrast, the three populations of herring in the North Sea spawn at different times as well as places (Buchan, north, August; Dogger, central, October; Downs, south, December). The fish all look the same, but the mean values of certain characteristics (e.g., vertebral count and fecundity) differ significantly, although the distributions overlap. More important is that the three populations have exhibited markedly different long-term dynamics. The central and southern populations began to decline from about 1940 onwards, accelerating after 1950 to reach a very low level by 1970. In contrast, the northern population increased substantially during the 1960s until drastically depleted by fishing in the early 1970s (14). Although, as in plaice, fish from the three populations mix to a considerable extent during the feeding season, lumping the three populations to comprise a single stock produces an artificially smoothed picture which masks what really happened, both in the decline and in the subsequent recovery.

**Spatial Aspects of Single-species Dynamics**

We concluded our discussion of this topic by reviewing the general question of the spatial (geographical) aspects of the dynamics of fish populations and their influence on assessments and management. Little theoretical study seems to have been done of this topic since Beverton and Holt (9) and Beverton and Gulland (8) explored the response to fishing of a population only part of which was fished, but between which (the "stock" as defined above) and the unfished part there was mixing at exponential rates. They used a conventional yield per recruit model with all parameters other than the fishing mortality rate, F, uniformly distributed. Even so, there were marked differences in the pattern of response of the "stock" (both yield and catch per unit effort) according to whether changes in the amount of fishing were made by varying the true intensity (i.e., effort per unit area) or by varying the fished proportion of the total population.

In practice, of course, depending on the zoogeographical characteristics of the species, it is usual to find that populations covering an extensive latitudinal range have well-defined gradients in density and population parameters such as growth. Of particular scientific interest, and potentially of great practical importance, is to know what happens to the pattern of those gradients when the total population abundance changes, depending on whether these are due to fishing or to natural causes.

There are, in fact, a number of documented cases (particularly in clupeids, e.g., (64)) in which expansion or contraction of the geographical range of a fish population has occurred in parallel with changes in its abundance. When the California sardine (Sardinops sagax caerulea) was at its peak, its density decreased and its size-at-age increased from its center northwards to its limit off British Columbia; when its collapse set in, the decline started at the northern extremity and spread southward (63). Again, when the Hokkaido herring (Clupea harengus) declined after 1940, the area occupied by the population shrunk while the density at the center tended to stay more nearly constant. As a consequence, although the total population (and fleet size) was declining, the overall catch per unit effort from the residual stock changed much less until the final collapse of the fishery (89).

In the opposite direction, the increase in abundance of the Pacific sardine (S. sagax sagax) in Peruvian and Chilean waters in the last decade or so has been accompanied by a pronounced southward extension of its latitudinal range (81). Long-term trends in the spawning of the Atlanto-Scandinavian herring during a period of population increase have been recorded by Devold (29). From the beginning of the present "herring period" at the beginning of the century to peak abundance in the 1950s, there appears to have been a progressive shift in the onset of spawning from October to February, rapidly until about 1920 and then more slowly. In the years after 1946, when the detailed distribution of the fishery was recorded, there was also a pronounced northward shift of the spawning area along the Norwegian coast.

Expansion and contraction of the geographical range of a population clearly interacts with many aspects of dynamics which determine its response to exploitation. Thus MacCall (53) has developed a model linking the density-dependence of cannibalism by adults of the northern anchovy (Engraulis mordax) on their eggs to changes in the distribution of the spawning population. One consequence of introducing the spatial dimension into the cannibalism model is to make the stock-recruitment relationship less dome-shaped than it would otherwise be (50).

In discussion, MacCall developed the concept of a distributional model of an exploited fish population based on the premise that where density is the highest (i.e., at the center), density-dependent compensation for removals by harvesting should be most effective. Conversely, at the extremities, density-independent (i.e., environmental) processes would be expected to dominate and removals to be correspondingly less well compensated. He suggested that while the center is the main productive area of such a population, it should not be depleted to the point at which those natural compensatory responses are impaired. These ideas are developed further by MacCall (54).

In summing up, we concluded that the spatial structure and dynamics of populations was an important topic deserving further attention. In cases where the fishery covers only part of the geographical range of the population, the response of the fished stock, even in the short term, may depart significantly from that predicted for a fully fished population. Evidence of events at the limits of the range of a population may give valuable early warning of general trends. Of particular interest for future research are the mechanisms, behavioral or otherwise, by which reproductive centers and spatial gradients of parameters are maintained, and how those patterns are liable to be affected by fishing and by environmental (e.g., climatic) factors.

## NATURAL REGULATION IN SINGLE-SPECIES POPULATIONS

### Introduction – Evidence of Regulation

Variability in fish populations, especially that due to year-to-year fluctuations in recruitment, has traditionally received much attention in fisheries dynamics and is well reviewed by Sissenwine (this volume). We therefore decided to focus our discussion on the state of knowledge about the occurrence and causes of natural regulation of population size in fish, and its implications to management.

It is difficult, if not impossible, to conceive that the upper extreme of population size is not limited ultimately by density-dependent processes, which is one manifestation of self-regulation (homeostasis) in the sense used here. Less clear is the role of density-dependence in the dynamics of populations in their usually observed size range; this is of more practical relevance, but its significance is likely to vary greatly between species.

It is essential here to distinguish conceptually between true density, i.e., numbers or biomass per unit of habitat space, and total population size. Density-dependent processes typically occur on a local scale, much influenced by spatial heterogeneity both of the population in question and of environmental factors. It usually has to be assumed for practical purposes that total population size and the intensity of density-dependent processes are proportional. There could, however, be circumstances in which this assumption does not hold; as, for example, if changes in total population size are accompanied by expansion or contraction of its geographical range (see section on Spatial Aspects of Single-species Dynamics).

Natural regulation can be manifest through a variety of density-dependent mechanisms - predation (including cannibalism), infection, competition for food, and reproductive processes. Sissenwine (this volume) reviews the key evidence on this subject. The same pattern of regulative mechanisms is unlikely to hold throughout the range of population size. For example, those whose action is strongly mediated by limitations of habitat space may have little effect until a certain level of density is approached and the carrying capacity of the available habitat is becoming saturated. In consequence, regulative processes may well not be symmetrical above and below the long-term average population size, either in kind or degree.

For further analysis of natural regulation it is necessary to identify where in the life cycle density-dependent processes are most effective. The first-order classification is into the size of the parent population (e.g., as biomass of mature fish) and the number of progeny which survive to reach maturity in the next generation, resulting in the conventional "stock-recruit" data arrays of the fisheries literature.

There are not many populations from which a clear stock-recruit pattern has yet emerged, owing mainly to the high variability of the recruit data (some of which is error variation) and the limited range of adult population size usually covered by the observations. The search for a functional relationship is further complicated by the fact that, except where the mature population consists of a single age-group only (e.g., in certain species of Pacific salmon and very short-lived tropical species), population values are bound to be serially correlated, if for no other reason than that most of the age-groups contributing to the biomass in year X are also present in year X + 1.

Recruit values also may be serially correlated if environmental conditions affecting survival persist for two or more years. The Georges Bank haddock data array shown in Fig. 1 of Sissenwine (this volume) are a dramatic demonstration of this phenomenon, since the ten points closest to the X-axis refer to the ten consecutive years 1965–1974 and yet cover the whole range of population size. The presumption is that environmental factors affecting survival changed markedly for the worse during this period, although a shift of fishing pressure onto the smallest sizes was a contributory factor.

In contrast to the Georges Bank haddock, the variability of recruitment in the North Sea plaice is relatively low (less than ± 50%) and environmental disturbances apparently small (except for the cold winter of 1962/63, see below). This population would therefore be expected to provide a better test of regulative capacity; but the value of the long series of data (back to 1922, in principle) is qualified by changes in the pattern of the fishery and fishing methods over the decades, which makes it difficult to obtain accurately comparable abundance indices throughout. Figure 1 shows the latest estimates of stock (as total biomass of mature females) and recruitment for the years 1946–1979, from which the conclusion may reasonably be drawn that recruitment has been effectively independent of stock size over the range of biomass covered by this data set. A similar picture emerges for some other flatfish, e.g., for North Sea sole (Solea solea) (44) and for the Dover sole (Microstomus pacificus) on the Pacific West Coast (36).

This form of stock-recruit relationship, or indeed any pattern in which recruitment changes less than proportionately with adult stock (i.e., the stock-recruit curve is concave downwards), implies that regulation (compensation) exists somewhere between the parents of one generation and progeny surviving to maturity in the next generation. There are, in theory, two possible places where this could occur: in the adult phase

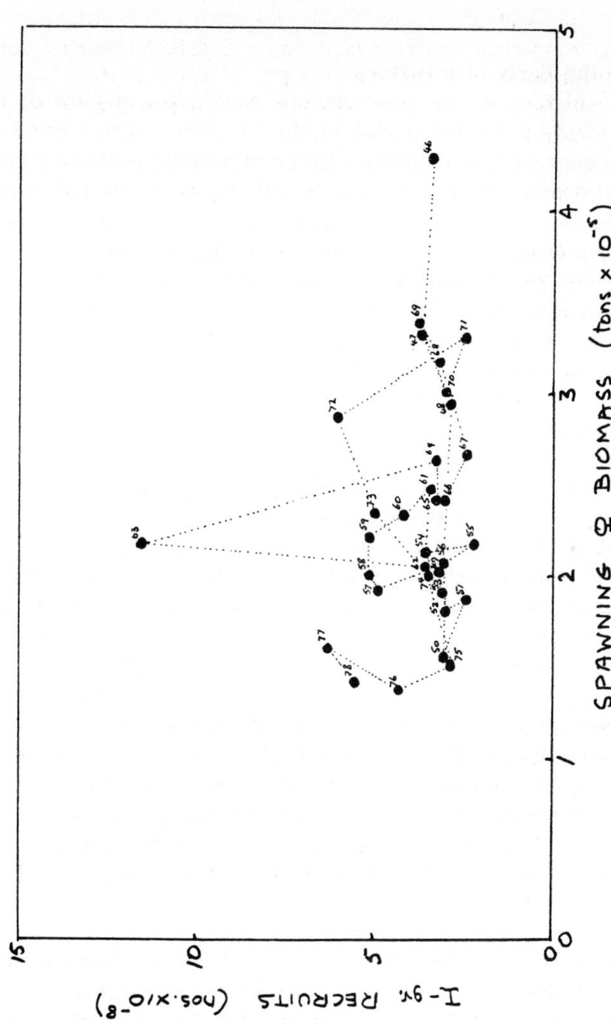


FIG. 1 – Stock-recruit data for North Sea plaice, 1946–1978. Compiled by Beverton from ICES Flatfish Working Group Report CM 1982 (1960–79) (44), and from Garrod (31), Bannister (1), and Gulland (33) for earlier years.

in the course of production of fertilized eggs by the parent population, and during the early life history of the next generation from eggs through to recruitment. If the adults are direct predators on their own offspring, this between-generation cannibalism could be a third regulative mechanism.

**Regulation in the Early Life History**

The early life history is the most obvious candidate, in view of the very high mortality rate over that period of the life span. Direct proof requires empirical evidence of the in situ mortality rate being positively correlated with the local density of fish, such as would happen if the fish were themselves grazing their food supply down to the point at which their death resulted from starvation. Positive evidence of death of larval fish from lack of food has been obtained in a few cases, notably by O'Connell (67) for the northern anchovy, E. mordax (see also Sissenwine, this volume), but it has not yet been shown in any pelagic larval or post-larval stage that such mortality is density-dependent in the sense defined above.

Lack of such evidence is not, of course, proof that regulation is absent in the pelagic larval phase, especially in view of the practical difficulties involved in measuring accurately events in this phase of the life history. Nevertheless, the balance of circumstantial evidence is pointing to that conclusion, at least for some species. The Lowestoft Laboratory's plaice egg and larval studies over the period 1961-1971 (2) were intended to test this hypothesis. However, Beverton (6), reviewing in 1974 the results published up to that time, noted that the density of pelagic larvae, calculated from catches in individual oblique tow-net hauls in well-mixed water rarely exceeding 50m depth, was seldom more than one per $m^3$ and was usually one or two orders of magnitude lower. As these larvae did not exceed 10mm length and were capable of feeble movements only, it was difficult to conceive that they were actively grazing down their food supply even at the highest of these densities.

Beverton cited in this connection the view expressed verbally to him by Michael Graham some years earlier that density-dependence was more likely to be found in the early demersal stages when the juvenile fish have migrated to shallow water close to the shore. Graham's prognosis for plaice has recently been substantiated by Lockwood (51), who brought together data from studies of 0-group plaice during their first summer as early demersal juveniles on various British sandy coasts; he showed that a strongly positive relationship exists between monthly mortality rate and density (in absolute units of numbers per $m^2$).

Dutch investigations (Zijlstra et al. (95)) on the early demersal stages of plaice in the Wadden Sea, one of the main nurseries for the North Sea plaice, provide confirmatory evidence. This is shown in Fig. 2 where data from both Lockwood and Zijlstra et al. are plotted together using the same units of density. The effect of the above-average values of the natural mortality coefficient M in years 1963 and 1969 operating over two to three months would be quite sufficient to reduce numbers by the 5- to 10-fold amount necessary to account for the relatively small differences in final recruit values (1963, 3 times average; 1969, about average).

It is difficult to compare absolute densities in meaningful ecological terms between a three-dimensional pelagic habitat and a two-dimensional seabed habitat. Nevertheless, it is not unusual to find local concentrations of 0-group plaice in the shallow littoral zone on the order of 10 or more per $m^2$. These are active foragers for their preferred food (small worms and siphons of young stages of molluscs) and are easy prey for predators. Other North Sea species of flatfish pass through a similarly confined littoral phase in their early life history (78); like plaice, they are also characterized by low variability of recruitment and, at least in sole,

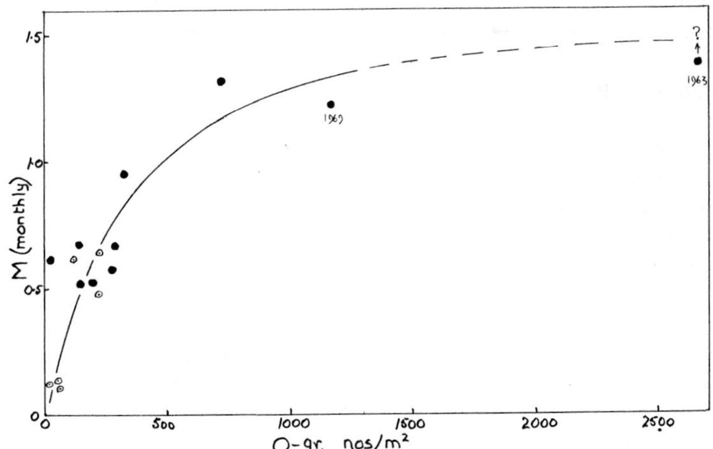


FIG. 2 - Mortality and density of 0-group plaice during their first four months of demersal life (ages 4-8 months), in their coastal nursery areas. Data (•) from Lockwood (51) for various British coastal regions; data (⊙) from Zijlstra et al. (95) for the Dutch Wadden Sea.

by strong regulative capacity.

Ursin (92) drew attention to the marked contrast between this life-history pattern and that for North Sea haddock (Melanogrammus aeglefinus), the young of which remain widely distributed in the northern North Sea when they become demersal towards the end of their first year of life. He pointed out that the North Sea haddock have probably the highest recorded variability of recruitment (over 100-fold, at the extremes (47)). Beverton reported to the group that he had found that if other well documented northeast Atlantic species (e.g., cod and herring) are ranked according to the degree of "concentration" in the early juvenile phase of their life history, the order corresponds roughly inversely with the variability of their recruitment and, so far as can be judged, with their natural regulative capacity. This question is considered again in the Introduction to the section on IMPLICATIONS FOR MANAGEMENT.

We concluded from these discussions, and from the reasoning developed by Sissenwine (this volume), that it would be rewarding to extend the search for regulative processes to the juvenile phase of the early life history. Clearly, this approach would need to be applied with due regard to the life-history pattern of the species in question and to the practical considerations of representative sampling, although the latter may not necessarily be more demanding than for the pelagic phase.

Although, as explained earlier, the group did not deal in detail with the influence of environmental changes on single-species dynamics, in discussion several instances arose in which physical factors interacted with density-dependent mechanisms in the early life history, either enhancing or moderating their effect. These included:

1 - the spread of warm surface water due to El Niño, concentrating the spawning anchovies (E. ringens) close to the Peruvian coast. This would be expected both to increase the cannibalistic predation of eggs and larvae by their parents and to intensify any subsequent density-dependent mortality of the survivors (18).

2 - the effect of the cold winter of 1962-1963 on the North Sea plaice, which slowed down their pelagic development, causing the larval stages to be carried nearer to their coastal nursery grounds (11). It also killed predators on the nursery grounds and favored the meiobenthic food suitable for juvenile plaice (R. Boddeke, personal communication). It is therefore possible that the mortality rate of the

1963 year class as O-group (see Fig. 2) would otherwise have been higher and its eventual size closer to the long-term average.

3 - the spread and contraction of relatively warm (1° - 2° C) Atlantic water in the Barents Sea having a corresponding effect on the extent of the tolerable nursery grounds available for the Arcto-Norwegian cod (79).

We concluded that interactions of environmental and intrinsic density-dependent mechanisms are of great importance in fish population dynamics and would be rewarding for further study.

**Regulation in the Adult Phase**

The significance of regulative processes in the adult phase is not easy to assess quantitatively on present evidence but certainly cannot be discounted, especially at extremes of density. The phenomena typically associated with a pronounced decrease in abundance of a fish population, whether due to fishing or natural causes, are a) increase in growth rate, and b) decrease in age (and usually in size) at first maturity (see (17, 88) for recent reviews). Since fecundity is roughly proportional to weight, these changes are in a direction which tends to compensate for the loss of total egg production of the population due to its depletion. The timing of sex-reversal in those species of fish and crustaceans with this feature of their life history may also be affected by population density and so have regulatory significance (15, 73).

The contribution of these processes to the overall regulative capacity of the population is largely dependent on the slope of the stock-recruit relationship in the operative range. If there is a near-horizontal part, such as shown in Fig. 1, the expected restriction of the change of adult biomass due to density-dependent growth would have a negligible effect on recruitment. However, on the ascending limb of the stock-recruitment curve (through which all populations must pass if depletion becomes severe enough), an additional compensation of biomass of possibly up to twofold could be instrumental in halting a collapse or in hastening a recovery once the primary cause of depletion has disappeared (see below).

The compensatory effect of changes in the adult population at very high densities, i.e., depression of growth rate and hence of specific fecundity, is in the opposite direction, tending to offset such further increases in total egg production as might otherwise have occurred. In certain circumstances such regulatory mechanisms might prevent or at least modify what might otherwise have been a catastrophic collapse due to overcrowding.

Evidence of pronounced changes associated with "overcrowding," such as stunted growth and precocious (by size) maturation is rare in marine fish populations - but it has to be remembered that most of those investigated had already been depleted by fishing. The best examples come from protected freshwater populations in confined habitats (e.g., (28) for perch (Perca fluviatilis L.) in Dutch lakes).

We concluded that while the conventional view that regulatory processes in the adult phase are of minor significance compared with those in the early life history is still broadly true, at least over the main part of the population size range, the circumstances in which this view does not hold are more important than is usually thought. These include populations in which compensation in the early stages of the life history is weak (e.g., in some clupeids and engraulids), and at both lower and upper extremes of density. The topic is certainly deserving of further study, not least in respect of the genetic implications (see section on Ultra-low Densities).

**Ultra-low Densities**

The last forty years or so have seen the well documented "collapse" of around ten of the world's major fisheries, "collapse" here being defined as a decline of population size to 1/50th or less of its size when abundant. Beverton (7) gives a recent summary. The Norwegian spring-spawning herring, the Icelandic summer-spawning herring, the North Sea herring, and the Pacific mackerel (Scomber japonicus) (69) all began to recover, albeit at varying speeds, within a few years of fishing on them being stopped by decree. Of the populations on which fishing was never stopped by legislation, the Japanese sardine (Sardinops melanosticta) disappeared for thirty years but has recently recovered strongly (49); the Hokkaido herring and the Icelandic spring-spawning herring have not yet shown signs of recovering after periods of forty and twelve years, respectively. The California sardine was fished to extreme depletion (to 1/1000th or less (52)) some years before fishing was stopped (effectively, in 1970); it has appeared in small numbers since 1980 but this is in no sense a "recovery."

There is little doubt that intensive fishing, accentuated by an escalation of catchability (q) at low stock sizes (e.g., (91)) was the prime cause of most of these collapses, particularly of those that recovered following cessation of fishing. However, it is characteristic of clupeid and engraulid species to have a record of long-term fluctuations, judging from historical and other evidence (e.g., see (20)), and natural causes have undoubtedly been contributory factors. This is particularly so in the earlier collapses (Hokkaido herring, Japanese sardine, Bohuslän herring, and the central and southern North Sea herring populations), when vessels were not equipped with the modern gear and fish detection devices now associated with the rapid escalation of q at low stock sizes.

This uncertainty serves to underline the need to understand more about the dynamics and ecology of fish populations which have declined, for whatever the cause, to these ultra-low densities. Of special interest, scientifically and for management, are the circumstances which cause the population's natural regulative capacity to break down irretrievably, which evidently happened in some of the above examples but not in others.

It seems hardly possible to predict the population size at which breakdown will happen, knowing only the properties of the system at abundances two or even three orders of magnitude higher; yet for obvious reasons, there is little direct evidence about events at these ultra-low densities. However, some possible factors can be envisaged, namely:

1 - Assembly of the mature fish in the chosen place and time, with the two sexes represented in numbers sufficient to begin successful reproduction, is in many species the culmination of a year's migration cycle, possibly covering hundreds or even thousands of miles. There are some clues about migration mechanisms in marine fish and evidence to show that they have considerable "homing" ability; there is, however, little to indicate whether, if numbers are too low, the continuity and integrity of the annual migration and spawning assembly could be broken.

2 - Effective mating and fertilization of eggs by fish having reached the general spawning grounds could well be a function of the square or higher power of total numbers. Species living in specialized and confined habitats, such as rocky shores and reefs, and many freshwater species typically have elaborate courtship and mating behavior and exhibit various forms of parental care. Little is known about these characteristics in the major deep-sea commercial species, although observations of spawning in aquaria of cod and haddock (e.g., (37)) suggest that mating in these species may be a less prosaic affair

than is usually assumed - and its success perhaps at risk if densities are too low.

3 - While some of the density-dependent mechanisms responsible for regulation at high densities, e.g., intraspecific competition causing shortage of food, are presumably weak or nonexistent at ultra-low densities, this may well not be true for the effect of predators. This could change in either direction (38). If the depleted species is the preferred or only food of specialized feeding predators, the latter may intensify their hunting as their prey becomes scarce. If, on the other hand, the predators are generalized opportunistic feeders, they are likely to turn their attention to other, more abundant (albeit less attractive) species if their original, preferred prey becomes too scarce. The implications for the capacity of the population to recover in the two cases are obvious.

A special aspect of population dynamics at very low densities arises in the management of coastal marine habitats for conservation purposes. A valid management objective in such a situation is to preserve as much species diversity as possible, and to this end steps need to be taken to prevent any one species being reduced to near-extinction levels. Possible procedures include creating (or not destroying) refuges which would give protection to a section of the population at risk, encouraging colonization by fugitive species from elsewhere, and (where possible) identifying the threat and removing it.

Finally, there is the question of the genetic consequences when a population declines to an extremely low level and then recovers. Whether the major fish populations referred to above which have recovered ever became small enough for genetic diversity to be diminished by the classical Seawal-Wright effect is problematical. More likely is that the recovering population derives wholly or mainly from the particular spawning group which was best able to maintain its reproductive performance in the face of whatever factor caused the depletion. It is then presumably better equipped to deal with a recurrence of the cause of the previous decline but, with a restricted genetic repertoire, not necessarily with new adversities that might arise in the future. We are not aware of any published information bearing on this question.

## SELECTIVE ACTION OF FISHING AND ITS GENETIC IMPLICATIONS

A great deal of research has been done on the selectivity of fishing from the point of view of establishing the probability that fish of different sizes will be released or retained by the gear in question. The implications of this process to the genetic makeup of the exploited populations, some of which have now been fished for many tens of generations (of fish), have received much less attention. Miller (61) considered the evidence then available for selection-induced genetic change to be weak. In contrast, Kirpichnikov (48) considers that the adverse genetic effects of the selectivity of fishing could outweigh the conventional advantages of preventing the capture of smaller or immature fish by using selective gear. We therefore decided to inquire into the selective effects of fishing on population characteristics and their genetic implications.

**Selectivity of Fishing**

A useful review of various kinds of selectivity by fishing is given by Nikolskii (66). Although selection for the size of the individual is the most obvious, the precise mechanism is not necessarily straightforward. For example, as Nikolskii points out, nets select not directly for length or weight but for girth, which can be much affected by the condition of the fish at the time. Size selection can occur, however, irrespective of the gear used. Thus, since it is usual for the fastest growing members of a year class to mature first, those will be the individuals first to be recruited to any fishery based on spawning aggregations of mature fish (as many are), even if the gear itself is unselective. Indeed, fisheries which are not in some way selective for the larger and faster growing fish are probably the exception rather than the rule; one such could be the "interception" capture of a migrating population in which the size of fish need have no bearing on whether it is caught or allowed subsequently to escape.

It is not infrequently stated that a decrease in average size of fish in the catch as a fishery develops is a demonstration of the selective removal of the fastest growing individuals (e.g., (62)). The diagnostic test is, of course, not that the average size decreases, which is almost bound to happen as the older fish become scarcer, but that the average growth rate (i.e., size-at-age) of the fully recruited cohorts decreases as the fishery develops. Yet, paradoxically, as noted in the section on IDENTITY OF SINGLE-SPECIES POPULATIONS, the almost invariable direction of change observed in a heavily depleted population is the opposite, with growth rate increasing and age at first maturity decreasing. One of the few recorded exceptions is in some species of Pacific salmon, but in this case both growth and size at maturity decreased (77).

One explanation may be that the relaxation of competition for food in the much depleted population enhances the growth to such a degree that it masks the reduction in intrinsic growth rate that has actually happened and would otherwise have been visible. There is another explanation that may apply in some cases, namely, that having entered the exploited phase, the fastest growing members of the cohort become thereafter progressively better able to evade capture or occupy less intensively fished areas, and so are less rapidly depleted in later life than slower growers. In this case, selectivity and density effects would work in the same direction to enhance the size-at-age of the older cohorts. It is intriguing, though hardly conclusive, that two of the most striking examples of an increase of growth rate following intensive fishing are both ones in which the fastest growing and earliest maturing individuals would have been the first to escape from intense to less intense exploitation; these are the Arctic cod, which are most heavily fished in their immature phase (45), and a coregonid gillnet fishery in which fish above a certain size are not retained by the gear (35).

Fishing must be presumed to be selective for or against other characteristics than size as such, also with genetic implications. Fishing, like predators, will presumably tend to eliminate those individuals least able to evade capture, including those weakened through illness or injury.

Differences between distribution and behavior of the two sexes are often observed in finfish, crustaceans, and marine mammals, particularly at spawning time, which can cause fishing to affect the populations of males and females differently. Data on the sex-ratios of commercial catches of various species show wide variations. Thus, the sex-ratio of catches of gadoids is usually close to 1:1; this is true also for immature flatfish, but for a few years after reaching maturity males dominate in catches of North Sea plaice, especially at spawning time when they are more active and more easily caught than females (26). Local aggregations consisting predominantly of one sex are also found at spawning time in engraulids (42) and Sebastes spp. (55). Atlantic salmon (Salmo salar) caught off West Greenland consist predominantly of females (43).

A departure from 1:1 in the sex-ratio of catches cannot necessarily be equated with a corresponding difference in fishing pressure on the two sexes (i.e., in terms of the fishing mortality rate, F), if only because their natural mortality rates may also differ. In rather few cases have mortality rates for the two sexes been estimated and carried through to stock assessments. Again, North Sea plaice is an example (5), and the ICES North Sea Flatfish Working Group now regularly prepares assessments for each sex of plaice and sole (Solea solea) separately.

These assessments have so far been on a "per recruit" basis only. The effect of sex-selective fishing on the reproductive capacity of a population and hence on its longer-term dynamics is more speculative. Mathison (56) observed that the size-selectivity of gill nets caused male red salmon (O. nerka) to be caught in preference to the females. However, he found with experimental populations that a proportion of males to females as low as 1:15 caused a decrease of only 5% in the production of fertilized eggs compared with a 1:1 sex-ratio. He therefore concluded that in wild populations there was normally a surplus of males which could be harvested without detriment to the reproductive capacity of the population. Later, Reed (76) extended this concept and calculated the optimum sex-differential harvesting regime.

The same concept of a surplus of males has long been assumed to apply to sperm whales, and intensive exploitation of those males has been justified on that basis. Recently, however, the validity of that reasoning has been questioned, on two grounds (40). For one thing, it has not yet been demonstrated that mature males not associated with a "harem" of females necessarily have no current reproductive function. For another, the data on change (or lack of change) in the pregnancy rate in populations from which a high proportion of females has been removed are inconclusive. The effects of sex-selectivity are important in the exploitation and assessment of baleen as well as sperm whales (41), but they have only recently been recognized in management advice. In neither case have predictions or tests been possible of the long-term effects of sex-selection, partly because of insufficient knowledge about the social behavior of either sperm or baleen whales and partly, of course, because of their very long generation time.

**Genetic Implications**

All the above characteristics are likely to be genetically based, although it is difficult to establish to what extent observed changes in fish populations are truly genetic because of the absence of controls to distinguish them from environmental or density-dependent effects. Genetic differentiation in respect of quantitative traits such as growth and age at maturation has, however, been established experimentally by techniques such as transplantation and hybridization.

The rate of genetic change, which is potentially very rapid, depends on two factors: the intensity of selection and the "genetic architecture" of the traits under selection. An extreme example is the demonstration that the weight at maturation of the platyfish, Xiphophorus maculatus, can be changed fourfold by substitution of a single allele at the "p" locus (58), although environmental variation can produce even larger differences. Such rapid changes are possible only when selection is intense and genetic heritability is reasonably high. More usually, genetic selection for growth (i.e., size-at-age) is less dramatic and in the region of 20%-30% in one generation; examples are Atlantic salmon (Salmo salar) (32) and channel catfish (Ictalurus punctatus) (12).

It seems that very little research has been carried out to test directly under controlled conditions the genetic implications of size-selective fishing. An exception is the work of Silliman (85) who found that, after three generations of removing the larger individuals from an experimental population of Tilapia mossambica, the male members of the residual population showed a marked decrease in growth capacity compared with those from the unselected control population under the same feeding conditions. There was no detectable change, however, in the growth capacity of females. Although fairly large numbers of fish were involved, the interpretation of these results must be qualified by the lack of replication in the experimental design, Tilapias being notoriously capricious as experimental animals.

The intensity of artifical selection associated with various fisheries management procedures can, however, be calculated and is often, in fact, rather high. The procedure for doing this has been outlined by Doyle (30). In general, selection intensities will be large if fishing is intense and/or selective for specific sizes or behavior patterns. Variable development rate, partial recruitment, and a strong correlation between size and fecundity will enhance selection intensities. Geographically localized stocks (e.g., salmon) may be especially responsive to selection.

As to the second factor which affects the rate of evolution, the genetic architecture of the characteristics in question, a certain amount is known about cultivated freshwater species such as carp, catfish, and salmonids (94), but virtually nothing for commercially important marine fish. However, the plasticity of growth and maturation in marine fish is comparable to that of freshwater species, and there are no a priori grounds for believing that their heritability of growth, fecundity, behavior, and development generally is lower. It is therefore reasonable to proceed on the assumption that marine and freshwater species respond genetically in the same way.

On that basis, calculation of selection intensities, coupled with plausible guesses about genetics, could show to what extent changes in quantitative characteristics observed in the major marine fish populations under exploitation are consistent with the expected rate of genetic change. Clearly, therefore, fishery-induced genetic changes of an essentially evolutionary nature are a real possibility and an important field for future research.

## IMPLICATIONS FOR MANAGEMENT

### Introduction

Our aim here is to bring together our conclusions on the adequacy of the knowledge about the dynamics of single-species populations needed for their management. For this purpose we mean by "management" not only the legislative control of a fishery and the associated resource, at the national or international level, to achieve some desired objective; the term includes also decisions by individuals or groups of fishing operators or of the whole industry, on whether to develop, expand, or contract the level of activity and investment in exploiting the stock in question. A brief analysis of the kind of questions arising under these headings is therefore first required.

To this end it seemed that three "levels" of questions could be distinguished, each associated with different kinds of scientific information and relating to different time scales, namely:

1) How reliable is the resource as a long-term investment?

2) What order of magnitude of fishing effort and resulting yield can it be expected to support in the medium to long term?

3) What measures, on what time scales, are needed to manage the fishery to meet medium- to long-term objectives arising from 2)?

Obviously, not all these questions will formally be asked in every case, nor in the above sequence. A great deal will depend on whether the fishery is an established one with good records or a newly developing exploitation with little or no data about it. Nevertheless, all three levels arise at some stage, implicitly in the minds of some decision-makers if not explicitly on formal management agendas, and they form a useful framework for evaluating the scientific requirements.

## Qualitative Assessments from General Biology and Fishery Characteristics

The level 1) questions have considerable significance these days in view of what is now known about the variability of fish populations. Certainly, those contemplating the funding of new or expanding fishery developments are much exercised by the degree of risk involved over various time horizons in committing capital and manpower to what may prove to be an unsound investment. The more perspicacious among the decision-makers, with a wary eye on the collapses of long-established fisheries in the last two decades, may well ask a supplementary question to 1) above, namely, is the stock likely to be vulnerable to unrestrained fishing, i.e., is it "collapse-proof" if management is ineffective?

As the basis for discussion about the scientific knowledge needed to answer level 1) questions, our group considered the hypothesis examined by Beverton (7), namely, that even if the conventional quantitative fisheries data are lacking, general biological and environmental information about a species, and of its habitat and the kind of fishing activity likely to be used, can still provide useful guidance. For this purpose, the following categories of information are considered:

1 - Stability of habitat: does the species live in an upwelling area or near a major current system (i.e., at an oceanic front)?

2 - General ecology: are the adults pelagic or demersal? Is the habitat near the center or the edges of the general geographical range of the species? Where is the species located in the food chain?

3 - Life-history characteristics: is there a marked "concentration" phase in the early life history? Does it have a long premature phase during which it is potentially exploitable? Is it short- or long-lived?

4 - Catchability: is it a shoaling fish? Is it readily detectable, e.g., by sonar or from the air? Does the species have access to habitat refuges (i.e., rocky ground) where it can lessen or avoid the risk of capture?

Table 1 summarizes information grouped in this way from some well documented fisheries which can from their history be classified according to whether they have proved steady or erratic, and hence reliable or uncertain as an investment, together with an indication of their capacity to withstand heavy fishing. Entries A - G are based on Beverton (7); entries F - K were added by the group (blue whiting by Bailey, penaeid prawn by Garcia, southern fin whale by Cooke).

The general, though not universal, view of the group was that this was an instructive and potentially useful way of codifying the main kinds of general information which might be available before a fishery has started, or before it has generated enough quantitative data to permit more rigorous analysis. Reservations were expressed about its predictive value in view of the evident association between some of the characteristics and the dominance of the catchability factor in the shoaling species. Nevertheless, we agreed that the approach was worth pursuing further, for example, by examining the experience of other fisheries whose characteristics differ from those shown in Table 1. We noted also that the approach links up with the estimation of population parameters from general biological information, which is discussed in the section on Parameter Estimation (below).

**Quantitative Assessments on Various Time Scales**

It was proposed by Hempel and agreed by the group that the interest of the fishing industry typically centers around the following quantitative objectives:

1 - high total catch (particularly for pelagic fish which are subsequently processed in bulk);

2 - high catch per unit effort (particularly for demersal fish which are destined for the table);

3 - high price (e.g., desired size, quality, and species composition of catch); and

4 - long-term stability (or advance warning of impending changes).

Shepherd's analysis (this volume) of the data required for such assessments, short- and long-term, was taken as the basis for our discussion. The starring of parameters in his Table 1 was accepted with the inclusion of three additional factors, namely, catchability, shifts of relative distribution of fish and fishing, and environmental stability. The expanded table is shown here as Table 2, to be read in conjunction with Shepherd's Table 1. Among the more significant features of Table 2 is the importance of information on recruitment; specifically, the number of incoming recruits for short-term predictions and the factors determining future recruitment levels (including stock-recruitment relationships) for long-term assessments.

TABLE 1 - Fishery characteristics and general ecology/biology. Entries A - G adapted from Beverton (6); entries H, J, and K contributed by Bailey, Garcia, and Cooke, respectively (see text for further explanation).

| | Species | Fishery Characteristics | Physical Environment | Ecology | Life History | Catchability |
|---|---|---|---|---|---|---|
| A | North Sea plaice | Reliable<br>Steady<br>Robust | Stable | Demersal<br>South part of range<br>Near top of food chain | Marked concentration phase as juveniles<br>Exploitable premature phase<br>Long-lived | Non-shoaling<br>Non-detectable<br>Some habitat refuge |
| B | North Sea sole | Reliable<br>Fairly steady<br>Robust | Stable | Demersal<br>North end of range<br>Near top of food chain | Marked concentration phase as juveniles<br>Expl. as prematures<br>Medium-long life | Non-shoaling<br>Non-detectable<br>Some habitat refuge |
| C | North Sea haddock | Fairly reliable<br>Erratic<br>Robust | Stable | Demersal<br>South end of range<br>Near top of food chain | No concentration phase<br>Expl. premature phase<br>Medium life span | Weakly shoaling<br>Weakly detectable<br>Good habitat refuge |
| D | North Sea herring | Uncertain<br>Fairly steady<br>Vulnerable | Stable | Pelagic<br>South-Central part of range<br>Middle of food chain | Moderate concentration phase as juveniles<br>Expl. premature phase<br>Medium life span | Strongly shoaling<br>Easily detectable<br>No habitat refuge |
| E | Atlanto-Scandian herring | Uncertain<br>Episodic<br>Very vulnerable | Moderately stable; in major current system | Pelagic<br>Central part of range<br>Middle of food chain | Slight concentration in juveniles<br>Expl. premature phase<br>Medium-long life | Strongly shoaling<br>Very easily detectable<br>No habitat refuge |

| | Species | Fishery Characteristics | Physical Environment | Ecology | Life History | Catchability |
|---|---|---|---|---|---|---|
| F | California sardine | Unreliable<br>Episodic<br>Very vulnerable | Unstable<br>(upwelling area) | Pelagic<br>Central part of range<br>Middle of food chain | No concentration phase<br>Expl. premature phase<br>Medium life span | Shoaling<br>Easily detectable<br>No habitat refuge |
| G | Peruvian anchovy | Very unreliable<br>Episodic<br>Very vulnerable | Very unstable<br>(upwelling area) | Pelagic<br>Central part of range<br>Low in food chain | No concentration phase<br>Not normally exploit-able as prematures<br>Short life span | Shoaling<br>Easily detectable<br>No habitat refuge |
| H | Blue whiting (N. Atlantic) | Reliable<br>Fairly steady<br>Probably robust? | Moderately stable; in major current system | Bathypelagic<br>Central part of range<br>Middle of food chain | No concentration phase<br>Expl. as prematures<br>Medium life span | Loose aggregations<br>Detectable<br>No habitat refuge |
| J | Penaeid shrimp (brown) | Reliable<br>Rather erratic<br>Robust | Variable | Demersal<br>Central part of range<br>Low in food chain | Marked concentration phase (post-larval)<br>Expl. as prematures?<br>Short life span | Non-shoaling<br>Non-detectable<br>? |
| K | Southern fin whale | Reliable<br>(hard to measure)<br>Vulnerable | Moderately stable | Pelagic<br>Migratory<br>Top of food chain | No concentration phase<br>Expl. as prematures<br>Long life span | Aggregations<br>Detectable<br>(searching)<br>No habitat refuge |

TABLE 2 - The relevance of various sorts of information to short-term and long-term assessments (Section A from Shepherd (this volume), Section B added by group).

| | Information | Short-term catch forecast | Long-term strategic assessment |
|---|---|---|---|
| A | Natural Mortality | - | ** |
| | Current Fishing Mortality or Stock Size | * | * |
| | Exploitation Pattern | * | ** |
| | Imminent Recruitment | *** | - |
| | Factors Determining Recruitment in the Long Term (including stock-recruit relationships) | - | *** |
| | Recent Catch Levels | *** | * |
| | Stock Identification | - | ** |
| | Discards | - | ** |
| | Biological Interactions | - | ** |
| B | Catchability | ** | *** |
| | Shifts in Relative Distribution of Fish and Fishing | ** | ** |
| | Environmental Stability | * | *** |

Key: - of little importance; * moderately important; ** very important; *** vital.

We spent little time discussing the long-term objectives of fisheries management per se, beyond noting that Shepherd's analysis of information was designed to deal with the question: if exploitation proceeds in a particular direction, whether by design or mischance, what is the expected response of the stock in the long term? We reiterated that the prime task of the fisheries scientist must continue to be to emphasize how the natural characteristics of the population and its environment determine

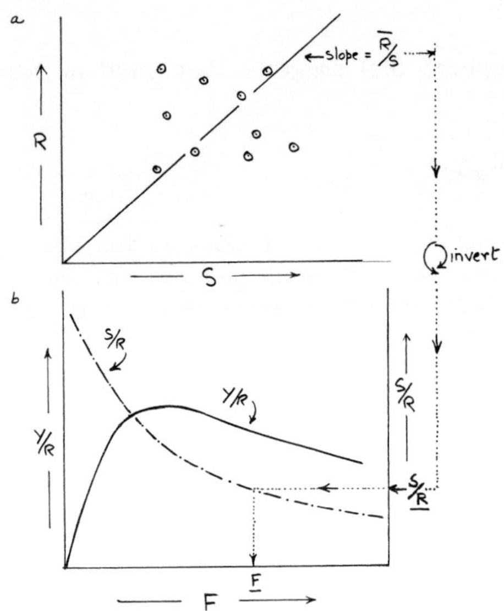


FIG. 3 - Method of Shepherd and Sissenwine (see text) for estimating F for persistence when the stock-recruit relationship is indeterminate.

and constrain its productivity as a resource - whatever the economists, financiers, and administrators may think they want to know. It seems, however, that the advice of the fisheries biologists is still too often either disregarded (as in the European Commission's fisheries policy) or taken too literally (as in the Canadian East Coast fisheries). The problem of communication has clearly not been solved, for the most part; and for this reason the various endeavors being made to improve things - e.g., in the U.S. Regional Fisheries Councils and the Dialogue Meetings convened by the International Council for the Exploitation of the Sea (see ICES Cooperative Research Reports) - are important developments.

Short-term assessments, where the aim is essentially to maintain the fishery at or about its present level (status quo policy), are not demanding of data - provided that the assumptions necessarily made are explicit and compatible with the general characteristics of the resource. These short-term methods may generally be described as limited auto-regressive/ moving average (ARIMA) methods (13), in which analysis of more complex models is used to guide the inclusion or exclusion of various terms in the ARIMA model.

For example, Shepherd (84) suggests that yield in year X + 1 may be estimated as

$$Y_{X+1} = aY_X + bR_X , \tag{1}$$

provided that fishing effort has not changed too much between the pair of years in question. Here $R_X$ is an index of new recruits to the fishery and a and b are coefficients whose magnitude is not completely free (a must be between O and 1, for example).

Equation 1 can be fitted to such data as are available by the usual techniques, provided that a reasonable and reliable index of recruitment is available. The latter might be obtained most economically by special surveys or by sampling a particular segment of the commercial fishery in which the recruiting fish are caught most representatively. This method can be applied quite soon after a fishery first develops and should become more precise as data accumulate, thus bridging the gap until more comprehensive analytical assessments can be attempted.

For the reasons outlined in the section on Spatial Aspects of Single-species Dynamics, care needs to be taken in applying this method to a fishery which grows by exploiting a small segment of the population and rapidly extends its spatial coverage as it builds up. More generally, of course, simplified, short-term methods such as these depend heavily on the presumption that the dynamics of the system in the period covered by the data will continue essentially unchanged for at least the immediate future. They are not intended to replace more complete and thorough methods of assessment but may provide valuable guidance where the costs of research and data collection are prohibitive.

A technique for bridging the gap between short- and long-term assessments, for the management objective of at least minimizing the risk of the fishery "collapsing" from excessive fishing, was put forward by Shepherd and Sissenwine during the meeting. To safeguard against this, there must be "adequate" spawning, i.e., to provide sufficient recruitment to replace losses. In most cases, however, as noted in the section on NATURAL REGULATION IN SINGLE-SPECIES POPULATIONS, the form of the relationship between stock and recruitment is indeterminate, owing to the high variability of the number of recruits surviving from a given parent stock, i.e., of the ratio R/S. Nevertheless, the mean or other central value of the R/S array can serve as a reference for establishing the highest level of fishing mortality coefficient (F) which is compatible with persistence of the population, in the following way.

Suppose that the mean value of recruits per mature biomass, R/S, is calculated from a stock-recruit array such as that shown in Fig. 3a. In addition, the equilibrium relationships between fishing mortality coefficient (F) and yield per recruit (Y/R) and mature biomass per recruit (S/R) are calculated by the conventional methods, giving the curves shown in Fig. 3b. The inverse of the observed mean R/S from Fig. 3a is then entered on the S/R curve of Fig. 3b and the value of F read off.

This resulting estimate, F, corresponding as it does to a value of recruit per stock (R/S) as high or higher than half the values that have so far been observed, satisfies reasonably well the requirement of persistence without undue caution. It is not, of course, a substitute for a properly evaluated target F taking yield, CPUE, and socioeconomics into account. However, if combined with an estimate of the current value of F, it will at least give some idea of how the present level of exploitation stands in relation to the requirement of safeguarding the future of the stocks. If some knowledge is available of the biomass of the stock before exploitation began, the above approach could be used in conjunction with the techniques developed by Beddington and Cooke (4) for evaluating the potential yield of unharvested fish stocks.

As the stock/recruit data series builds up, some semblance of order may emerge in the resulting array which enables more sophisticated analysis to be used. For example, a trend in R/S values with time may become detectable; in that case, a weighted mean giving more prominence to recent values can be calculated, although it would be advisable to have some idea of the reasons for such a trend before placing too much reliance on its forward projection. If a tendency for the R/S ratio to increase with decreasing stock size becomes detectable (i.e., there is evidence of compensation), then it would be possible to introduce an appropriate form of stock-recruitment relationship into the yield equation and use established methodology (e.g., (9, 82)).

In summing up on the question of short-term vs. long-term approaches, the group concluded that reliance on year-to-year assessments alone was unlikely to constitute a viable management strategy. For one thing, it would mean that management would be unprepared to respond to warnings of decline or impending collapse - or of improved productivity - even if the scientists had been able to forecast them and management had been disposed to listen. So, unless the scientists bear the longer-term perspectives, however uncertain, very much in mind and keep the decision-makers fully informed, the prospects for improving the communication between the two are not bright. The concepts of adaptive management (39) may point the way ahead in the appropriate circumstances.

**Parameter Estimation**

The question of the information required for assessment leads directly to the methodology of parameter estimation, which is reviewed in that context by Shepherd (this volume). Our discussion of this subject was limited to some general observations on the present state of the art of parameter estimation for analytical models.

Conventional fisheries data consist of catch and fishing effort, disaggregated in varying degrees by time, location, and method of capture; and with fish of a given species being sampled for size, age, and perhaps sex and maturity. Experience over the last two decades reveals that considerable confusion exists about the intrinsic information content of such data and the appropriate mathematical methods required to extract it. Much of that confusion centers around the capability of the Virtual Population Analysis (VPA) technique, which was reintroduced in fisheries assessment in the mid-1960s by Gulland (34) as a development of the original method of De Lury (25).

The VPA method, providing that certain requirements are satisfied, is a valuable way of reconstructing the past history of a moderately or heavily fished stock and deriving estimates of past stock sizes, recruitment, and fishing mortality rates. Those requirements have, however, proved to be more demanding than was earlier thought. One set of them concerns the natural mortality rate, M. This parameter and its age-specificity must be estimated (or, more usually, guessed) and assumed as remaining unchanged over the period in question - or else changed on the basis of independent evidence.

Equivalent assumptions are also required about the way in which the relative fishing mortality rate changes with the age of fish. If that age-specificity is assumed constant over the period of analysis, then the "separable" version of the VPA technique developed by Pope and Shepherd (75), used in conjunction with effort data, is a more powerful method.

In reality, few fisheries will properly satisfy all these requirements. A high (and probably variable) natural mortality, imperfectly estimated, remains a serious weakness. The difficulties are further compounded if fishing practices change with time in an unknown way, or if the stock is fished by fleets using different gears and deploying their effort over the stock in different and variable ways. In these circumstances there are too many degrees of freedom, and the VPA method may not give reliable results unless additional independent information can be obtained. This can be of any of three kinds, namely, a) on catchability and the selectivity of fishing operations, b) on stock size, and c) on the natural mortality rate, M.

Changes in selectivity from some causes may be detectable by ancillary data or experiments in favorable circumstances: examples include tests of the relative catching power of gear developments; tagging experiments on fish of different sizes; and changes in the distribution of fishing relative to that of the target species. It is unlikely, however, that all possible causes of variation in selectivity can be detected by such means, so that this source of information, though valuable, is not in itself sufficient.

More powerful are independent measures of stock size to "anchor" the estimates obtained from the VPA. Research vessel surveys, using standard gear and a consistent sampling pattern, can give estimates of relative stock size free of some of the sources of bias which are usually present in commercial data. Better still are absolute measures of stock size obtained either by egg (or, less reliably, larval) surveys from which the biomass of the spawning female stock can be deduced (see (87) for review), or by acoustic surveys of the fished stock (e.g., (3)). Both are expensive and each presents technical and logistic problems. Fecundity estimation can be troublesome and hence seasonal egg production difficult to measure with the required precision, but recent techniques of histological analysis (42) and of egg sampling (68, 72) reduce some of the problems involved. Again, with acoustic survey, the relationship between signal strength and true fish abundance is complex even in pelagic fish with good sound-reflecting swim bladders, attributes possessed by only some of the major commercial species.

The need in long-term assessments for as reliable an estimate of M as possible therefore remains an important requirement. Developments in deducing at least an approximate estimate of the average level of M from the general population biology of the species are encouraging. Thus, recent evidence (e.g., (17, 71)) has tended to confirm the positive association between M and the parameter K of the von Bertalanffy growth equation, and hence that the ratio M/K (e.g., as used in the dimensionless form of the yield equation (10)) may vary less and be more predictable than either parameter alone.

The further development of this approach is likely to lie in searching for broader associations between parameters or groups of parameters hitherto treated as independent. One promising line of research derives from the concept of the Evolutionary Stable Strategy (ESS) due to Maynard Smith and Price (57), in which the trade-off between age-specific reproduction and natural mortality rate is assumed to be optimized by natural selection. This approach has been applied to fish populations by Tanaka (90) and by Myers and Doyle (65); it is likely to be most appropriate when the population either is fairly close to its "primordial" state or is able effectively to track environmental change by suitably rapid selection.

While these general approaches could be of considerable value in establishing the general level of natural mortality, they cannot deal with the year-to-year variation or medium-term changes in M that could be a source of serious bias in the VPA method. To detect such variability, direct evidence of changes in the causal factors (e.g., environmental conditions or predation, see section on Empirical Evidence of Species Interaction) is required.

## LIMITATION OF THE SINGLE-SPECIES APPROACH

The reader should perhaps be reminded at this point that this concluding section of our report is intended as no more than a link with the substantive discussion of multispecies dynamics and management undertaken by Sugihara et al. and Larkin et al. (both this volume). The aim here is limited to identifying those limitations of single-species dynamics which are most likely to be overcome by extension to some form of multispecies analysis, and the most promising lines of research to that end.

### Operational Interaction

One potential source of weakness in treating species individually when they are a component of a multispecies fishery arises not from biological interaction between them but from the operational response of the fleet to the mixed-species resource. In such circumstances it is to be expected that the fishermen would try to maximize their economic return per trip by seeking out the best contemporary "mix" of species, depending on their relative abundance, market value, and distribution. To the extent that this happens, so will the relationship between the catch of any one species per unit of nominal effort and its true abundance be affected by the relative economic attractiveness of the other species in the fishery.

There are in principle several ways of minimizing, or at least detecting, the existence of this source of bias, depending on the detail and reliability of the statistical record of fishing activities. If the local concentrations of the various species within the general fishing area are fairly distinct, haul-by-haul records may make it possible to detect bias due to species preference and to make some allowance for it. Some fishing skippers may be willing to report which species had been their prime target; others may be specialists in fishing for one particular species and concentrate on it irrespective of changes in the abundance of other species.

Conceptionally, the most satisfactory approach in such circumstances is to analyze and attempt to model the tactical response of the fleet as a whole to variations in the composition of the mixed-species resource. This would require taking account of the various factors that determine how the vessels distribute their effort in relation to the component species, including market demand and the extent (and reliability) of contemporary information available to each vessel on fishing success elsewhere in the general area.

**Empirical Evidence of Species Interaction**

We first considered the current status of empirical evidence for interspecific biological interaction at the population level, based on the correspondence or otherwise between trends in stock size of individual species in the same general area. Some of the difficulties are illustrated by Daan (23) and by several authors in (60). The North Sea fish populations, especially in view of the dramatic decline in herring and mackerel, might be expected to show pronounced interactions, but the evidence is less conclusive than appears at first sight. Thus Cushing (19) concluded that the gadoid outburst in the North Sea seemed neither in its ecology, location, nor timing to bear much relationship to the decline in herring and mackerel. In the most recent review of the North Sea evidence by Jones (46), the most convincing changes that could be associated with the decline of herring and mackerel were the increases in sprat (Sprattus sprattus) and sand eel (Ammodytes marinus), respectively. Even here, however, there are some doubts about the reliability of stock estimates for the two latter species in the earlier years from which their subsequent increases have been measured, while the precise mechanisms remain speculative. The fish community of the Georges Bank region is ecologically similar to the North Sea and has also experienced major changes. However, a detailed analysis of data for fifteen species of fish and two of squid by Sissenwine et al. (86) failed to reveal valid evidence of interaction.

Both of these regions are ecologically complex systems, and it could be argued that this is partly why convincing empirical evidence of interaction at the population abundance level is difficult to find. On the other hand, the increase of squid in the Gulf of Thailand following depletion by fishing of their fish predators (70) is almost the only convincing evidence known to the group of predator-prey interaction at the population level in a major sea area. Yet the Gulf of Thailand is generally regarded as among the most ecologically complex of major marine ecosystems.

The changes in the much simpler Antarctic marine ecosystem following the drastic depletion of the great whales are therefore of special interest. However, the supposed increase in pregnancy rate and changes in maturation of whales since their decline, which would have constituted strong though indirect evidence of improved food supply, have now been shown to be largely artefactual (Cooke, personal communication). One possible explanation is that there was no shortage of krill for whales even before their postwar depletion. Alternatively, other krill consumers such as squid, penguins, and crab-eater seals may have increased in abundance more rapidly than the whales and utilized the surplus krill as fast as it was produced. Unfortunately, the key evidence - particularly changes in abundance of krill, but also of krill consumers other than whales - is lacking.

Our conclusion has therefore to be that clear and unambiguous evidence of interspecific interaction in major marine ecosystems, which could be used directly to achieve a significant improvement in single-species assessments, as yet hardly exists. This does not, of course, mean that such interaction does not occur or even that it is unimportant; rather limitations of data and the influence of other factors have so far rendered the empirical approach alone of little diagnostic value.

## Mechanisms of Interaction between Species

The way ahead must therefore be to gain a better quantitative understanding of the biological mechanisms of interaction so that their effects at the population level can be deduced and incorporated into assessment methodology. These biological mechanisms can be of two main kinds: a) competition for food (or, conceivably, habitat space), and b) predation. Of these, predation is likely to be by far the most significant, but intensive competition for food may enhance predatory interaction by slowing the growth of prey and thus increasing its vulnerability while searching for food.

Qualitative evidence of what eats what is among the oldest of naturalists' records. The infinitely more difficult task, logistically at least, is to turn this into an estimate of the component of the natural mortality rate in the prey species due to predation. One approach suggested by Doyle is to test whether any statistical significance emerges from a multiple regression of stock-to-recruit "survival" coefficient (i.e., log R/S) on abundance of suspected predators, i.e.,

$$\log R/S = aS + bP_1 + cP_2 + \ldots, \text{etc.}, \tag{2}$$

where $P_1$, $P_2$, etc., are the abundances of predator populations and b, c, etc., are the predation coefficients. This could be a useful exploratory technique for mapping the predation surface in the region of the states covered by the observations. If significance emerges, the predator effect could then be formulated more explicitly in the appropriate model of the system. It would be important in this connection to establish at what stage or stages predation operates in the life history of both the prey and predator populations. Allowance should also be made for shifts in predator selectivity and satiation, without which predator-prey models may be misleading except for small changes close to the observed state.

Considerable emphasis has been given in recent years to obtaining quantitative data on predation in fish species, and some important results are emerging. We noted, for example, the evidence of cross-predation (as well as cannibalism) on eggs and larvae by spawning adults of the Pacific sardine (S. sagax) and Peruvian anchovy (E. ringens) (80), which should throw further light on the interaction between these two major commercial stocks. Following Daan's (21, 22) preliminary estimates of the potentially very large amount of fish consumed by the North Sea cod population, the International Council for the Exploration of the Sea (ICES) has organized a six-nation survey of the quantity of food organisms consumed by the cod population in the North Sea. The results are about to be published, the indications being that they will confirm Daan's earlier conclusions as to the importance of predation by cod, especially on the juveniles of their own and other species. Shepherd's (83) exploratory study of the effect of predation by cod on haddock for assessments of the North Sea fisheries for these species was noted in this connection.

Important though cod as the classical large predator seems to be in a fish community such as the North Sea, the crucial predatory role may not always be confined to species occupying this top position in the conventional food chain. Thus, Christensen (16) has shown from experimental studies that adult sand eels (Ammodytes marinus), normally regarded as a typical prey species, are voracious predators on herring larvae, which they much prefer to the copepods normally comprising their stomach contents in field samples. Christensen also points out that herring larvae, unlike copepods, are rapidly digested by sand eels, being undetectable after 15-30 minutes in stomach samples: this illustrates the value of experimental studies to supplement field data collected for quantitative estimation of fish consumption.

Significant predation on fish is not, of course, confined to other fish predators. The consumption of fish by birds and sea mammals in some major fisheries is undoubtedly large, perhaps the same order of magnitude as the commercial catch, though accurate measures of predation mortality by either group of predators are scarce. Coelenterates also are emerging as a potentially important predator on larval and juvenile fish. For example, Van der Veer et al. (93) discuss the influence that the scyphomedusa Aurelia aurita, the hydromedusa Sarsia tubulosa, and the ctenophore Pleurobranchia pileus may have on survival of larvae and early juveniles of plaice, sole, flounder, herring, and sprat in the nursery grounds of the Dutch North Sea Coast.

These are a few examples of many recent and contemporary studies which, in due course, are likely to advance substantially our understanding of the dynamics of fish populations. While it would be premature to speculate in detail, it seems safe to anticipate that the main impact of predation will often be in the larval or juvenile phase of the life history. This again emphasizes the central role of recruitment and of its relationship to parent stock, in both single- and multispecies fish population dynamics.

## SUMMARY OF PROMISING LINES FOR FURTHER RESEARCH

In discussion of the adequacy of present knowledge and methodology concerning single-species dynamics, the group endeavored to identify where further research would be most rewarding. The topics are referred to at or near the end of each section and are listed here for ease of reference.

**Sections on "Definition of Population" and "Spatial Aspects of Single-species Dynamics"**

(a) Response to exploitation of fished stocks which consist of various proportions of the biological population.

(b) Significance of spatial (geographical) distribution of population processes and parameters to dynamics of single-species fisheries.

(c) Biological and ecological mechanisms for the maintenance of spatial gradients in population structure, and their stability under environmental and exploitation stress.

**Section on "Regulation in the Early Life History"**

(d) Density-dependent processes in early life history, with special reference to the juvenile phase.

(e) Interaction of environmental and density-dependent processes.

**Section on "Regulation in the Adult Phase"**

(f) Regulation in the adult phase through interaction of growth, maturation, and mortality under different population densities.

(g) Spawning stock per recruit assessments, with rigorous definition and measurement of recruitment.

**Section on "Ultra-low Densities"**

(h) Dynamics and ecology of single-species populations at ultra-low densities, with special reference to breakdown of regulative mechanisms and effect on genetic diversity.

**Section on "Selective Action of Fishing and Its Genetic Implications"**

(i) Genetic effects due to selectivity of fishing operations and their evolutionary significance.

**Section on "Qualitative Assessments from General Biology and Fishery Characteristics"**

(j) Assessment of "reliability" of a fishery to be expected from general ecology and biology of the species and its habitat.

**Section on "Quantitative Assessments on Various Time Scales"**

(k) Further development of assessment techniques when data are limited.

**Section on "Parameter Estimation"**

(l) Clarification and further development of the VPA technique.

(m) Estimation of M, or groups of associate parameters, from general biological considerations, including the Evolutionary Stable Strategy approach.

**Section on "Mechanisms of Interaction between Species"**

(n) Analysis of fleet tactics to improve reliability of catch per unit effort data in mixed-species fisheries.

(o) Mechanisms of predatory interaction within and between species, with special reference to stage of life history.

**Acknowledgements.** The above account is the outcome of three days of intensive debate at the workshop, to which every member of our group and some from other groups made significant contributions. It is hardly to be expected that all these varied ideas and experience could be reflected accurately in a single text. Nevertheless, the aim of this rapporteur has been to produce a synthesis which he hopes is a reasonable consensus of how the group responded to their remit.

It will be appreciated that much remained to be done after the workshop by way of putting flesh on bare bones. Arguments had to be developed in an orderly way and supported by properly documented source references. These later stages proceeded by correspondence, and the rapporteur gratefully acknowledges the help he received from his group on topics and literature with which he was unfamiliar.

At the risk of seeming individious by mentioning names in what was essentially a team effort, the rapporteur would nevertheless wish to record his special thanks to A.D. MacCall, J.B. Csirke, and G. Hempel for their careful scrutiny of the first draft of the final report and, in so doing, for filling many lacunae in the rapporteur's knowledge of Pacific fisheries and recent literature. The sections on genetics and other less conventional aspects of fishery dynamics owe much to the patient help of R.W. Doyle and D.J. Policansky. A special word of appreciation is due to S.J. Holt for his able and stimulating leadership as the group's moderator.

Finally, the rapporteur wishes to record his special thanks to J. Lupp and K. Geue of the Dahlem staff for their patient and unflagging help, both at the workshop and in the subsequent editing.

## REFERENCES

(1) Bannister, R.C.A. 1977. North Sea plaice. In Fish Population Dynamics, ed. J.A. Gulland. London: John Wiley.

(2) Bannister, R.C.A.; Harding, D.; and Lockwood, S.J. 1974. Larval mortality and subsequent year-class strength in the plaice (Pleuronectes platessa, L.). In The Early Life History of Fish, ed. J.H.S. Blaxter. Berlin, Heidelberg, New York: Springer-Verlag.

(3) Bazigos, G.P., ed. 1981. A manual on acoustic surveys. CECAF/ECAF Ser. 80/17: 137.

(4) Beddington, J.R., and Cooke, J.G. 1983. The Potential Yield of Fish Stocks. FAO Fish. Tech. Paper 242.

(5) Beverton, R.J.H. 1964. Differential catchability of male and female plaice in the North Sea, and its effect on estimates of stock abundance. Rapp. P.-v. Reun. Cons. Int. Explor. Mer 155: 103-112.

(6) Beverton, R.J.H. 1974. Epilogue. In Sea Fisheries Research, ed. F.R. Harden-Jones, pp. 431-440. London: Elek Science.

(7) Beverton, R.J.H. 1983. Science and decision-making in fisheries regulation. In Proceedings of the Expert Consultation to Examine Changes in Abundance and Species Composition of Neritic Fish Resources, eds. G.D. Sharp and J. Csirke. FAO Fish. Report 291(3): 919-938.

(8) Beverton, R.J.H., and Gulland, J.A. 1958. Mortality estimation in partially fished stocks. In Some Problems for Biological Fishery Survey and Techniques for Their Solution. ICNAF Spec. Pub. No. 1, pp. 51-66.

(9) Beverton, R.J.H., and Holt, S.J. 1957. On the dynamics of exploited fish populations. Fish Invest. Ser. II, XIX, H.M.S.O., pp. 533.

(10) Beverton, R.J.H., and Holt, S.J. 1964. Tables of Yield Functions for Fishery Assessment. FAO Fish. Tech. Paper 38.

(11) Beverton, R.J.H., and Lee, A.J. 1965. Hydrographic fluctuations in the North Atlantic and some biological consequences. In The Biological Significance of Climatic Changes in Britain, eds. C.G. Johnson and L.P. Smith, pp. 79-107. New York: Academic Press.

(12) Bondari, K. 1983. Response to bidirectional selection for body weight in channel catfish. Aquaculture 33: 73-81.

(13) Box, G.E.P., and Jenkins, G.M. 1976. Time Series Analysis: Forecasting and Control. San Francisco: Holden-Day.

(14) Burd, A.C. 1978. Long-term changes in North Sea herring stocks. Rapp. P.-v. Reun. Cons. Int. Explor. Mer 172: 137-153.

(15) Charnov, E.L. 1982. The Theory of Sex Allocation. Princeton, NJ: Princeton University Press.

(16) Christensen, V. 1983. Predation by sand eel on herring larvae. ICES C.M. 1983/L: 27 (Mimeo).

(17) Craig, J.F. 1982. Population dynamics of Windermere perch. In 50th Annual Report of the Freshwater Biological Association, pp. 49-59. Ambleside: FBA.

(18) Csirke, J. 1980. Recruitment in the Peruvian anchovy and its dependence on the adult population. Rapp. P.-v. Reun. Cons. Int. Explor. Mer 177: 307-313.

(19) Cushing, D.H. 1980. The decline of the herring stocks and the gadoid outburst. J. Cons. Int. Explor. Mer 39(1): 70-81.

(20) Cushing, D.H. 1982. Climate and Fisheries, p. 373. London: Academic Press.

(21) Daan, N. 1973. A quantitative analysis of the food intake of North Sea cod, Gadus morhua. Neth. J. Sea Res. 6: 479-517.

(22) Daan, N. 1975. Consumption and production in the North Sea cod, Gadus morhua: an assessment of the ecological status of the stock. Neth. J. Sea Res. 9: 24-55.

(23) Daan, N. 1980. A review of replacement of depleted stocks by other species and the mechanisms underlying such replacement. Rapp. P.-v. Reun. Cons. Int. Explor. Mer 177: 405-421.

(24) de Ligny, W., ed. 1971. Special meeting on the biochemical and serological identification of fish stocks, Dublin, 1969. Rapp. P.-v. Reun. Cons. Int. Explor. Mer 161: 179.

(25) De Lury, D.B. 1947. On the estimation of biological populations. Biometrics 3(4): 145-167.

(26) de Veen, J.F. 1964. On the merits of sampling spawning fish for estimating the relative abundance of different year-classes in plaice. Rapp. P.-v. Reun. Cons. Int. Explor. Mer 155: 94-98.

(27) de Veen, J.F. 1978. On selective tidal transport in the migration of North Sea plaice (Pleuronectes platessa L.) and other flatfish species. Neth. J. Sea Res. 12(2): 115-147.

(28) Deelder, C.L. 1951. A contribution to the knowledge of stunted growth of perch (Perca fluviatilis L.) in Holland. Hydrobiologia 3(4): 357-378.

(29) Devold, F. 1963. The life-history of the Atlanto-Scandian herring. Rapp. P.-v. Reun. Cons. Int. Mer 154: 98-108.

(30) Doyle, R.W. 1983. An approach to the quantitative analysis of domestication selection in aquaculture. Aquaculture 33: 167-185.

(31) Garrod, D.J. 1982. Stock and recruitment - again. Fish. Res. Tech. Report 68. Lowestoft: MAFF.

(32) Gjedrem, T. 1979. Selection for growth rate and domestication in Atlantic Salmon. Z. Tierzücht. Zuchtgsbiol. 96: 56-59.

(33) Gulland, J.A. 1964. The reliability of the catch per unit effort as a measure of abundance in North Sea trawl fisheries. Rapp. P.-v. Reun. Cons. Int. Explor. Mer 155: 99-102.

(34) Gulland, J.A. 1965. Estimation of mortality rates. Annex to Report of the Arctic Fisheries Working Group. ICES C.M. 1965/3 (Mimeo).

(35) Handford, P.; Bell, G.; and Reimchen, T. 1977. A gillnet fishery considered as an experiment in artificial selection. J. Fish. Res. Board Can. 34: 954-961.

(36) Hayman, R.A., and Tyler, A.V. 1980. Environment and cohort strength of Dover sole and English sole. Trans. Am. Fish. Soc. 109: 54-70.

(37) Hawkins, A.D., and Rasmussen, K.J. 1978. The calls of gadoid fish. J. Mar. Biol. Ass. UK 58: 891-911.

(38) Holling, C.S. 1965. The functional response of predators to prey density. Mem. Entomol. Soc. Can. 48: 1-86.

(39) Holling, C.S. 1978. Adaptive Environmental Assessment and Management. Wiley IIASA International Series on Applied Systems Analysis, 3. London: John Wiley.

(40) Holt, S.J. 1980. On the protection levels for sperm whales. Rep. Int. Whal. Comm. 30: 177-181.

(41) Holt, S.J. 1984. Assessment of Area II minke stocks. 1984 Meeting of the International Whaling Commission, Doc. IWC SC/36/Mi 10.

(42) Hunter, J.R., and Goldberg, S.R. 1980. Spawning incidence and batch fecundity in northern anchovy, Engraulis mordax. U.S. Fish. Bull. 77: 641-652.

(43) ICES. 1981. Report of the ICES Working Group on North Atlantic Salmon, 1979 and 1980. Coop. Res. Report 104: 46.

(44) ICES. 1982. Report of the North Sea Flatfish Working Group. ICES C.M. 1982/Assess. 1 (Mimeo).

(45) ICES. 1983. Report of the Arctic Fisheries Working Group. ICES C.M. 1983/Assess. 2 (Mimeo).

(46) Jones, R. 1983. The decline in herring and mackerel and the associated increase in other species in the North Sea. In Proceedings of the Expert Consultation to Examine Changes in Abundance and Species Composition of Neritic Fish Resources, eds. G.D. Sharp and J. Csirke, pp. 507-520. FAO Fish. Report 291.

(47) Jones, R., and Hislop. 1978. Changes in North Sea haddock and whiting. Rapp. P.-v. Reun. Cons. Int. Explor. Mer 172: 58-71.

(48) Kirpichnikov, V.S. 1981. Genetic Bases of Fish Selection. Berlin, Heidelberg, New York: Springer-Verlag.

(49) Kondo, K. 1980. The recovery of the Japanese sardine - the biological basis of stock-size fluctuations. Rapp. P.-v. Reun. Cons. Int. Explor. Mer 177: 332-354.

(50) Lasker, R., and MacCall, A.D. 1983. New ideas on the fluctuations of the clupeid stocks off California. In Proceedings of the Joint Oceanographic Assembly, 1982 - General Symposia, pp. 110-120. Ottawa: Canadian National Committee/Scientific Committee on Ocean Research.

(51) Lockwood, S.J. 1981. Density-dependent mortality in 0-group plaice (Pleuronectes platessa, L.) populations. J. Cons. Int. Explor. Mer 39(2): 148-153.

(52) MacCall, A.D. 1979. Population estimates for the waning years of the Pacific sardine fishery. CA Coop. Ocean. Fish. Invest. Report 20: 72-82.

(53) MacCall, A.D. 1980. The consequences of cannibalism in the stock-recruitment relationship. Intergovt. Oceanogr. Comm. Workshop Report 28: 201-220. UNESCO.

(54) MacCall, A.D. 1984. Population models of habitat selection, with application to the northern anchovy. U.S. Natl. Mar. Fish Service Admin. Report LJ-84-01.

(55) Magnusson, J. 1983. The Irminger Sea oceanic stock of redfish "spawning" and "spawning areas". ICES C.M. 1983/9: 56 (Mimeo).

(56) Mathison, O.A. 1962. The effect of altered sex ratio on the spawning of red salmon. In Studies of Alaska Red Salmon, ed. T.Y. Koo. Seattle, WA: University of Washington Press.

(57) Maynard Smith, J., and Price, G.R. 1973. The logic of animal conflict. Nature 246: 15-18.

(58) McKenzie, W.D.; Crews, D.; Kallman, K.D.; Policansky, D.; and Sohn, J.J. 1983. Age, weight and the genetics of sexual maturation in the platyfish, Xiphophorus maculatus. Copeia (3): 770-774.

(59) Meng, H.J., and Stocker, M. 1984. An evaluation of morphometrics and meristico for stock separation of Pacific herring (Clupea harengus pallasi). Can. J. Fish. Aquat. Sci. 41: 414-422.

(60) Mercer, M.C., ed. 1982. Multispecies approaches to fisheries management advice. Can. Spec. Publ. Fish. Aquat. Sci. 59: 169.

(61) Miller, R.P. 1957. Have the genetic patterns of fishes been altered by introductions or by selective fishing? J. Fish. Res. Board Can. 14(6): 797-806.

(62) Moav, R.; Brody, T.; and Hulata, G. 1978. Genetic improvement of wild fish populations. Science 201: 1090-1094.

(63) Murphy, G.I. 1965. Population biology of the Pacific sardine (Sardinops caerulaea) - Prosp. CA Acad. Sci 34(1): 1-84.

(64) Murphy, G.I. 1977. Clupeids. In Fish Population Dynamics, ed. J.A. Gulland. London: John Wiley.

(65) Myers, R.A., and Doyle, R.W. 1983. Predicting natural mortality rates and reproduction - mortality trade-offs from fish life-history data. Can. J. Fish. Aquat. Sci. 40: 612-620.

(66) Nikolskii, G.V. 1969. Theory of Fish Population Dynamics. London: Oliver and Boyd Ltd.

(67) O'Connell, C.P. 1980. Percentage of starving northern anchovy (Engraulis mordax) larvae in the sea. U.S. Fish. Bull. 78: 475-489.

(68) Parker, K. 1980. A direct method for estimating northern anchovy Engraulis mordax, spawning biomass. U.S. Fish. Bull. 78: 541-544.

(69) Parrish, R.H., and MacCall, A.D. 1978. Climatic variation and exploitation in the Pacific mackerel fishery. CA Dept. Fish Game, Fish Bull. 167: 110.

(70) Pauly, D. 1979. Theory and management of tropical multi-species stocks: a review, with emphasis on the Southeast Asian demersal fisheries. ICLARM Studies and Reviews No. 1, p. 35. Manila: International Center for Living Aquatic Resources Management.

(71) Pauly, D. 1980. On the interrelationships between natural mortality, growth parameters, and mean environmental temperature in 175 fish stocks. J. Cons. Int. Explor. Mer 39(2): 175-192.

(72) Picquelle, S.J., and Hewitt. 1983. The northern anchovy spawning biomass for the 1982-83 California fishing season. CA Coop. Ocean. Fish. Invest. Report 24: 16-28.

(73) Policansky, D. 1982. Sex change in plants and animals. Ann. Rev. Ecol. Syst. 13: 471-495.

(74) Policansky, D. 1983. Sex, age and demography of metamorphosis and sexual maturation in fishes. Am. Zool. 23: 57-63.

(75) Pope, J.G., and Shepherd, J.G. 1982. A simple method for the consistent interpretation of catch-at-age data. J. Cons. Int. Explor. Mer 40: 176-184.

(76) Reed, W.J. 1982. Sex-selective harvesting of Pacific salmon: a theoretically optimal solution. Ecol. Mod. 14: 261-271.

(77) Ricker, W.E. 1981. Changes in average size and average age of Pacific salmon. Can. J. Fish. Aquat. Sci. 38: 1636-1656.

(78) Riley, J.D.; Symonds, D.J.; and Woolner, L. 1981. On the factors influencing the distribution of 0-group demersal fish in coastal waters. Rapp. P.-v. Reun. Cons. Int. Explor. Mer 178: 223-228.

(79) Saetersdal, G., and Loeng, H. 1983. Ecological adaptation of reproduction in Arctic Cod. PINRO/IMR Symposium on Arctic Cod, Leningrad (Mimeo).

(80) Santander, H.; Alheit, J.; MacCall, A.D.; and Alamo, A. 1983. Egg mortality of the Peruvian anchovy (Engraulis ringens) caused by cannibalism and predation by sardines (Sardinops sagax). In Proceedings of the Expert Consultation to Examine Changes in Abundance and Species Composition of Neritic Fish Resources, eds. G.D. Sharp and J. Csirke, pp. 1011-1026. FAO Fish. Report 291.

(81) Serra, J. 1984. Changes in abundance of pelagic resources along the Chilean coast. In Proceedings of the Expert Consultation to Examine Changes in Abundance and Species Composition of Neritic Fish Resources, eds. G.D. Sharp and J. Csirke, pp. 255-284. FAO Fish. Report 291.

(82) Shepherd, J.G. 1982. A versatile new stock-recruitment relationship for fisheries and the construction of sustainable yield curves. J. Cons. Int. Explor. Mer 40(1): 67-75.

(83) Shepherd, J.G. 1984a. A promising method for the assessment of multi-species fisheries. ICES C.M. 1984/G: 4 (Mimeo).

(84) Shepherd, J.G. 1984b. Status quo catch estimation and its use in fishery management. ICES C.M. 1984/G: 5 (Mimeo).

(85) Silliman, R.P. 1975. Selective and unselective exploitation of experimental populations of Tilapia mossambica. U.S. Fish. Bull. 73(3): 495-507.

(86) Sissenwine, M.P.; Brown, B.F.; Palmer, J.E.; Essig, R.J.; and Smith, W. 1982. Empirical examinations of population interactions for the fishery resources off the north-eastern U.S.A. In Multispecies Approaches to Fisheries Management Advice. Can. Spec. Publ. Fish. Aquat. Sci. 59: 82-94.

(87) Smith, P.E., and Richardson, S.L. 1977. Standard techniques for pelagic fish egg and larvae surveys. FAO Fish. Tech. Paper 75: 100.

(88) Stearns, S.C., and Crandall, R.E. 1984. Plasticity for age and size at sexual maturity. A life-history response to unavoidable stress. In Fish Reproduction: Strategies and Tactics, eds. G.W. Potts and R.J. Wootton. London: Academic Press.

(89) Tanaka, S. 1960. Studies on the dynamics and management of fish populations. Bull. Tokai Reg. Fish. Res. Lab. 28: 1-200.

(90) Tanaka, S. 1983. A mathematical consideration on the adaptation strategy of marine fishes. Res. Pop. Ecol. Suppl. 3: 93-111.

(91) Ulltang, Ø. 1980. Factors affecting the reaction of pelagic fish stocks to exploitation and requiring a new approach to assessment and management. Rapp. P.-v. Reun. Cons. Int. Explor. Mer 177: 489-504.

(92) Ursin, E. 1982. Stability and variability in the marine ecosystem.

Dana 2: 1-50.

(93) Van der Veer, H.W.; Van Garderen, H.; and Zijlstra, J.J. 1983. Impact of coelenterate predation on larval fish stocks in the coastal zone of the Southern North Sea. ICES C.M. 1983/L: 8 (Mimeo).

(94) Wilkins, N.P., and Gosling, E.M., eds. 1983. Genetics in Aquaculture. Amsterdam: Elsevier.

(95) Zijlstra, J.J.; Dopper, R.; and Witte, J.J. 1982. Settlement, growth and mortality of post-larval plaice (Pleuronectes platessa) in the Western Wadden Sea. Neth. J. Mar. Res. 15(2): 250-272.

Exploitation of Marine Communities, ed. R.M. May, pp. 59-94. Dahlem Konferenzen 1984. Berlin, Heidelberg, New York, Tokyo: Springer-Verlag.

# Why Do Fish Populations Vary?

M.P. Sissenwine
National Marine Fisheries Service
Northeast Fisheries Center, Woods Hole Laboratory
Woods Hole, MA 02543, USA

**Abstract.** Fish populations vary because of density-dependent and -independent processes that determine recruitment, growth, and natural mortality, and in response to fishing. Most of the natural (non-fishing) variability is associated with recruitment, presumably the density-independent effect of fluctuating environmental factors.

Numerous empirical models have been used to explain recruitment variability. While "statistically significant" correlations are plentiful, most empirical studies are flawed because they a) use an inappropriate proxy for recruitment, b) are data exploration exercises that are not based on a plausible a priori hypothesis, c) do not consider, simultaneously, physical variables and spawning biomass, and/or d) fail to test predictions on independent data (i.e., not used to establish correlation). As a result, many empirical models fail to predict post-publication events. Furthermore, fundamentally different empirical models may be indistinguishable because they account for virtually the same proportion of variability in recruitment.

Process-oriented studies of recruitment variability have focused on starvation, particularly of first-feeding larvae. Starvation may be related to the amount of suitable food, the contagious nature of its distribution, currents which transport eggs and larvae, the match or mismatch of the annual reproductive cycles with the annual production cycle of prey, and the stability of the environment.

Only recently has a new hypothesis emerged, namely, that predation is the major cause of prerecruit mortality. It is supported by a) evidence of prey concentrations which are adequate for a high survival of larvae as indicated by laboratory and modeling studies, b) lack of evidence of starvation for field-collected larvae, c) a high survival rate of larvae in large, predation-free enclosures, d) the high mortality rate of eggs and yolk-sac larvae which are not subject to starvation, and e) the identification of fish and invertebrates as predators of egg, larval, and post-larval stages.

At least for some systems, fish consume most of their own production. Two important implications are that fish populations are probably able to compensate for fishing, and post-larval mortality must be high, thus affecting recruitment. The lack of correlation between larval abundance and recruitment also implies that year-class strength is not established until the post-larval stage.

Physical factors are presumed to be responsible for recruitment variability. There are numerous possible mechanisms if recruitment is determined by starvation of larvae. It is less apparent how physical factors are related to variability in predation mortality of post-larvae.

## INTRODUCTION

The dynamics of a fish population are determined by the balance between increases due to recruitment and growth, and losses due to fishing and natural mortality. This paper concerns the biological mechanisms and physical factors that control recruitment, growth, and natural mortality. Broadly speaking, this is most of fishery science. Herein, I only attempt to highlight ideas from some of the most important and recent literature.

Population size may vary in either a density-independent or density-dependent manner. Density-independent means that population growth per unit size (i.e., numbers or biomass) is unrelated to population size; density-dependent means that there is a relationship. A density-dependent relationship is referred to as compensatory if population growth per unit size is inversely related to population size (all other factors remaining constant), or depensatory if there is a direct relationship. The terms density-dependence and -independence are often taken literally to imply numbers or biomass per unit area or space. While this may be the case, these terms are used in the context of population size without regard to local density.

Physical factors are usually considered to have a density-independent influence since their potential effect on each member of the population is the same regardless of population size. On the other hand, biological mechanisms are usually density-dependent. For example, the constraining effect of competition increases with the size of the population competing for a limiting resource. Yet, physical factors may act indirectly as density-dependent (e.g., physical changes may affect the environment's carrying capacity, thus exacerbating competition). So, too, may biological processes be density-independent. Predators may consume a constant proportion of their prey independent of prey population size. Depensation occurs when predators consume a constant amount (or increasing proportion) of their prey as prey population size decreases.

The theory of fishing depends on the existence of compensation. The theory is that an unexploited fish population remains in a quasi-equilibrium with no surplus production (i.e., recruitment plus growth minus natural mortality). Fishing reduces population size, resulting in a compensatory response with a surplus production available for harvest. Compensation may involve an increase in recruitment or growth rate as population size decreases, or a decrease in natural mortality rate. Unfortunately, density-dependent responses of fish populations (particularly involving recruitment) are often obscured by variability, presumably due to the density-independent effect of fluctuating environmental factors.

**RECRUITMENT, GROWTH, AND NATURAL MORTALITY**

The growth rate of a fish depends on its consumption rate and the efficiency of the conversion of food energy to body tissue (i.e., growth efficiency). Compensation in growth occurs when fish reduce the abundance of their prey as their own population size increases. When this happens, there is less prey available for consumption and more energy is required to search for and/or capture it. Compensatory growth may also result from competition for habitat space. In this case, the "interference" effect of an increase in population size per unit space reduces consumption rate or growth efficiency. Growth rate is also affected by physical factors. Temperature has a density-independent effect. Physiological processes (e.g., metabolism) that determine growth efficiency are a function of temperature.

There is ample evidence that growth rate varies. Compensation in growth rate is indicated for Pacific halibut (26), Baltic Sea sprat (45), and Georges Bank haddock (15). Yet, there are also examples of remarkably little change in growth rate in spite of large changes in population size (e.g., North Sea haddock (46)).

Natural causes of death of fish are predation, starvation, lethal environmental conditions, and disease. Starvation of post-recruit fish has rarely been documented. Furthermore, the generally precise relationship between length and weight argues against the significant proportion of a population suffering from starvation. Fish on the verge of starvation would be expected to have an anomalously low weight at specific length, yet such anomalies are infrequent.

Disease is probably always a source of natural mortality, although routine monitoring of fish from the waters off the northeastern USA indicate that gross pathology and anomalies (e.g., ambicoloration, fin rot, lymphocystis, skeletal anomalies, and ulcers) are relatively infrequent (25). It is clear that epidemics occasionally devastate some populations. Northwest Atlantic herring populations were ravaged by periodic epizootics of a systemic fungus pathogen. Outbreaks occurred in the Gulf of Maine in 1932 and 1947, and in the Gulf of St. Lawrence in 1898, 1916, 1940, and 1955 (80, 81). The effect of contagious disease is probably density-dependent since transmission efficiency increases with density.

Disease and physical factors may act synergistically. Burreson (11) provides convincing evidence that virtually all of the juvenile summer flounder within the Chesapeake Bay were killed during the winter of 1980-81 by the combined effect of anomalously low temperatures and a hemoflagellate parasite. The parasite only became lethal when the fish were stressed by the low temperature.

There are numerous predators of fish, e.g., marine mammals, marine birds, and larger fish. As will be demonstrated later in this paper, predation is probably the major cause of natural mortality, although other factors (e.g., starvation) may increase vulnerability to predation.

Unfortunately, it is very difficult to estimate natural mortality rate on average, much less its variability. MacCall and Methot (64) attempt annual estimates of natural mortality of northern anchovy, but their results are confounded with variability in estimates of biomass. Nevertheless, an increase in estimated mortality in the late 1970s coincided with an increase in the abundance of Pacific mackerel, a known predator. Natural mortality is usually estimated by relating it to life span or other life-history characteristics such as growth and age at maturity (70).

Fisheries scientists have focused on the relationship between recruitment and spawning biomass (referred to as S-R relationship). They have done so because most production to the fishery results from recruitment, and ultimately population persistence requires replacement in numbers by the recruitment process.

There are numerous biological mechanisms that result in a compensatory S-R relationship. Ricker (72) noted several mechanisms including competition for breeding sites or living space, competition for food resulting in starvation, increased susceptibility to disease from crowding, cannibalism, and compensatory growth coupled with size-dependent predation mortality. Some of these mechanisms lead to characteristic mathematical relationships between spawning biomass and recruitment. The best known of these relationships are the Ricker (72) and Beverton and Holt (5) functions. The underlying differential equation model of the dome-shaped Ricker function assumes that mortality rate of offspring is proportional to the number of spawners. This relationship could result from cannibalism by spawners. According to Shepherd ((79), based on a personal communication from R. May), it may also result from depletion of a key food resource due to the stock itself. The asymptotic Beverton and Holt function results from a differential equation model with mortality rate of offspring proportional to their own number. This relationship results from intracohort cannibalism or competition.

Unfortunately, neither function nor other S-R functions proposed in the literature (e.g., (21)) account for the variation in typical stock-recruitment data (e.g., Figs. 1-2). It is usually hypothesized that the unexplained variability in recruitment results from the density-independent effects of environmental fluctuations.

In most cases, spawning biomass is estimated with the assumption that age at maturity and fecundity are constant and independent of density. This assumption is frequently invalid and it may contribute to the apparent lack of a relationship between spawning biomass and recruitment.

Recruitment variability is the central problem of fishery science and a major source of uncertainty in fisheries management (84). Only projections of abundance based on year classes already recruited to the fishery are possible until recruitment variability is understood. Sustainable yield strategies depend on quantifying density-dependence in the S-R relationship.

There are two general approaches to the variable "recruitment problem": empirical- and process-oriented. The empirical approach involves fitting statistical relationships between recruitment (or proxies for it) and

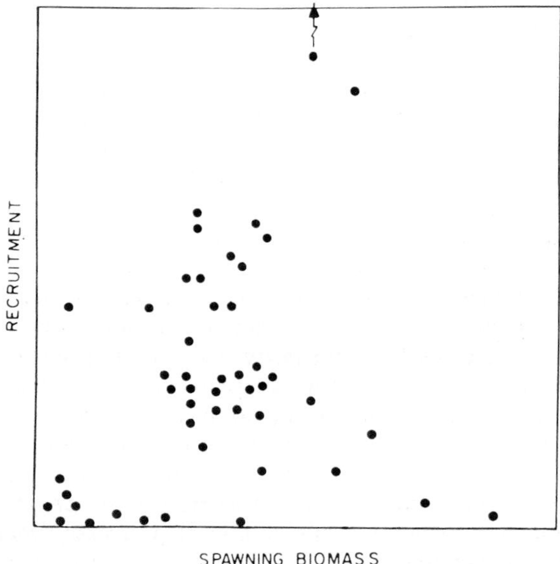


FIG. 1 - Stock-recruitment data for Georges Bank haddock (15).

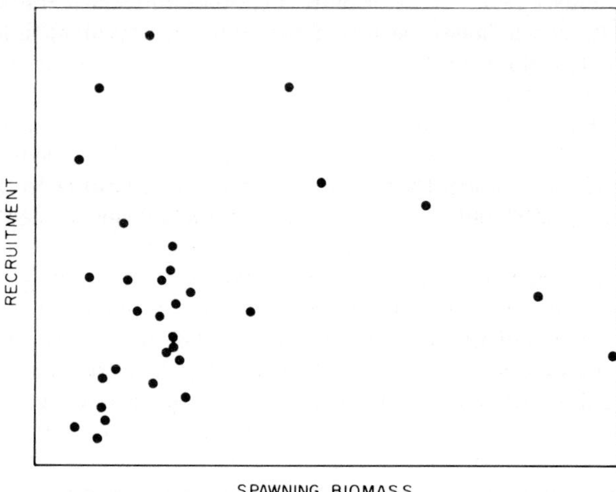


FIG. 2 - Stock-recruitment data for Northeast Arctic cod (44).

physical, and sometimes biological, variables. Presumably, these relationships can be used to predict recruitment. The process-oriented approach involves identifying the causes of prerecruit mortality (egg, larval, and post-larval, i.e., a portion of the juvenile stage) and the factors that cause prerecruit mortality to vary. The process-oriented approach can be applied holistically to several species or to functional units as an aggregate.

## EMPIRICAL APPROACH TO THE RECRUITMENT PROBLEM

Ultimately, the spatial distribution and abundance of fish is controlled by climate, e.g., there are no parrot fish in boreal waters, nor are there codfish in the tropics. Jones and Martin (47) found that the fisheries for demersal species relative to pelagic species became more prevalent in the Atlantic as bottom water temperature decreased. Therefore, it is only natural to try to correlate changes in abundance of fish with changes in climate.

Templeman and Fleming (92) considered long-term changes in hydrographic conditions and corresponding changes in abundance of fishery resources in the waters off Newfoundland and Labrador. They note similar trends in climate (as indicated by air temperature) and abundance (as indicated by catch or sightings) for mackerel, lobster, squid, billfish, capelin, and cod. Taylor et al. (91) examined climatic trends and the distribution of fishery resources off New England. They noted a general warming trend from 1900 to 1940. They relate landings statistics for mackerel, lobster, whiting, menhaden, and yellowtail flounder to air temperature and water temperature records. They also relate range extensions to the warming trend.

Cushing and Dickson (23) examined variations in atmospheric and marine climate ranging from changes over a century in duration to changes occurring within a year. They speculate that long-term changes in fisheries (e.g., herring fisheries since 1400 A.D.) have been in response to climate. In particular, they cite the correlation between the periods of the Atlanto-Scandian herring and periods of ice cover north of Iceland (6). Cushing and Dickson (23) conclude that climatic changes occur on a worldwide scale and biological events in one region can be related to those in another. Likewise, Garrod and Colebrook (32) conclude that fish communities respond to climate on the "pan-Atlantic scale."

It is important to understand the nature of changes in fish communities that occur over a time scale of decades so that long-term fisheries expectations are not unrealistic. Yet the success of fisheries management also depends on understanding inter-annual variability in fish abundance. The empirical approach has been used frequently to examine this scale of variability (see (1) for a more complete review than is given herein).

Many of these attempts have demonstrated correlations between catch and environmental factors (usually temperature). Dow (27) compared landings of lobster, scallop, and shrimp with annual average water temperature at Boothbay Harbor. Lags between variables were inspected until a relationship was found. Similarly, Dow (28) considered Maine lobster, in more detail, for the period from 1905. He found that the optimum center of catch moved southward as temperature decreased. Flowers and Saila (30) and Dow (29) correlated several water temperature records (both surface and bottom) with lobster catch from the Gulf of Maine region and found a direct relationship.

Gunter and Edwards (36) found a correlation between the catch of white shrimp from the Gulf of Mexico and rainfall in Texas that year and two years earlier. Sutcliffe et al. (90) correlated the catch of seventeen species of fish from New England with water temperature. The correlations were improved by considering fishing effort and select time lags. Ulanowicz et al. (93) found that an appreciable portion of the annual variation in catch of seven commercially important fisheries of Maryland could be linked to fluctuations in the past physical environment.

These studies use catch as a proxy for abundance and, presumably, as a proxy recruitment a short time prior. The validity of this substitution depends on the degree of variability in fishing effort and the degree of dependence of the fishery on annual recruitment. In the case of the lobster, the fishery is almost totally dependent on annual recruitment, and fishing effort is relatively stable (30). Therefore, catch is a reasonable proxy for recruitment. On the other hand, substitution of catch for recruitment is questionable for several of these correlations.

There are many examples where physical factors are correlated directly with recruitment (instead of with a proxy). Carruthers (12) correlated computed winds with recruitment of haddock, herring, cod, and hake of the northern North Sea. Sissenwine (82) correlated recruitment of southern New England yellowtail flounder with air temperature. In order to minimize the chances of a spurious relationship, Sissenwine tested correlation coefficients fit to each half of the data base separately and determined that there was no statistically significant difference between them. In this case, air temperature was shown to be a reasonable proxy for surface and bottom water temperature. Hayman (38) modeled Dover sole and English sole recruitment of populations off the Oregon Coast of the USA using numerous environmental variables (i.e., temperature, sea level, barometric pressure, index of offshore and alongshore transport, offshore divergence index, wind speed, solar radiation, Columbia River discharge, water quality, number of wind direction shifts, storm frequency, and storm duration).

A major shortcoming of most of the studies cited above are that they are primarily examples of data exploration exercises. The correlations cannot be viewed as a test of a previously stated hypothesis, and nominal levels of statistical significance are misleading due to multiple testing. At best, they serve as a basis for establishing a future hypothesis.

Some empirical studies have been based on a plausible reason for expecting a correlation between recruitment and a physical variable in advance of the analysis. Walford (96) correlated Pacific sardine recruitment for 1934-1941 with salinity. Salinity was taken as an index of upwelling which would affect production of focd suitable for larval fish.

Chase (13) hypothesized that the loss of haddock pelagic larvae from Georges Bank should be related to northwesterly winds as derived from pressure measurement. He achieved a relatively good correlation ($r^2$ =0.59) by taking account of the effects of temperature fluctuations on spawning time.

Bakun (2) and Bakun and Nelson (3) used longshore wind stress data to calculate monthly upwelling indices and surface layer divergence indices for the California Current area. Coastal upwelling is always associated with positive nearshore divergence, but the intensity and sign of offshore divergence depends on the rotation of the wind field, i.e., wind stress curl (Fig. 3, (4)). These indices of upwelling and divergence have been used to examine fluctuations in recruitment of several species (including Dover and English sole, cited above). Bakun and Parrish (4) review several of these applications (e.g., for coho salmon, dungeness crab, rockfish, bonito, hake, sardine, mackerel, and anchovy). In particular, poor upwelling in 1947-1952 was associated with recruitment failures of anchovy, mackerel, sardine, and bonito, although some of these analyses were primarily data exploration exercises.

So far, we have focused on empirically derived relationships between

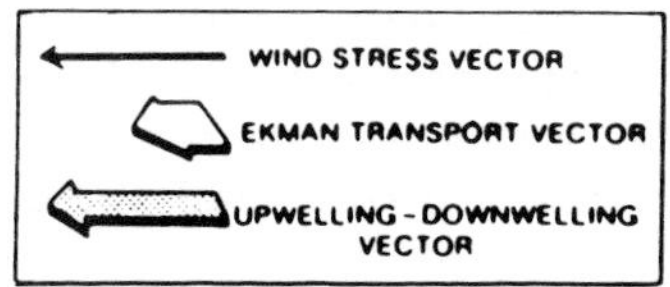


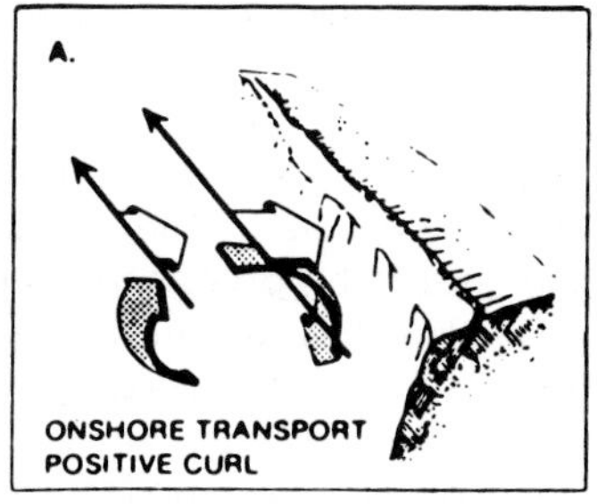


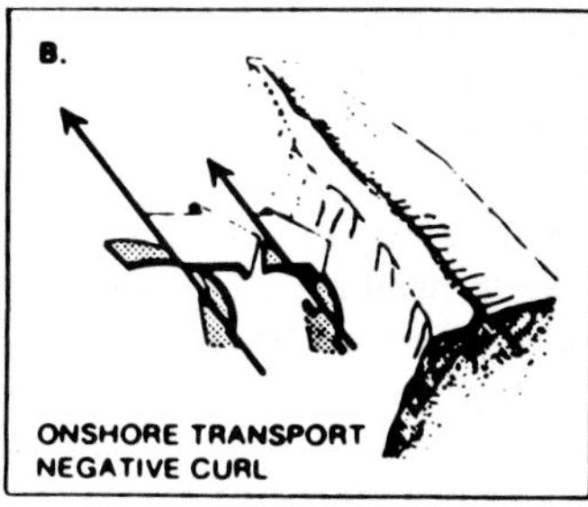


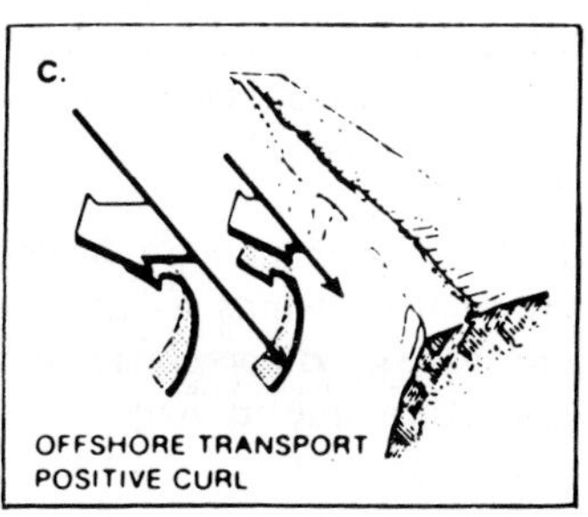


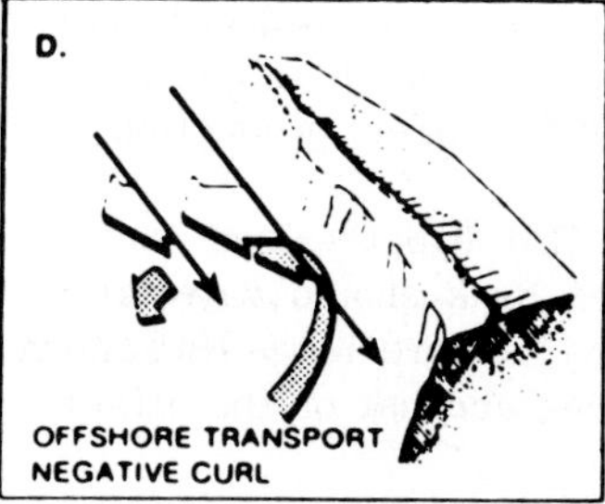


FIG. 3 - Types of couplings between surface convergence or divergence at the coast (due to the onshore-offshore component of Ekman transport) and that (due to wind stress curl) occurring offshore of the coastal boundary zone. A. Convergence and downwelling at the coast, divergence and upwelling offshore. B. Convergence and downwelling at the coast, continued convergence offshore. C. Divergence and upwelling at the coast, continued divergence offshore. D. Divergence and upwelling at the coast, convergence offshore (after Bakun and Nelson (3)).

recruitment and physical variables, ignoring the relationship with spawning biomass. There are some noteworthy investigations that consider both.

Bakun and Parrish (4) model Pacific sardine recruitment per unit spawning biomass as a multiplicative function of indices of upwelling and divergence at 30°N and 39°N, spawning stock size, and a linear trend. There was a statistically significant fit, with a positive dependence on upwelling, a negative dependence on divergence, a negative dependence on stock (implying compensation), and a negative trend. They speculate that the negative trend is due to competition with anchovy.

Parrish and MacCall (69) present a number of empirical models of Pacific mackerel recruitment. They consider several compensatory stock recruitment functions (i.e., Ricker, Beverton and Holt, Cushing), physical variables (i.e., sea surface temperature at two locations, barometric pressure, wind speed, sea level height, upwelling index, and surface layer divergence index), and a variety of combinations of monthly values of the physical variables. In some cases, they excluded a portion of the data time series from parameter estimation procedures for use in testing the predictive value of the models. They were able to explain 76% of the variability in recruitment for the period 1946-1968, and 59% for 1931-1968 (Fig. 4).

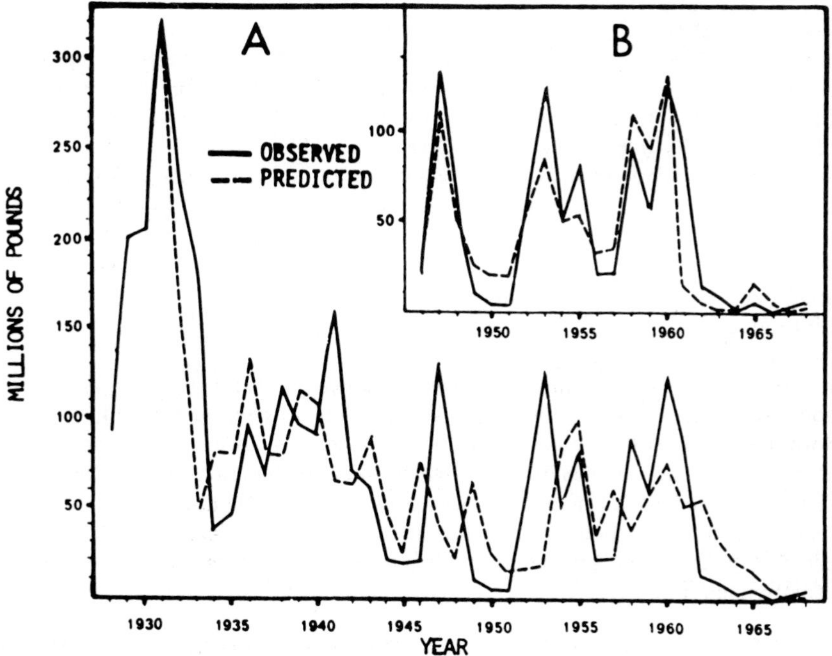


FIG. 4 - Observed and predicted recruitment for Pacific mackerel (69).

Nelson et al. (66) fit a Ricker stock recruitment model to Atlantic menhaden year-class strength data for 1955-1970. Deviations from the relationship were used to calculate a "survival index," and these were correlated with several environmental variables; 84% of the variations in the survival index was explained. Zonal Ekman transport, which acts to transport larvae from offshore spawning grounds to inshore nursery grounds, was the most statistically significant physical variable.

Csirke (19) hypothesized that recruitment of the Peruvian anchovy was dependent on the density of spawners, as well as their absolute biomass. Presumably, the amount of suitable spawning area varies as a result of environmental fluctuations, and this affects density, which in turn affects the rate of cannibalism. Csirke modified the Ricker S-R function to include a variable "concentration index." He calculated the index by comparing commercial catch-per-unit data with spawning biomass estimates. His analysis provides a family of S-R functions depending on the concentration index (Fig. 5).

Sissenwine (83) developed simulation models of the southern New England yellowtail flounder fishery. Catch was simulated based on age- and size-structured models, with variable fishing effort, and the effects of temperature variations (82) superimposed on underlying S-R functions. Reminiscent of the classic debate between Clark and Marr (14), recruitment was modeled as independent of the spawning biomass (Marr's position) and linearly dependent on spawning biomass (quasi Clark's position). The models explained 83 and 85% variability in catch, respectively (Figs. 6-7) and appear to have predictive value, since only information up to 1965 was used to fit parameters, yet they continued to predict after that date.

While it appears that much progress has been made using the empirical approach, the approach has some serious limitations. Empirical methods do not demonstrate "cause and effect." There is always a concern about spurious correlations, particularly when several combinations of variables are considered (35). In only a few cases have some data been excluded from the fitting procedure for use in model testing (e.g., (69, 82)). The predictive value of empirical models beyond the date of publication is usually unknown.

I am aware of some models, which were based on plausible hypotheses about the effects of physical factors, which have failed to predict post-publication events (e.g., (13, 66)). I suspect that post-publication failure

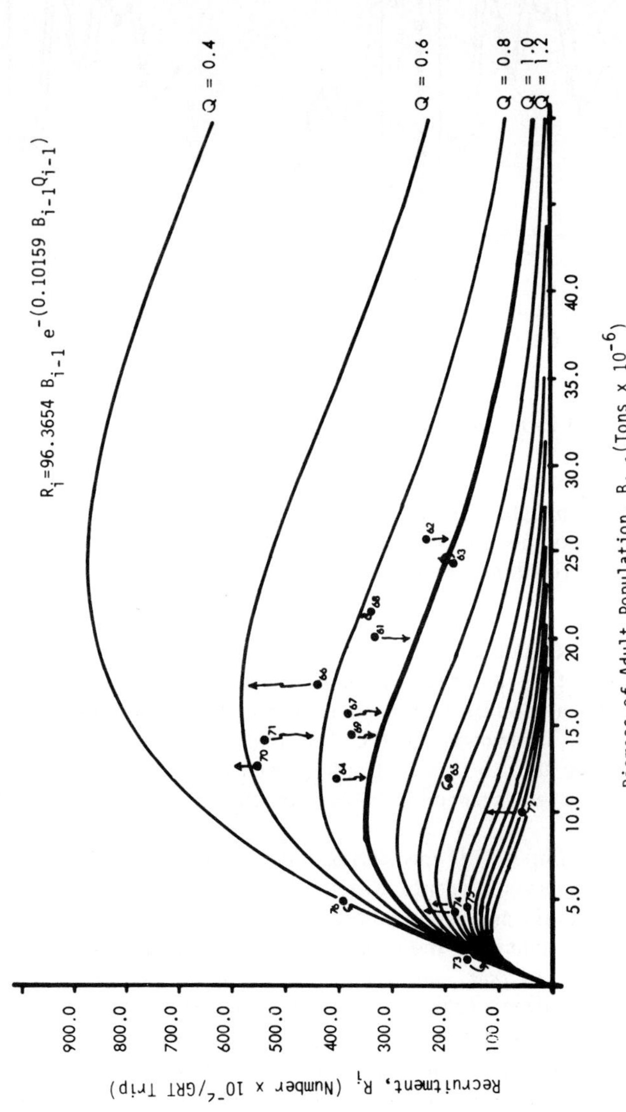


FIG. 5 – Family of recruitment functions for Peruvian anchovy depending on density of spawners (19).

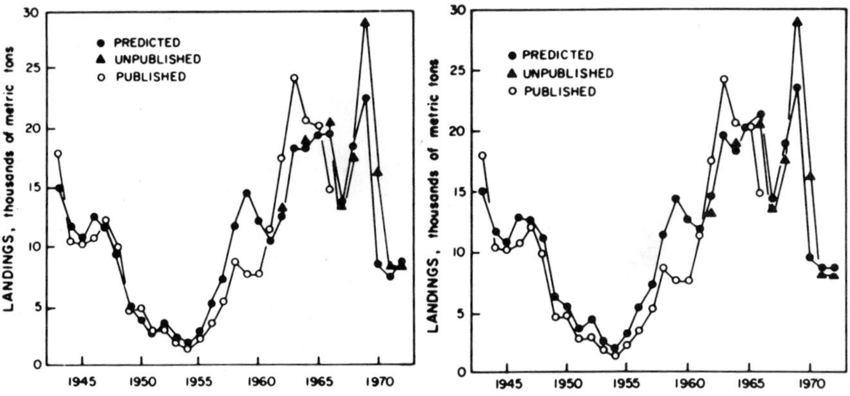


FIGS. 6-7 - Simulation of Southern New England yellowtail flounder with recruitment independent (left) or linearly dependent (right) on spawning biomass (83).

of empirical models is not infrequent.

Perhaps an even more important limitation of empirical models is that they may not be useful for quantifying density-dependence. The influence of physical variables may be so dominant that it may not be possible to distinguish between fundamentally different S-R models (e.g., Figs. 6-7, (83)).

**PROCESS-ORIENTED APPROACH**

Hjort (40) hypothesized that Atlanto-Scandian herring recruitment fluctuated according to the availability of food to early stage (first-feeding) larvae. This hypothesis is known as the "critical period hypothesis." Although Hjort was never able to test it, the hypothesis remains attractive today. Most process-oriented research is focused on starvation of larvae.

Since Hjort's work, two important corollaries have evolved. One is that recruitment depends on water currents which maintain pelagic eggs and larvae in, or transport them to, areas where food is concentrated. This hypothesis is the basis of many of the empirical studies reviewed earlier (e.g., (13, 66)). The second corollary is referred to as the "match-mismatch" hypothesis (20). According to this hypothesis, recruitment depends on the match or mismatch of the annual reproductive cycle of fish and the annual production cycle of larval fish prey. Cushing and Dickson (23) give several examples of unique events that could be explained by the match-mismatch hypothesis. They note that the 1963 year class of southern North Sea plaice was the largest ever recorded. They speculate that an unusually cold winter delayed egg and larval development, resulting in a better match with production of prey.

One of the first attempts to investigate the "critical period hypothesis" was described by Sette (78). Sette began egg and larval surveys off New England in 1926. By 1932 he had developed the necessary quantitative techniques to estimate larval growth and mortality rates and describe egg and larval drift. The 1932 year class turned out to be poor. Mortality was high (10-14% per day) throughout the egg and larval stages, but not especially high for the small yolk sac and first-feeding larvae. The highest mortality appeared to have been connected with the transition from larval to post-larval stages. Sette attributed this to the particular drift pattern caused by dominant winds, which left the larvae further from their nursery grounds along the southern New England Coast than usual. Unfortunately, Sette was unable to continue his study and compare the 1932 year class with more successful year classes that followed. The research vessel ... "ALBATROSS II was withdrawn from service as a government economy measure."

Lasker (49-51) qualitatively compared year-class strength to prey density for three year classes (1975, 1976, 1978) of northern anchovy of the California current area. Larval anchovy required a "threshold" concentration of prey particles in order for them to feed and survive.

This concentration is substantially higher than the average, and Lasker hypothesized that necessary concentrations occur in small-scale layers that form in inshore waters during periods of water column stability. He observed that these layers could be dispersed by storms and by intense upwelling. These observations led Lasker (51) to hypothesize that anchovy year-class strength depended on the stability of the environment. A complicating factor in 1975 was that the dominant food particle in the environment of larval anchovy was a dinoflagellate which is nutritionally inadequate (77). The 1976 spawning season was characterized by stable conditions.

Consistent with Lasker's "stability" hypothesis, the 1975 year class was poor, that of 1976 was outstanding. The 1978 year class was predicted by Lasker (50) to be of intermediate strength due to the onset of stable conditions and food aggregations during the latter part of the spawning season. Methot (65) showed that juveniles of the 1978 year class were, in fact, survivors from the latter half of the spawning season. He used daily growth rings of otoliths to determine birth date. In 1979, the birth date distribution of juveniles corresponded more closely with the seasonal distribution of larval production (Fig. 8). In fact, the 1978 year class seems to be one of the largest ever (64). In general, water column stability has not proved to be a good predictor of year-class strength.

Lasker's investigation of the relationship between anchovy year-class strength and larval prey density is supplemented by a simulation model. Vlymen's (95) model showed the importance of the microstructure of larval prey. He treated the spatial distribution of prey as negative binomial. His simulation indicated that extremely different larval growth rates could occur at the same average concentration of prey as a function

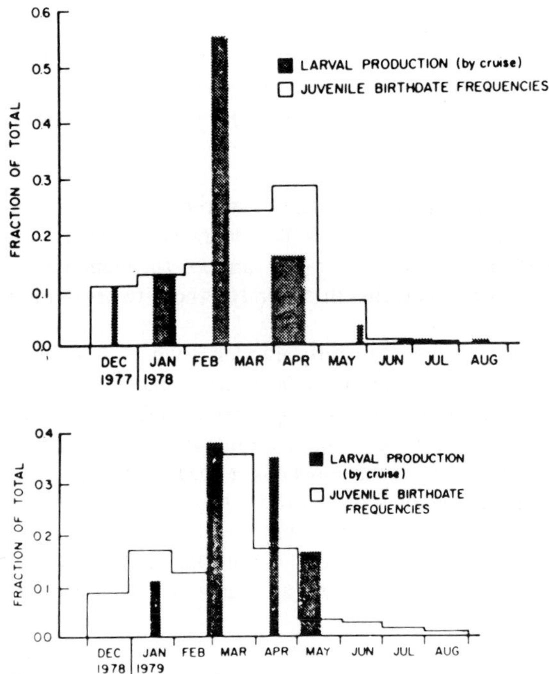


FIG. 8 - Birth date frequencies of juvenile anchovies compared to the larval production. Hatched bars indicate each of seven cruises (after (65)). Top, December 1977-August 1978; Bottom, December 1978-August 1979.

of contagion. The highest growth rate was at an intermediate level of contagion.

Lasker and Zweifel (53) modified Vlymen's model to simulated growth and survival of first-feeding larvae on different prey sizes and concentrations. Their simulation was based on laboratory and field data. They showed that at nominal capture efficiency of 20-30%, there is a threshold of 30 to 50 small particles (45-50 microns in diameter) per ml necessary for substantial survival and growth. This threshold concentration is found in inshore areas during periods of environmental stability.

The relationship between larval survival and prey concentration and distribution has also been the focus of process-oriented research concerning coastal waters off the northeastern USA. Georges Bank haddock and cod are the target species. Laurence (54, 55) determined their laboratory growth, respiration, and survival rates in relationship to prey concentration and temperature. Laurence (56) and Lough (59) described a sampling strategy for determining fine to microscale vertical distribution of fish larvae. Their initial results are pertinent to Lasker's stability hypothesis. In 1981, they observed the vertical distribution of cod and haddock larvae and copepod prey on Georges Bank, before and after a storm (Figs. 9-10). Before the storm, a thermocline was established and fish larvae and copepods were found in greatest density at about the thermocline depth. After the storm, the water column was well mixed (isothermal) and copepods were dispersed throughout the water column. The peak density of larvae was at the mid-depth. The implication is that the concentration of prey in the immediate environment of the fish larvae was nearly an order of magnitude lower following the storm. The actual impact of the storm on larval growth and survival is unknown.

There have been a series of stochastic larval growth simulation models (7, 8, 57) used to evaluate the importance of prey concentration. The most recent version of the model treats the following as stochastic: a) prey encounter frequency conditional on prey concentration (poisson distribution), b) prey size frequency (poisson distribution), c) larval size frequency at hatching (normal distribution), and d) daily variation in prey concentration (uniform distribution). Individual larvae are simulated and their growth and survival is determined based on their stochastic feeding rate compared to laboratory-determined energy requirements.

Laurence (56, 57) concludes that haddock require a mean prey concentration of 50 per liter for 50% survival through their 42-day larval period,

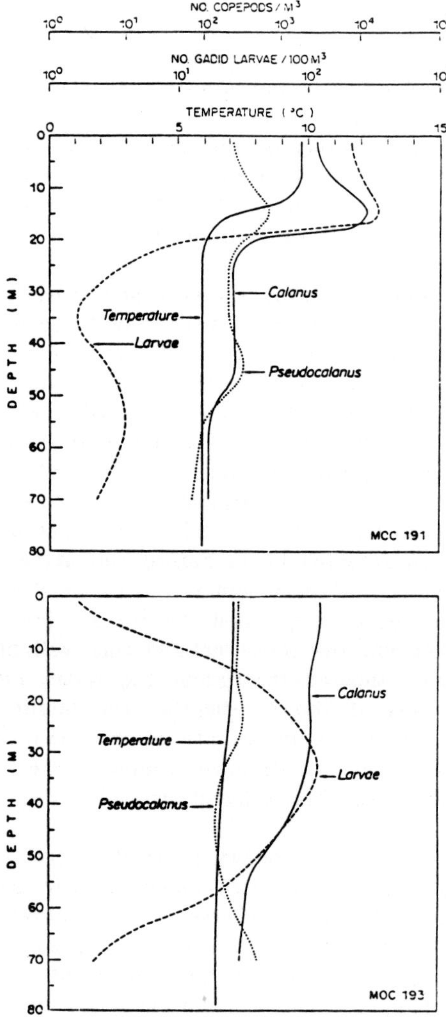


FIGS. 9-10 - Vertical distributions for Georges Bank before (21 May 1981, top) and after (24 May 1981, bottom) storm (59).

while cod larvae require about 20 per liter for 50% survival. Lough (59) reports prey concentrations of 10-65 per liter on Georges Bank in the vicinity of larval cod and haddock. Laurence (57) concludes that, although starvation mortality is important, it does not appear to be the single controlling mortality factor under normal ranges of prey density.

This conclusion is supplemented by RNA/DNA ratios of field-collected larval fish. Buckley (10) showed that the RNA/DNA ratio increases with growth rate of laboratory-reared larval fish. Since then, the Northeast Fisheries Center (National Marine Fisheries Service, Woods Hole, Massachusetts, USA) has monitored RNA/DNA ratios of larval fish collected in the field. To date, these studies indicate that most field-collected larval fish have a high growth rate and are not on the verge of starvation.

What about the effects of larvae on their prey? Laurence (56, 57) and Cushing (22) address this question. Laurence concludes that larvae are too dilute to affect the concentration of their prey. Cushing bases his analysis on several of Laurence's results and confirms that at least early stage larvae are too dilute to affect the concentration of their prey, although this may not be the case as larvae approach metamorphosis.

The effect of larvae on their prey is an important aspect of the recruitment problem. If they have no effect, then starvation mortality is density-independent since the density of prey does not change in response to larval density. Thus, even if fluctuations in recruitment are a result of varying amounts of starvation mortality, additional processes are necessary if the existence of a compensatory stock-recruitment relationship is to be confirmed.

Cannibalism is a source of density-dependent mortality (72). Hunter and Kimbrell (43) estimate that cannibalism by adults accounts for 32% of the daily egg mortality of northern anchovy. By comparing the number of eggs in anchovy stomachs and the density in the same area, they estimate that the consumption rate of eggs increases as the 1.6 power of egg concentration. One hypothesis is that at low concentration eggs are consumed incidentally during filter feeding, whereas at higher concentrations eggs contribute to stimulating feeding (63).

MacCall (63) and Lasker and MacCall (52) considered density-dependent effects of cannibalism for a population which expands and contracts its range with changes in abundance. This behavior is clearly demonstrated by the distribution and abundance of anchovy larvae off southern California. MacCall (63) developed a model that indicated that this geographic behavior is a consequence of density-dependent habitat selection under the following assumptions: a) spawning habitat is most favorable near the center of the range, b) local spawning habitat becomes less favorable as spawning intensity increases, and c) fish individually attempt to spawn in the most favorable location. Cannibalism is one mechanism that could make local spawning habitat less favorable as spawning intensity increases, as in assumption b.

An important consequence of the geographic expansion-cannibalism model is that the stock-recruitment relationship will be less dome-shaped than would result from cannibalism alone. This conclusion is consistent with stock-recruitment data for the California sardine and anchovy (61, 62). It is also consistent with Csirke's ((19), described earlier in this paper) stock-recruitment model of Peruvian anchovy, which took account of density of spawners.

Hunter (42) considers, in general, the importance of predators as a cause of egg and larval mortality. He rejects the hypothesis that starvation is the largest cause of mortality. He cites evidence that growth rates "in situ" are about the same as growth rates when larvae are fed a high ration in laboratories. He cites histological studies that indicate that the starvation rate, based on the frequency of detection of starving larvae, is much lower than observed mortality rates. He notes that high mortality rates of eggs and yolk sac larvae could not result from starvation and concludes that predation is the major cause of mortality.

There are numerous potential invertebrate and vertebrate predators of fish eggs and larvae. Hunter (42) speculates that pelagic fish are the most important predator. There are numerous examples of fish predation on eggs and larvae (e.g., (37, 43)). One reason for hypothesizing that fish are the most important predator is that egg and larval mortality remains high over a broad size range. Fish are capable of consuming prey over a broader size range than are small invertebrate predators. Hunter notes that a serious problem that limits research on predation is that eggs and particularly larvae are digested beyond recognition in the gut of fish within a short period (about an hour).

Results of enclosure studies support Hunter. Øiestad (67) conducted experiments with cod, herring, capelin, and plaice larvae in enclosures in Norway and found a very high survival potential to metamorphosis in systems without predators (e.g., 70% for herring, 50% for cod), even at marginal feeding conditions where growth rate was reduced. Survival was markedly reduced when predators were introduced to the system. He notes that post-larval fish can be important predators of larval fish of their own and other species.

It should now be apparent that the recruitment problem is extremely complex. Recruitment is probably influenced by both starvation and predation, and the modifying effects of physical factors. This was the conclusion of a meeting of experts during 1982 (i.e., Fish Ecology III (73)). In fact, several of the important (42, 56) works cited above were presented during Fish Ecology III. The meeting represents a turning point in the scientific attack on the recruitment problem. Several plans for future research were developed.

Fish Ecology III gave little consideration to processes that occur during the post-larval (but prerecruit) stage. I next review a more holistic approach which reflects on the importance of the post-larval stages.

## A HOLISTIC APPROACH

The holistic approach has been used to examine the roles of starvation and predation for Georges Bank. Georges Bank and the surrounding region have been the focus of intensive ecological studies (31, 33, 85). The results of these investigations have been summarized in the form of an energy budget of Georges Bank (86). The energy budget is most certain for primary productivity and fish production. Estimates of primary productivity are based on a three-year study with samples collected throughout the year (68). Estimates of fish abundance, production, and consumption are based on extensive bottom trawl surveys and fisheries statistics. While estimates of other components (e.g., macrozooplankton) are less certain, conclusions based on the entire energy budget are generally robust. This robustness results from the bounding effect of the more precise information at the lowest (primary productivity) and the higher (fish) trophic levels.

Average biomass, annual consumption, and annual production estimates of finfish of Georges Bank for the periods 1964-1966 and 1973-1975 were calculated by Grosslein et al. (34). The former period was one of increasing fishing pressure from distant-water-fleets in response to high fish abundance. The latter period was one of low abundance, following a decade of excessive fishing pressure. Pre-exploitable fish (individuals which were either too small or young to be captured by commercial or research vessel bottom trawl survey gear) are not represented in these traditional estimates of abundance. Nevertheless, they are important components of the ecosystem.

Relatively little is known about the population dynamics of these small fish, particularly after the late larval stage and before they grow large enough to be captured in trawls. However, some valuable information is available. The initial number and biomass of the cohort of pre-exploitable fish can be estimated from the abundance of adults, the proportion of the total adult production used for reproduction, and the average size of an egg. As the young fish reach exploitable size, their number and biomass can be determined for traditional stock assessments based on trawl surveys and fisheries statistics. With these beginning and end points known for the pre-exploitable fish, a simple model can be used to calculate estimates of average biomass, production, and consumption. The model assumes that growth of individual fish and of the entire cohort of pre-exploitable fish is exponential. While pre-exploitable fish are only 10% of the biomass of exploitable fish, their consumption is nearly as great and their production is two and a half times as high (86).

Extensive stomach content investigations indicate that fish consume macrozooplankton (e.g., by herring, mackerel, and redfish), benthos (e.g., haddock and flounder), and fish (e.g., by cod and silver hake). Sissenwine et al. (86) noted that consumption of fish by silver hake and cod accounted for 40 to 50% of the total consumption by the demersal component of the fish community. For the purposes of this discussion, we assume that approximately 50% of demersal consumption is fish.

The Georges Bank energy budget is summarized in Table 1. Estimates of particulate phytoplankton, zooplankton, and benthic production are from Cohen and Grosslein (16).

Fish production was 1.2% to 2.1% of particulate primary productivity. Considering the complexity and number of trophic levels of the food web, trophic efficiency must be high relative to traditional thinking (10% (88)). One implication of the result is that the energy budget is "tight," and fish production is ultimately limited by their food resource. In fact, Table 1 indicates that fish consume from 30 to 50% of the production of suitable prey types. This is remarkably high considering that microzooplankton (about 60% of the zooplankton production) is a suitable food for only a brief period during the life cycle of fish, and that consumers

TABLE 1 - Components of Georges Bank energy budget. Based on Sissenwine et al. (86), Powers (71), Scott et al. (76), Cohen and Grosslein (16), average catch for 1968-1982. Range of values is for 1973-1975 (low) and 1964-1966 (high).

| | Kcal/m$^2$yr | |
|---|---|---|
| Production | | |
| Phytoplankton (particulate) | 3780 | |
| Macrozooplankton | 496 | |
| Benthos | 106 | |
| Fish (exploitable) | 13-17 | |
| (pre-exploitable) | 29-52 | |
| (total) | 42-69 | (1.3-2.1% of Phytoplankton Production) |
| Potential Fish Prey | 644-671 | |
| Fish Consumption | | |
| All Prey | 197-344 | (31-50% of Potential) |
| Of Fish | 39-42 | (73-61% of Their Own Production) |
| Consumption of Fish | | |
| By Fish | 39-42 | |
| By Birds | 2.0 | |
| By Mammals | 5.4 | |
| By Large Pelagics | 2.0 | |
| By Humans | 6.1 | |
| Total | 54.6-57.6 | (130 and 83% of Fish Production, Respectively) |

other than finfish and squid are dependent on the components of the ecosystem which have been labeled in Table 1 as potential fish prey.

Table 1 also indicates that fish consume most of their own production (61-93%). Other consumers are marine mammals, birds, large pelagic migratory fish (e.g., sharks), and humans. Estimates of their consumption are also included in Table 1. The estimated total consumption of fish ranges from 83 to 130% of production for the two periods considered. The deviations of these estimates from 100% are considered within the level of precision of available data.

Although this shows that predation is the major cause of mortality of fish in terms of biomass, it does not necessarily follow that most fish (in terms of numbers) are victims of predation. Most mortality occurs during the egg and larval stage, but these stages account for relatively little biomass. With respect to starvation, there is generally enough food for the fish community in aggregate, but food probably limits production of some species or life stages, particularly when fish biomass was higher prior to overfishing. Density-dependent growth (as observed in juveniles and adult fish) seeems likely.

It is noteworthy that most of the predation of fish was by silver hake. The primary prey of silver hake are post-larval, pre-exploitable fish. In fact, silver hake consume most of the production of the pre-exploitable component. The implication is that predation must cause a high post-larval mortality. This predation probably affects recruitment. If post-larval mortality is high, then there exists a potential that year-class strength is not established until this life stage since only small variations in mortality would be necessary to account for large changes in recruitment.

There is clearly a biological basis for expecting compensation in production by the Georges Bank finfish community. They modify their own abundance by predation (cannibalism at the community level). It seems likely that this mechanism operates at the species as well as community level. Some species are cannibalistic (e.g., silver hake); furthermore, an abundance of fish as prey may enhance production of predators, ultimately resulting in compensation.

There is also the potential for depensation. Sissenwine et al. (87) speculate that the marine mammals contributed to the demise of herring on Georges Bank, although overfishing was certainly the primary cause. The estimated biomass of marine mammals in the range of the Georges Bank herring is sufficient to have consumed the entire population that was estimated to remain after the last commercial harvest in late 1977. There has been virtually no evidence of the population of herring since that time.

Cohen and Grosslein (16) compared Georges Bank to other Continental Shelf ecosystems. These comparisons indicate that Georges Bank is not unique in its efficient conversion of primary productivity to fish production. For example, fish production is about 1% of particulate primary productivity for the North Sea and about 3% for the East Bering Sea. These ecosystems are also "tight" like Georges Bank, and probably food limits their overall production as well. Furthermore, North Sea (24) and East Bering Sea (58) fish also consume a significant proportion of their own production.

**DISCUSSION**

The focus of this paper has been variability. While variability is generally the case, recruitment of some populations is relatively stable (e.g., North Sea plaice (94)). Ursin (94) hypothesized that this stability results from space limitation during the post-larval stage. Competition for space is, potentially, a strong compensatory mechanism that will stabilize recruitment. The mechanism appears to be particularly important for species depending on coastal areas where space is measured in a linear scale.

It is clear that there should be more emphasis on predation mortality of early life stages, including post-larvae. It is ironic that post-larvae have been nearly ignored since there is an obvious clue that year-class strength may not be established until this stage. It is widely recognized that year-class strength is not correlated with larval abundance (e.g., (4, 31, 41, 74, 75, 89)). For this reason, stock assessment working groups of the International Council for the Exploration of the Sea (ICES) use egg and larval surveys to back-calculate spawning biomass of certain species (e.g., (18)) and use young fish (primarily age one) surveys to predict recruitment (e.g., (17)). Apparently, year-class strength is established somewhere between. Saville and Schnack (75) argue that the lack of correlation between larval abundance and year-class strength probably indicates a problem of quantitative sampling, not the time at which year-class strength is established. If this were the case, then larval abundance should not correlate with spawning stock biomass.

One reason why the egg and larval stages have received so much emphasis is that mortality rates (M) are very high, and presumably only a small change is necessary to account for large changes in year-class strength. But year-class strength is determined by the product of M and the duration of the time period (t) during which M applies (assuming the initial number is constant). It is more logical to compare Mt between life stages, in order to evaluate the sensitivity of year-class strength to prerecruit mortality, than it is to compare M alone. Sissenwine et al. (86) compared egg and early larval Mt to late larval and post-larval Mt for several species of Georges Bank. They found that the latter was always higher, but since they did not examine size composition data, the distinction between early and late stage larvae is unclear.

Mt for the entire prerecruit stage is $-\log_e(R/E)$, where R is the number of recruits and E is the number of eggs spawned. For values of R and E reported in Sissenwine et al. (86), Mt is 12.6 and 13.5 for prerecruit herring and haddock, respectively. Lough et al. (60) applied catch-curve analysis to larval herring length frequencies from Georges Bank. By comparing day and night catches, they were able to take account of the bias that would result from an increase in net avoidance with length. Using the catch-curve analysis and larval growth data, Lough et al. estimated that M ranged from 0.022 to 0.047/day for the first 150 days of life (to about 30 mm length) for 1971-1978. This amounts to an egg and larval mortality of about 5.2 (0.35 x 150). By subtraction, the post-larval, prerecruit Mt is estimated as 7.4.

Unpublished larval length frequencies and larval growth data for Georges Bank haddock indicate a mortality rate of 0.094 to 0.131/day for a 50-day period for 1977-1982. The average was 0.108/day (Morse, personal communication, National Marine Fisheries Service, Sandy Hook, New Jersey, USA). Assuming that pelagic eggs suffer the same mortality rate for 14 days, Mt for eggs and larvae is 6.9. Therefore, the post-larval Mt is 6.6. In the case of haddock, larval mortality rate was certainly overestimated since no correction was made for an increase in net avoidance with length.

Thus, post-larval mortality is at least comparable to egg and larval mortality for two of the principal species on Georges Bank. This is not surprising in light of the energy budget evidence that most production of post-larval fish is consumed by larger fish. We suspect that Georges Bank is not unique. If predation mortality of post-larval fish has a major role in the determination of year-class strength, then why is year-class strength so variable? The number of recruited fish that are predators of post-larval fish is certainly less variable than recruitment. Fogarty et al. (31) noted that although year-class strength is highly variable, it is auto-correlated for several species of Georges Bank. This indicates that the factor(s) controlling year-class strength has a tendency to persist between years. This would be characteristic of predation by recruited fish.

Although the number of fish predators of post-larvae may not be very variable (in a relative sense), their diet composition may be. For example, the percentage (by weight) of fish in the stomachs of silver hake collected on Georges Bank during spring 1973-1979 was 23.7, 57.9, 88.2, and 68.6%, respectively (9). Variability in diet composition could have a significant effect on post-larval mortality, and predator switching could be a compensatory mechanism.

It is almost axiomatic that variability in recruitment is caused by fluctuations in physical factors. In spite of the pitfalls of empirical studies, there are enough correlations to indicate that physical factors are certainly important. There are numerous possible mechanisms associated with physical factors that could influence starvation of early life stage larvae. It is less apparent how predation, particularly on post-larval fish, is influenced by the physical environment, although growth rate is affected by environmental factors such as temperature, and this in turn affects vulnerability to predation.

Future hypotheses about the effects of physical factors on recruitment should be compatible with certain statistical characteristics of the frequency distribution of year-class strength. The distribution is usually skewed towards poor recruitment (39). The implication is that research should focus on physical factors that explain occasional "bonanzas," not "disasters."

Recruitment is frequently auto-correlated. As noted above, this could be associated with the effects of predation by recruited fish or auto-correlation of spawning potential. If it is associated with a physical variable, then the physical variable must also be auto-correlated. The average of physical variables is more likely to be auto-correlated the longer the period included in the average (e.g., annual average temperature is more likely to be auto-correlated than average temperature of a specific day of the year). There is a coherence in recruitment between stocks over broad geographic areas (32, 48). Therefore, if a physical factor determines recruitment, it must occur over a broad area.

There are numerous reasons why fish populations vary. Predation plays a much greater role than had been hypothesized until recently. As a result of predation by fish on their own and other species, and a limited amount of food potentially available to them, fish populations are able to compensate for some fishing pressure. Recruitment is likely to be a multiplicative function of highly variable processes occurring throughout the first year of life, including the post-larval stage. The more we learn about it, the better able we are to ask the right questions in the future.

## REFERENCES

(1) Austin, H.M., and Ingham, M.C. 1979. Use of environmental data in the prediction of marine fisheries abundance. In Climate and Fisheries, pp. 93-118. Kingston, RI: University of Rhode Island.

(2) Bakun, A. 1973. Coastal upwelling indices, west coast of North America, 1946-71. NOAA Tech. Rept. NMFS SSRF-671. Washington, D.C.: U.S. Dept. of Commerce.

(3) Bakun, A., and Nelson, C.S. 1977. Climatology of upwelling related processes off Baja, California. CA Coop. Oceanic Fish. Invest. Rept. 19: 107-127.

(4) Bakun, A., and Parrish, R. 1980. Environmental inputs to fishery population models for eastern boundary current. In Workshop on the Effects of Environmental Variations on the Survival of Larval Pelagic Fishes, ed. G.D. Sharp, pp. 67-104. Intergovernmental Oceanographic Commission, UNESCO.

(5) Beverton, R.J.H., and Holt, S.J. 1957. On the dynamics of exploited fish populations. UK Min. Agric. Fish., Fish. Invest. (Ser. 2) 19.

(6) Beverton, R.J.H., and Lee, A.J. 1965. Hydrographic fluctuations in the North Atlantic Ocean and some biological consequence. In The Biological Significance of Climatic Changes in Britain, eds. C.G. Johnson and L.P. Smith, pp. 79-107. Symposia No. 14. London: Institute of Biology.

(7) Beyer, J., and Laurence, G.C. 1980. A stochastic model of larval fish growth. Ecol. Mod. 8: 109-132.

(8) Beyer, J., and Laurence, G.C. 1981. Aspects of stochasticity in modelling growth and survival of clupeoid fish larvae. Rapp. P.-v. Reun. Cons. Int. Explor. Mer 178: 17-23.

(9) Bowman, R.E. 1980. Silver hake's regulatory influence on the fishes of the Northwest Atlantic. NMFS, NEFC Woods Hole Lab. Ref. Doc. 80-05.

(10) Buckley, L.J. 1979. Relationships between RNA-DNA ratio, prey density and growth rate in Atlantic cod (Gadus morhua) larvae. J. Fish. Res. Board Can. 36: 1497-1502.

(11) Burreson, E. 1981. Effects of mortality caused by the hemoflagellate (Trypanoplasma bullocki) on summer flounder populations in the Middle Atlantic Bight. Int. Cons. Explor. Sea, C.M. 1981/G: 61.

(12) Carruthers, J.N. 1951. An attitude on "fishery hydrography." J. Mar. Res. 10(1): 101-118.

(13) Chase, J. 1955. Winds and temperatures in relation to the brood strength of Georges Bank haddock. J. Cons. Perm. Int. Explor. Mer

21: 17-24.

(14) Clark, F.N., and Marr, J.C. 1955. Population dynamics of the Pacific sardine. CA Coop. Ocean Fish. Invest. Progr. Rept.: 11-48.

(15) Clark, S.H.; Overholtz, W.J.; and Hennemuth, R.C. 1982. Review and assessment of the Georges Bank and Gulf of Maine haddock fishery. J. Northw. Atl. Fish. Sci. 3: 1-27.

(16) Cohen, E.B., and Grosslein, M.D. 1984. Total ecosystem production on Georges Bank: comparison with other marine ecosystems. In Georges Bank, ed. R. Backus. MIT Press, in press.

(17) Corten, A. 1980a. Report of ICES young fish survey, 1980: herring data. Int. Cons. Explor. Sea, C.M. 1980/H: 35.

(18) Corten, A. 1980b. The use of larval abundance data for estimating the stock size of North Sea herring (II). Int. Cons. Explor. Sea, C.M. 1980/H: 37.

(19) Csirke, J. 1980. Recruitment in the Peruvian anchovy and its dependence on the adult population. Rapp. P.-v. Reun. Cons. Int. Explor. Mer 177: 307-313.

(20) Cushing, D.H. 1973. Recruitment and parent stock in fishes. University of Washington, Seattle.

(21) Cushing, D.H. 1973. The dependence of recruitment on parent stock in different groups of fishes. J. Cons. Int. Explor. Mer 33(3): 340-362.

(22) Cushing, D.H. 1983. Are fish larvae too dilute to affect the density of their food organisms? J. Plankton Res. 5(6): 847-854.

(23) Cushing, D.H., and Dickson, R.R. 1976. The biological response in the sea to climatic changes. In Advances in Marine Biology, eds. F.S. Russell and M. Younge, vol. 14. New York: Academic Press.

(24) Daan, N. 1983. The ICES stomach sampling project in 1981: aims, outline and some results. Northw. Atl. Fish. Organ. Sci. Res. Doc. 83/IX/93.

(25) Depres-Patanjo, L.; Ziskowski, J.; and Murchelano, R.A. 1982. Distribution of fish diseases monitored on stock assessment cruises in the Western North Atlantic. Int. Cons. Explor. Sea, C.M. 1982/E: 30.

(26) Deriso, R.B. 1983. Management of North Pacific halibut fishery. II. Stock assessment and new evidence of density dependence. Inter. Pac. Halibut. Comm., unpublished.

(27) Dow, R.L. 1964. A comparison among selected species of an association between sea water temperature and relative abundance. J. Cons. Perm. Int. Explor. Mer 28: 425-431.

(28) Dow, R.L. 1969. Cyclic and geographic trends in sea-water temperature and abundance of American lobster. Science 146: 1060-1063.

(29) Dow, R.L. 1977. Effects of climatic cycles on the relative abundance and availability of commercial marine and estuarine species. J. Cons. Perm. Int. Explor. Mer 37(3): 274-280.

(30) Flowers, J.M., and Saila, S.B. 1972. An analysis of temperature effects on the inshore lobster fishery. J. Fish. Res. Board Can. 29: 1221-1225.

(31) Fogarty, M.J.; Sissenwine, M.P.; and Grosslein, M.D. 1984. Fisheries research on Georges Bank. In Georges Bank, ed. R. Backus. MIT Press, in press.

(32) Garrod, D.J., and Colebrook, J.M. 1978. Biological effects of variability in North Atlantic Ocean. Rapp. P.-v. Reun. Cons. Int. Explor. Mer 173: 128-144.

(33) Grosslein, M.D.; Brown, B.E.; and Hennemuth, R.C. 1979. Research, assessment, and management of a marine ecosystem in the Northwest Atlantic: a case study. In Environmental Biomonitoring, Assessment, Prediction, and Management - Certain Case Studies and Related Quantitative issues, eds. J. Cairns, Jr., P. Patil, and W.E. Waters, pp. 289-357. Fairland, MD: International Co-operative Publishing House.

(34) Grosslein, M.D.; Langton, R.W.; and Sissenwine, M.P. 1980. Recent fluctuations in pelagic fish stocks of the Northwest Atlantic, Georges Bank region, in relationship to species interaction. Rapp. P.-v. Reun. Cons. Int. Explor. Mer 177: 374-404.

(35) Gulland, J.A. 1952. Correlations on fisheries hydrography. Letters to the editor. J. Cons. Perm. Int. Explor. Mer 18: 351-353.

(36) Gunter, G., and Edwards, J.C. 1967. The relation of rainfall and freshwater drainage to the production of the Penaeid shrimps (Penaeus anviatillis and P. aztecus) in Texas and Louisiana waters. Proceedings of the World Science Conference on Biology and Culture of Shrimps

and Prawns. FAO Fish Repts. 57(3): 875-892.

(37) Harding, D.; Nichols, J.H.; and Tungate, D.S. 1978. The spawning of plaice (Pleuronectes platessa L.) in the southern North Sea and English Channel. Rapp. P.-v. Reun. Cons. Int. Explor. Mer 172: 102-113.

(38) Hayman, R.A. 1977. The relationship between environmental fluctuation and year-class strength of Dover sole (Microstomus pacificus) and English sole (Parophrys vetulus) in the fishery off the Columbia River. Master's Thesis, Oregon State University, Corvallis, OR.

(39) Hennemuth, R.C.; Palmer, J.E.; and Brown, B.E. 1980. A statistical description of recruitment in eighteen selected fish stocks. J. Northw. Atl. Fish. Sci. 1: 101-111.

(40) Hjort, J. 1914. Fluctuations in the great fisheries of northern Europe viewed in the light of biological research. Rapp. P.-v. Reun. Cons. Perm. Int. Explor. Mer 20: 1-228.

(41) Hunter, J.R. 1981. Feeding ecology and predation of marine fish larvae. In Marine Fish Larvae, ed. R. Lasker, pp. 33-79. Seattle: University of Washington press.

(42) Hunter, J.R. 1982. Predation and recruitment. Fish Ecology, III CIMAS, University of Miami, September 7-10, 1982.

(43) Hunter, J.R., and Kimbrell, C.A. 1980. Egg cannibalism in the northern anchovy, Engraulis mordax. Fish. Bull. (USA) 78: 811-816.

(44) ICES. 1983a. Report of the Arctic Fisheries Working Group. Int. Cons. Explor. Sea, C.M. 1983/Assess: 2.

(45) ICES. 1983b. Report of the Working Group on Assessment of Pelagic Stocks in the Baltic. Int. Cons. Explor. Sea, C.M. 1983/Assess: 13.

(46) Jones, R. 1979. Relationship between mean length and year-class strength in North Sea haddock. Int. Cons. Explor. Sea, C.M. 1979/G: 45.

(47) Jones, R., and Martin, J.H.A. 1981. The relationship between demersal fish landings and bottom temperature. Int. Cons. Explor. Sea, C.M. 1981/G: 44.

(48) Koslow, A.J. 1984. Recruitment patterns in Northwest Atlantic fish stocks. Can. J. Fish. Aquat. Sci., in press.

(49) Lasker, R. 1975. Field criteria for survival of anchovy larvae: the relation between inshore chlorophyll layers and successful first feeding. Fish. Bull. (USA) 73: 453-462.

(50) Lasker, R. 1978. Ocean variability and its biological effects - regional review - Northeast Pacific. Rapp. P.-v. Reun. Cons. Int. Explor. Mer 173: 168-181.

(51) Lasker, R. 1981. Factors contributing to variable recruitment of the northern anchovy (Engraulis mordax) in the California current: contrasting years, 1975 through 1978. Rapp. P.-v. Reun. Cons. Int. Explor. Mer 178: 375-388.

(52) Lasker, R., and MacCall, A. 1982. New ideas on the fluctuations of the clupeoid stocks off California. In CNC/SCOR Proceedings of Joint Oceanographic Assembly 1982 - General Symposia, pp. 110-120. Ottawa, Ont.: Canadian National Committee/Scientific Committee on Ocean Research.

(53) Lasker, R., and Zweifel, J.R. 1978. Growth and survival of first-feeding northern anchovy larvae (Engraulis mordax) in patches containing different proportions of large and small prey. In Spatial Pattern in Plankton Communities, ed. J.H. Steele, pp. 329-354. New York: Plenum.

(54) Laurence, G.C. 1974. Growth and survival of haddock (Melanogrammus aeglefinus) larvae in relation to plankton prey concentration. J. Fish. Res. Board Can. 31: 1415-1419.

(55) Laurence, G.C. 1978. Comparative growth, respiration and delayed feeding abilities of larval cod (Gadus morhua) and haddock (Melanogrammus aeglefinus) as influenced by temperature during laboratory studies. Mar. Biol. 50: 1-7.

(56) Laurence, G.C. 1982. Nutrition and trophodynamics of larval fish - review concepts, strategic recommendations and opinions. Fish Ecology, III CIMAS, University of Miami, September 7-10, 1982.

(57) Laurence, G.C. 1983. A report on the development of stochastic models of food limited growth and survival of cod and haddock larvae on Georges Bank. NOAA, NMFS, NEFC, Narragansett Lab., unpublished.

(58) Laevastu, T., and Larkins, H.A. 1981. Marine fisheries ecosystems. Farnham Survey, England: Fishing News Books Ltd.

(59) Lough, R.G. 1984. Larval fish trophodynamic studies on Georges

Bank: Sampling strategy and initial results. In E. The propagation of Cod, Gadus morhua L., Fløvigen rapportser, eds. E. Dahl, D.S. Danielssen, E. Moksness, and P. Solemdal, vol. 1, pp. 395-434. Arendal, Norway: Institute of Marine Research, Fløvigen Biological Station.

(60) Lough, R.G.; Bolz, G.R.; Pennington, M.R.; and Grosslein, M.D. 1980. Abundance and mortality estimates for sea herring (Clupea harengus L.) larvae spawned in the Georges Bank-Nantucket Shoals area, 1971-1978 seasons, in relation to spawning stock and recruitment. NAFO SCR Doc. 80/IX/129 (Revised): 1-59.

(61) MacCall, A. 1979. Population estimates for the waning years of the Pacific sardine fishery. CA Coop. Oceanic Fish. Invest. Rept. 20: 72-82.

(62) MacCall, A. 1980. Population models for the northern anchovy (Engraulis mordax). Rapp. P.-v. Reun. Cons. Int. Explor. Mer 177: 292-306.

(63) MacCall, A. 1980. The consequences of cannibalism in the stock-recruitment relationships of planktivorous pelagic fishes such as Engraulis. Intergov. Oceanogr. Comm. Workshop Rep. 28: 201-220.

(64) MacCall, A.D., and Methot, R. 1983. Historical spawning biomass estimates and population model in the 1983 anchovy Fishery Management Plan. U.S. National Marine Fisheries Service, SW Fisheries Center, Adm. Rept. LJ-83-17.

(65) Methot, R.D., Jr. 1981. Growth rates and age distribution of larval and juvenile northern anchovy, Engraulis mordax, with influences on larval survival. Ph.D. Dissertation, University of California, San Diego, CA.

(66) Nelson, W.M.; Ingham, M.; and Schaaf, W. 1977. Larval transport and year class-strength of Atlantic menhaden, Brevoortia tyrannus. US Fish. Bull. 75(1): 23-41.

(67) Øiestad, V. 1983. Predation on fish larvae as a regulatory force illustrated in enclosure experiments with large groups of larvae. Northw. Atl. Fish. Organ. Sci. Res. Doc. 83/IX/73.

(68) O'Reilly, J.E., and Busch, D.A. 1984. The annual cycle of phytoplankton primary production (netplankton, nannoplankton and release of dissolved organic matter) for the Northwestern Atlantic Shelf (Middle Atlantic Bight, Georges Bank and Gulf of Maine). Rapp.

P.-v. Reun. Cons. Int. Explor. Mer 183, in press.

(69) Parrish, R.H., and MacCall, A.D. 1978. Climatic variations and exploitation in the Pacific mackerel fishery. CA Dept. Fish. Game, Fish. Bull. 167.

(70) Pauly, D. 1981. On the interrelationship between natural mortality, growth parameters, mean environmental temperature in 175 fish stocks. J. Cons. Perm. Int. Explor. Mer 39(2): 175-192.

(71) Powers, K.D. 1984. Estimates of consumption by seabirds on Georges Bank. In Georges Bank, ed. R. Backus. MIT Press.

(72) Ricker, W.E. 1954. Stock and recruitment. J. Fish. Res. Board Can. 11: 559-623.

(73) Rothschild, B.J., and Rooth, C. 1982. Fish ecology III. University of Miami Tech. Rept. No. 82008.

(74) Saville, A. 1977. Survey methods of appraising fishery resources. FAO Fish. Tech. Paper 171: 76.

(75) Saville, A., and Schnack, D. 1981. Overview - some thoughts on the current status of studies of fish eggs and larval distribution and abundance. Rapp. P.-v. Reun. Cons. Int. Explor. Mer 178: 153-157.

(76) Scott, G.P.; Kenney, R.D.; Thompson, T.J.; and Winn, H.E. 1983. Functional roles and ecological impacts of the cetacean community in the waters of the Northeastern U.S. continental shelf. Int. Cons. Explor. Sea, C.M. 1983/N: 12.

(77) Scura, E.D., and Jerde, C.W. 1977. Various species of phytoplankton as food for larval northern anchovy, Engraulis mordax, and relative nutritional value of the dinoflagellates Gymnodinium splendens and Gonyaulax polyedra. Fish. Bull. (USA) 75: 577-583.

(78) Sette, O.E. 1943. Biology of the Atlantic mackerel (Scomber scomberus) of North America. Fish. Bull. 50(38): 149-237. Fish and Wildlife Service of the U.S. Dept. of Interior.

(79) Shepherd, J.G. 1982. A versatile new stock-recruitment relationship for fisheries, and the construction of sustainable yield curves. J. Cons. Int. Explor. Mer 40(1): 67-75.

(80) Sindermann, C.J. 1963. Disease in marine populations. Trans. N. Am. Wildl. Conf. 28: 221-245.

(81) Sindermann, C.J. 1970. Principal Diseases of Marine Fish and Shellfish. New York: Academic Press.

(82) Sissenwine, M.P. 1974. Variability in recruitment and equilibrium catch of the southern New England yellowtail flounder fishery. J. Cons. Perm. Int. Explor. Mer 36: 15-26.

(83) Sissenwine, M.P. 1977. A compartmentalized simulation model of the southern New England yellowtail flounder (Limanda ferruginea) fishery. Fish. Bull. (USA) 73(3): 465-482.

(84) Sissenwine, M.P. 1984. The uncertain environment of fishery scientists and fishery managers. Mar. Resource Econ. 1(1): 1-29.

(85) Sissenwine, M.P.; Brown, B.E.; Grosslein, M.D.; and Hennemuth, R.C. 1984. The multispecies fisheries problem: a case study of Georges Bank. Lecture Notes Math. Biol., in press.

(86) Sissenwine, M.P.; Cohen, E.B.; and Grosslein, M.D. 1984. Structure of the Georges Bank ecosystem. Rapp. P.-v. Reun. Cons. Int. Explor. Mer 183: 243-254.

(87) Sissenwine, M.P.; Overholtz, W.J.; and Clark, S.H. 1984. In search of density dependence. In Proceedings of the Workshop on Marine Mammal Fishery. Interactions of the East Bering Sea, ed. B.R. Melteff. University of Alaska Sea Grant, Rept. 84-1: 119-139.

(88) Slobodkin, L.B. 1961. Growth and Regulation of Animal Populations. New York: Holt, Rinehart and Winston.

(89) Smith, P.E. 1978. Biological effects of ocean variability: time and space scales of biological response. Rapp. P.-v. Reun. Cons. Int. Explor. Mer 173: 117-127.

(90) Sutcliffe, W.H., Jr.; Drinkwater, K.; and Muir, B.S. 1977. Correlations of fish catch and environmental factors in the Gulf of Maine. J. Fish. Res. Board Can. 34: 19-30.

(91) Taylor, C.C.; Bigelow, H.B.; and Graham, H.W. 1957. Climate trends and the distribution of marine animals in New England. Fish. Bull. 57: 293-345.

(92) Templeman, W., and Fleming, A.M. 1953. Long term changes in the abundance of marine animals. Int. Comm. Northwest Atl. Fish. Annual Proc. 3: 78-86.

(93) Ulanowicz, R.E.; Ali, M.L.; Vivian, A.; Heinle, D.R.; Rickus, W.A.;

and Summers, J.K. 1982. Identifying climatic factors influencing commercial fish and shellfish landings in Maryland. Fish. Bull. (USA) 80(3): 611-619.

(94) Ursin, R. 1982. Stability and variability in the marine ecosystem. Dana 2: 51-67.

(95) Vlymen, W.M. 1977. A mathematical model of the relationship between larval anchovy (Engraulis mordax) growth, prey micro-distribution and larval behavior. Env. Biol. Fish. 2(3): 211-233.

(96) Walford, L. 1946. Correlation between fluctuations in abundance of the Pacific sardine (Sardinops caerulea) and salinity of the sea water. J. Mar. Res. 6(1): 48-53.

Exploitation of Marine Communities, ed. R.M. May, pp. 95-109. Dahlem Konferenzen 1984. Berlin, Heidelberg, New York, Tokyo: Springer-Verlag.

# The Availability and Information Content of Fisheries Data

J.G. Shepherd
Directorate of Fisheries Research
Fisheries Laboratory
Lowestoft, Suffolk NR33 OHT, England

**Abstract.** The availability of catch, age/size composition, effort, tagging, and survey data of various sorts is summarized according to the state of development and economic importance of the fishery. Fairly complete catch weight data are usually a prerequisite for any meaningful assessment or management. Data of all these types are available only in very rare cases, and even then the assessment of the stock may be ambiguous.

The information content of each type of data is reviewed. This is most often of a relative nature (e.g., current stock size relative to previous values). Absolute values can usually only be obtained by relatively complex analysis of several sorts of data simultaneously. Current absolute stock size in particular requires expensive data to be collected, and even then is difficult to determine.

The need for such information is, however, dependent on the type of management envisaged. Short-term status quo management can be carried out quite successfully, even without knowing absolute stock size at all accurately. Determination of appropriate long-term management strategies, however, requires much more information (especially of absolute stock sizes and the factors determining recruitment) which is difficult to obtain. Attempts to determine rational long-term management strategies in detail are therefore likely to be worthwhile only for stocks of major economic significance.

## INTRODUCTION

I have elsewhere likened the task of managing a fishery to that of rationally harvesting a forest in which the trees are all invisible and keep moving around (14). Various types of data can be collected to assist in this unenviable task, but in marine fisheries, at least, it is quite rare for the interpretation of these data to be unambiguous.

It is therefore necessary to consider rather carefully the information potentially available from various sorts of data which can be extracted using suitable interpretative techniques. This information would usually be expressed as estimates of the parameters entering into a model (or mathematical description) of the dynamics of the stock, such as fishing and natural mortality rates. We shall see that such estimates can rarely be derived from one type of data only. Usually it is necessary to utilize several types simultaneously in a combined interpretation.

However, it is also necessary to ask what information is necessary for the management task in hand. It has been suspected for some time, and is now becoming quite clear, that for certain management tasks, such as forecasting catches in the short-term, some information such as the level of natural mortality is not really very important at all (see section on INFORMATION REQUIREMENTS). On the other hand, if one seeks to estimate the likely effects of a certain level of exploitation in the long term, the natural mortality becomes very important.

There is thus in principle a correspondence between the management questions to be answered and the data required to do so, but this can only be elucidated by considering the information required by the questions and that obtainable from data of various types. The task of elucidation is quite difficult, and possibly controversial. I endeavor in this paper to set down my best understanding of the matter. I have not attempted to deal with the practical details of acquiring and processing the data, since these can be found in several standard texts and manuals (e.g., (5, 12)). Nor have I attempted to review the literature or provide complete citations for all the views expressed. To do so would require too much space and labor, both to write and to read, and in any case the investigations concerned are sometimes so recent that no relevant citation exists. I have, however, tried to set down as a basis for discussion a fairly wide-ranging set of opinions on both the information necessary for certain tasks and the information obtainable from certain data. These opinions are discussed in the text and summarized in tabular form. I hope that the synthesis of these may aid us as we seek to keep track of all those moving and invisible trees! It should be noted that I use "information" to mean something useful which can be extracted from data, and not in the technical sense of the measures proposed by (for example) Fisher or Shannon.

Before proceeding with the analysis of the information requirements of various sorts of assessment, it is useful to recall that these are generally required as a basis for management of the fisheries. The objectives of such management are not always clear - and sometimes (as stated) so diffuse and all-embracing as to be impossible of achievement. For the purposes of this discussion I have assumed that the objective as usually recognized in the UK apply. This is to manage the activities of the fishing fleets in such a way as to promote the maintenance of fish supplies at acceptable prices whilst allowing a reasonable return on the investments of capital and labor made by vessel owners and fishermen. Many elaborations of this are possible, but some key features should be noted. First, the emphasis is on maintenance of fish supplies, and therefore fishing fleets and fish stocks, for long periods of time. Second, biological factors are not paramount, but mingled with social and economic factors. Third, the objective is not easily quantifiable, since there is room for judgement over what may be considered acceptable and reasonable.

In practice fisheries tend to expand to the point where the returns or investments are inadequate, so that fishery management tends to become fishery regulation. In seeking to regulate the level and pattern of exploitation, it is usually necessary to have some idea of what level of exploitation would be appropriate in the long term to meet the objectives - at least whether the current level is too high or too low. Second, it is usually necessary to assess the level of catches or effort in the immediate future which would be appropriate to stabilize the level of exploitation, or to move towards the longer-term objective. Thus, ideally, management measures should be based on quantitative assessments of both the long-term and short-term prospects for the fisheries, together with analyses of the effects of various technical measures to control or modify the pattern of exploitation. This ideal is quite difficult to attain.

## INFORMATION REQUIREMENTS

The main purposes for which quantitative assessments of fish stocks are required are therefore: a) for short-term forecasts of catches and/or catch rates up to a few years ahead, b) for assessments of the likely long-term yields and catch rates at varying levels of exploitation, and c) for assessments of the effects of technical measures such as mesh size regulations.

For short-term forecasts it is usually possible to assume that the level of exploitation (expressed by measures such as fishing mortality or yield/biomass ratio, rather than catch in tonnes) does not change much from that of the recent past. Under these approximate "status quo" conditions it is now clear, especially from the work of Pope (8), that much of the information usually sought by assessment scientists is of little consequence. The principle determinants are the level of recent catches and the levels of imminent recruitment. This is confirmed by recent work of mine (15) using a much simpler methodology which also makes it clear that the current absolute stock size (roughly equivalent to the current fishing mortality) is not very important either. This is illustrated in Fig. 1 which shows for a real example how a variation of more than five times in the assumed yield/biomass ratio leads to only a variation of about 50% in the estimated status quo catch (which is the essence of a catch forecast (15)).

For the long-term assessments, on the other hand, the levels of recent catches and imminent recruitment are of little importance. The results

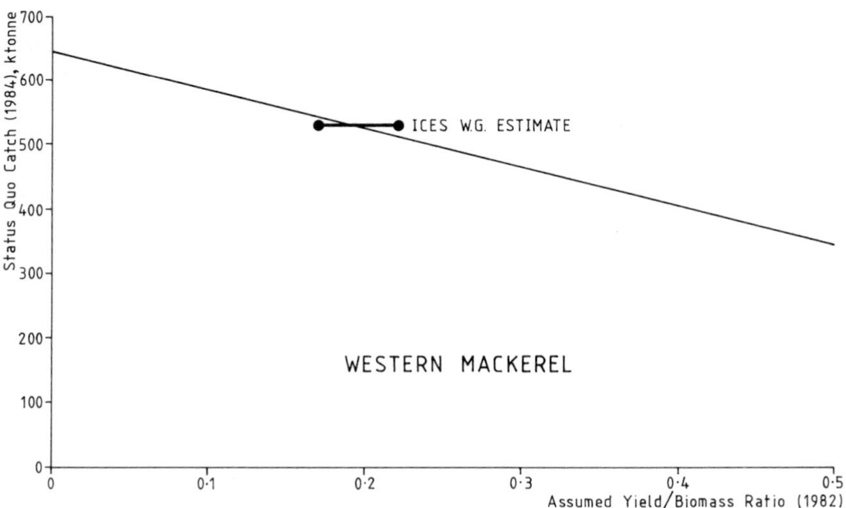


FIG. 1 - Example of the variation of the estimated status quo catch for a fish stock as a function of the assumed level of exploitation.

are, however, sensitive to the pattern of exploitation (i.e., fishing mortality as a function of age or size), to the level of natural mortality and its dependence on age or size, to the rate of growth (weight at age), and to the rates of discarding and discard survival. In particular, however, the traditionally highly valued pieces of information such as natural mortality and the factors determining recruitment (including stock-recruitment relationships) return to prominence. The importance of various sorts of information, as I see it, is summarized in Table 1. Indeed, the information requirements for the short and long term are almost orthogonal to one another. This means that, before seeking to gather data for fisheries management, it is important to spell out the goals. Much traditionally sought information may not be required if only short-term management is envisaged or likely to be feasible. Historically, of course, the assessment techniques embodied in calculations of yield-per-recruit and stock production models are geared implicitly to the long-term goals and strategies. These have, I suspect, been carried over uncritically to a world demanding more and more information about short-term prospects and more prepared to let the future look after

TABLE 1 - The relevance of various sorts of information to short-term and long-term assessments.

| Information | Short-term catch forecast | Long-term strategic assessment |
|---|---|---|
| Natural Mortality | - | ** |
| Current Fishing Mortality or Stock Size | * | * |
| Exploitation Pattern | * | ** |
| Imminent Recruitment | *** | - |
| Factors determining recruitment in the long-term (including stock-recruit relationships, environmental effects and fluctuations) | - | *** |
| Recent Catch Levels | *** | * |
| Stock Identification | - | ** |
| Discards | - | ** |
| Biological Interactions | - | ** |

Key: - of very little importance; * moderately important; ** very important; *** vital.

itself. It may be time to reassess our priorities and try to gather appropriate data at appropriate cost more deliberately than hitherto.

Nowadays assessments of the effects of technical measures, such as a change of mesh size, usually involve consideration of both short-term and long-term changes and thus are the most demanding of information. Similarly, the evaluation of closed areas and/or seasons usually requires information on the spatial and temporal distribution of stocks - which one may be able to do without for less specific assessments. Nevertheless, information on (for example) stock separation and identification is usually implicit in the way an assessment is carried out, even if the assumptions made are not explicit.

Finally, it should be noted that although assessment of many species together is still in its infancy, it is clear that the main effects of biological interactions are due to predation on juvenile fish and, therefore, fully apparent only in the long term. They may, however, be of great importance, particularly as they may influence the desirability of various management options.

## THE AVAILABILITY OF FISHERIES DATA

There are many sorts of data relevant to the assessment and management of fisheries. They may be summarized as follows: a) data on catches, such as catch weights of various species and their compositions by size (e.g., category or length) and/or by age; b) data on the operations of fishing fleets, such as fishing effort measured in various ways; c) data from tagging (mark/recapture) investigations; d) data from surveys by research vessels, notably fishing surveys, acoustic surveys, and surveys of eggs or larvae; and e) data on basic biological parameters such as growth (as length and/or weight at age) maturity, fecundity, etc.

These final data of type e) are often obtained as a by-product of other investigations, or by means of limited sampling at relatively low cost, and will not be considered further here. They are really basic "housekeeping" data which are very necessary but of little use on their own for assessment purposes, although they may be of considerable biological interest. It is regrettable that, perhaps because they are relatively cheap and straightforward to obtain, they are often the first to be considered in new investigations, whereas an early start on assembling time series of the other categories of data would probably be more useful. The data available on a long established fishery for a species of major commercial importance, in a region with a long history of scientific investigations (such as the North Atlantic), will probably include data of several of these main types, but not all. Examples may be found in the reports of the assessment working groups of ICES and ICNAF/NAFO. These, although not usually published, are generally available on request from the organizations concerned. In a typical case one might have: a) estimates of total international catch numbers at age for ten years or more (possibly broken down by nations or fleets); b) estimates of fishing effort for some (and possibly most) of the fishing fleets involved, possibly standardized for variations in fishing power (but probably not very well standardized); c) data from occasional tagging exercises and possibly also from regular exercises; and/or d) estimates of abundance of pre-recruits (and possibly adults also) from fishing surveys.

This represents the maximum that is likely to be available even in the most favorable case. Unfortunately, not all the data is likely to be of uniformly high quality. Coefficients of variation on the order of 10% on catch-at-age or abundance indices would be considered excellent. Thirty percent would be more usual, and coefficients of variation exceeding 50% are not uncommon. Not all of this variability is due to inadequate sampling - a substantial part may be due to natural variability.

The collection of fisheries data is not cheap. A single research vessel cruise may cost several hundred thousand dollars, and the collection of comprehensive catch statistics may require many man-years of effort and thus have a similar cost. Such expenditure can usually only be justified on fisheries of substantial commercial importance, even if the cost is spread by collecting data on many species at the same time. Thus, it is inevitable that fewer data will be available for minor fisheries. The more expensive items such as age compositions (which require sampling of catches, collection of otoliths or other material, and age determinations) are likely to disappear first. Tagging investigations are necessarily stock-specific, are difficult to scale down in cost, and are thus also rarely available for minor stocks. On the other hand, data on minor stocks from fishing surveys can be acquired as a by-product of other studies and may therefore survive.

For developing fisheries of any size, it is unlikely that extended time series of data will be available. This is serious because, as discussed below, many methods of interpretation require correlations between time series of relative indices (e.g., of recruitment) and absolute estimates which are only available post hoc. In addition, although there is no fundamental reason why age or size composition data should not be acquired early in the history of an emerging fishery, in practice the necessary sampling schemes usually take some time (several years) to establish and run in, so such data are usually also lacking. On the other hand, research vessel surveys can often be put in hand quite quickly, and acoustic surveys, in particular, are often carried out soon after (or even before) a significant fishery develops.

The gradation in availability with the size and age of the fishery for the various sorts of data is summarized in Table 2. In constructing the table it was assumed that the scientific investigations required develop in tandem with the fishery itself. If no investigations are carried out or no fishery statistics collected, there will obviously be no data. The position as given in the table may therefore be unduly optimistic.

## THE INFORMATION CONTENT OF FISHERIES DATA

I have also added to Table 2 my personal opinion of the utility of these data in a one- to three-star rating (this, of course, depends on the discussion which follows), together with a brief note of the type of information potentially available from the data in isolation. This can, of course, only be obtained by suitable data processing and analysis, which may not always be straightforward.

Total catch weight data on their own tell one little except the scale (order of magnitude) of the fishery, although if a time series is available some idea of possible variations may be obtained. The variations may, however, be due to changes of abundance or changes of exploitation rates, so that the interpretation is ambiguous. If the size composition of the catches is available, the relative contributions of pre- and post-recruits can be ascertained, so that some indication of imminent year-class strength may be deduced. It may also be possible if sufficient growth data are available to determine approximate age compositions and thus to deduce time series of year-class strengths. Usually, however, this requires extensive age composition data to have been assembled directly. There has been much work in the last decade on the analysis of catch-at-age data on its own (usually known as virtual population analysis), and this is discussed further below.

Effort data alone for a few indicator fleets is of little use for assessment purposes, and even if available for most fleets, it may be difficult to

TABLE 2 - The availability and information content of various types of fisheries data.

| Data | Utility | Availability | | | Basic Information Content |
|---|---|---|---|---|---|
| | | Major developed fishery | Minor developed fishery | Emergent developing fishery | |
| Catch Data | | | | | |
| Total Catch Weight | *** | Usually | Usually | Hopefully! | Scale of Fishery |
| Length Composition | ** | Usually | Often | Sometimes | Pre/Post-recruit Composition |
| Age Composition (NB: some species only) | * | Often | Sometimes | Rarely | Year-class Strength, etc. |
| Effort Data (with associated catches) | | | | | |
| Indicator Fleets | ** | Usually | Often | Sometimes | Relative Stock Size |
| Most Fleets | * | Sometimes | Rarely | Sometimes | Relative Exploitation Rate |
| Tagging Data | | | | | |
| Occasional Exercises | * | Often | Sometimes | Sometimes | Migration, Growth |
| Regular Exercises | ** | Sometimes | Rarely | Rarely | Fishing Mortality and Natural Mortality |
| Survey Data | | | | | |
| Fishing Survey | *** | Often | Often | Sometimes | Relative Stock Size<br>Year-class Strength |
| Acoustic Survey | * | Sometimes | Rarely | Often | Relative Stock Size<br>Absolute Stock Size?<br>Size Composition? |
| Egg/Larval Survey | ** | Sometimes | Rarely | Rarely | Absolute Stock Size |

deduce the time series of relative* exploitation rate because of the difficulty of reconciling different effort measures. When combined with the associated catch data, however, even effort data for minor fleets may sometimes be used to construct estimates of the relative size of the post-recruit population (strictly the exploited biomass - see (1, 3)), and the determination of relative exploitation rate is facilitated. This is only possible if there is a reasonably close relationship between catch rates and stock abundance (i.e., in general only for demersal species), and if mortality rates have been high enough that historic stock sizes can be estimated fairly unambiguously by virtual population techniques. This is the first example illustrating the general theme that more information can generally be obtained from the joint analysis of data of several types than from their analysis separately.

A related example is afforded by the analysis of catch-at-age data. It is now well-known that the usual method for analyzing such data on its own - virtual population analysis - leaves the terminal fishing mortality undetermined (7, 10). The method is, in fact, unusual in that it is capable of detecting details of the pattern of fishing mortality over time, but not the overall pattern.

Additional information is required, such as effort (or catch per unit of effort) data or trawl survey data. The utilization of effort data is (as discussed above) superficially straightforward, but so far as I am aware, there is even in 1983 no entirely respectable method of jointly analyzing catch-at-age and effort data with satisfactory assumptions about the error structure (1, 3), at least in the case of fisheries data where recruitment must be assumed to be highly stochastic. Furthermore, it is unfortunate that the utility of CPUE data rests on the assumption of constant catchability, yet one knows full well that there are good biological and technical reasons to allow that catchability may change with time. Even when catchability is held constant, the solutions for terminal F may not be very well determined (7, 11). When one allows catchability to change, even in a restricted manner, the precision with which F may be determined may be seriously degraded.

For this reason the use of indices of population size from research vessel

---

* Here and elsewhere "relative" is used to indicate that an index of the quantity in question can be constructed (usually as a time series), but that the absolute magnitude remains uncertain by a multiplicative constant.

fishing surveys, where the nature of the fishing operation and therefore (hopefully) the catchability can be well standardized, is attractive. Fairly well-founded methods for the analysis of such data exist (4), and more efficient methods for obtaining solutions are being developed (Pope and Shepherd, work in progress). These methods again involve the combined analysis of more than one sort of data. They use catch-at-age data which essentially determine differences between populations at adjacent times, together with survey indices which determine ratios of populations at adjacent times, in order to estimate absolute population sizes. They allow for errors in both sets of data and effectively use the rather variable survey indices mainly to control the magnitude of trends in the interpretation of the catch-at-age data. Such methods should be more productive than the simple interpretation of survey data alone, which can do little more than produce rather variable estimates of total mortality. They do, however, rely on survey indices having reasonably low variability (less than about 30%, say), else the computations become excessively difficult. The other main (and more traditional) use of fishing survey data is, of course, the construction of indices of the size of recruiting year classes.

The other main sort of fishery-independent data is that from tagging (mark-recapture) investigations. There is an extensive literature on the interpretation of such data; the technique is, of course, widely used in almost all branches of zoology (see, e.g., (13)). The information on migration and stock separation, as well as growth, which may be obtained from even a few isolated marking exercises, is unrivalled. The interpretation in terms of mortalities is, however, less straightforward, especially in fisheries. There are several reasons for this. First, fish are not easy to capture, mark, and release unharmed. Indeed, for some species (hake and megrim, for example) this is impossible. The initial survival after tagging is therefore usually in doubt, as is the subsequent representativeness of any mortality rates obtained. Second, the recapture returns are invariably mostly from commercial fisheries, and the quality of the information (particularly location of capture) is often doubtful. These difficulties are somewhat ameliorated when regular repeated marking exercises are carried out, since these provide cross-checks on some aspects of the investigation. Indeed, by using methods akin to virtual population analysis it is possible to set bounds on such parameters as natural mortality (6), which are otherwise difficult to obtain. The use of tagging data to determine natural mortality, particularly as a function of age, and the precision with which this can be done deserve further investigation. Such investigations are another example of the value of combining different data sets. Analysis of mark-recapture data alone can be analyzed in terms of mortalities but cannot be used to determine population size unless the associated catch data is also available. There is no doubt that mark-recapture investigations, when carried out with suitable rigor, can yield most valuable information. The main practical bar to more widespread use is simply that most fish populations of interest are so large and dispersed that the cost of marking sufficient numbers of individuals in anything approaching a random manner is very high, particularly since one can usually work with only one species at a time, and tagging is often only feasible in shallow waters where populations are not representative of the whole population.

Acoustic survey methods in principle should allow stock sizes to be estimated very rapidly, and without accumulating time series provided that target strength can be determined. However, in practice this has proved extraordinarily difficult (2). In most cases the target strengths are not known to better than a few dB, whilst an accuracy (not precision) of about $\pm 0.5$ dB is really desirable if the results are to be used directly to determine management options (9). There are also doubts about the precision obtainable using normal survey designs, particularly since acoustic methods are most applicable for pelagic species which tend to have a very patchy distribution. For the immediate future it seems that reliable information from acoustic surveys is most likely to be obtained if housekeeping procedures (such as calibration methods) are rigorously controlled, and the data are used in combination with other information (such as catch-at-age) in much the same way as data from fishing surveys which make no pretence of providing absolute stock size directly. An example of this approach is the work of Jakobsson reported in Anon (2).

Egg and larval surveys are also used to determine stock sizes, in the case of egg surveys without accumulating time series for "calibration" purposes (see, for example, the reports of 1980 and 1983 of the ICES Mackerel working group). Egg surveys are, in fact, the only reasonably viable technique for determining absolute (spawning) stock size without a long lead time. They are unfortunately usually very expensive exercises requiring several research vessel cruises and considerable inputs of skilled labor to sort and count eggs in the samples collected. Larval surveys are also used when eggs are not pelagic (e.g., with herring), but need to be "calibrated" against historic data because of the unknown egg and hatching mortality, which also introduces a substantial amount of natural variability.

## THE UTILIZATION OF FISHERIES DATA

Most methods for determining options for the management of fisheries assume that it is possible to determine the current absolute stock size (or fishing mortality), and that it is necessary to do so. However, recent work (8, 15) shows that if the short-term management strategy to make small proportional changes in the level of fishing mortality, this information is of little importance. This is perhaps fortunate, because the discussion above indicates that of all the information which might be required, current absolute stock size is extraordinarily difficult to determine with any confidence. The key pieces of information for short-term management turn out to be recent catch levels and a forecast of imminent recruitment (relative to the average). In the long term this is no longer true. However, practical (social, economic, political, and psychological) factors usually mean that it is virtually impossible to make major changes in a fishery at all rapidly. The principle information required from a long-term assessment is therefore an indication of sign - should the level of exploitation be controlled, reduced, or allowed to rise? Even if an estimate of a long-term target can be made, it is doubtful whether its magnitude could even be used to determine the rate at which it should be approached. This may also be just as well, because the biological information most important in a long-term assessment - the factors determining recruitment, the natural mortality, and the multispecies interactions - are precisely those which are most difficult to determine. Indeed, since environmental changes of significance undoubtedly occur on the (decadal) time-scale in question, one would in any case attempt to locate a moving target. Furthermore, the biological information (such as rates of growth, recruitment, etc.) has almost always been deduced from observations under conditions not far removed from present conditions - i.e., in the vicinity of the status quo. Extrapolations to conditions much different from the status quo are therefore of dubious validity, and a progressive approach is desirable.

These considerations lead one to ask whether traditional methods of fish stock assessment and management do not seek to do too much. They are too expensive to apply to stocks other than those of major economic significance, and simpler procedures would be more widely applicable. I therefore feel that we should consider seriously management strategies which attempt to do no more than make small proportional changes (of around ±20%, say) in levels of fishing mortality or yield/biomass ratio, with the sign of the change being determined by the interaction of biological and socioeconomic factors and updated regularly to allow for changes in the environment, surprises in the biology, and changes in the social and economic world around us.

## REFERENCES

(1) Anon. 1981. Report of the ICES ad hoc Working Group on the Use of Effort Data in Assessments. ICES CM 1981/G: 5.

(2) Anon. 1982. Report of the ICES Symposium on Fisheries Acoustics, Bergen, Norway.

(3) Anon. 1983. Report of the ICES Working Group on the Methods of Fish Stock Assessment. ICES CM 1983/Assess: 17.

(4) Collie, J.S., and Sissenwine, M.P. 1982. Estimating population size from relative abundance data. ICES CM 1982/G: 34.

(5) Gulland, J.A. 1983. Fish Stock Assessment: A Manual of Basic Methods. FAO/Wiley.

(6) Hamre, J. 1980. Biology, exploitation and management of the Northeast Atlantic mackerel. Rapp. P.-v. Cons. Int. Exp. Mer 177: 212-242.

(7) Pope, J.G. 1979. Population dynamics and management: current status and future trends. Invest. Pesqu. 43(1): 199-221.

(8) Pope, J.G. 1982. Short-cut formulae for the estimation of coefficients of variation of status quo TAC's. ICES CM 1982/G: 12.

(9) Pope, J.G. 1982. User requirements for precision and accuracy of acoustic surveys. ICES Symposium on Fisheries Acoustics, Contribution No. 55, Bergen, Norway.

(10) Pope, J.G., and Shepherd, J.G. 1982. A simple method for the consistent interpretation of catch-at-age data. J. du. Cons. Int. Exp. Mer 40(2): 176-184.

(11) Pope, J.G., and Shepherd, J.G. 1983. A comparison of the performance of various methods for tuning VPA's using effort data. ICES CM 1983/G: 9.

(12) Ricker, W.E. 1975. Computation and interpretation of biological statistics of fish populations. Bull. Fish. Res. Board Can. 191.

(13) Seber, G.A.F. 1973. The Estimation of Animal Abundance and Related Parameters. London: Griffin.

(14) Shepherd, J.G. 1981. Matching fishing capacity to the catches available: a problem in resource allocation. J. Agric. Econ. 32(3): 331-340.

(15) Shepherd, J.G. 1984. Status quo catch estimation and its use in fisheries managment. ICES CM 1984/9: 5.

Exploitation of Marine Communities, ed. R.M. May, pp. 111-128. Dahlem Konferenzen 1984. Berlin, Heidelberg, New York, Tokyo: Springer-Verlag.

# Dynamics and Evolution of Marine Populations with Pelagic Larval Dispersal

J. Roughgarden, S. Gaines, and Y. Iwasa
Hopkins Marine Station, Stanford University
Pacific Grove, CA 93950, USA

**Abstract.** A model for the demography of an open population with space-limited recruitment, together with field studies in the rocky intertidal zone of Monterey Bay in California, show that the settlement rate of larvae onto vacant substrate is a parameter that controls the structure of communities containing sessile marine organisms. Models of open populations can be combined, assuming a common larval pool is shared, into a model for a closed regional population. This regional model integrates processes in benthic ecology with coastal oceanographic processes, shows that local coexistence among species competing for space reflects the partitioning of habitats on a regional scale, and is the basis for a theory of life-history evolution pertaining to organisms with pelagic larvae and sessile adults. These new models may be relevant to the dynamics of fishes with demersal adults and pelagic larvae.

## INTRODUCTION

Understanding the population dynamics and evolution of marine populations, especially those with sessile adults and pelagic larvae, seems to require a new theoretical language. The theoretical vocabulary generated by traditional models in ecology and population genetics seems to lack terms that can refer to the processes affecting many marine populations. Moreover, it seems these populations cannot be understood by studies confined to the water column or to rocky or soft substrates; studies integrating oceanography and benthic ecology are needed. To justify these suggestions, we begin by noting unique features of marine populations that require a new theoretical vocabulary. Then we offer a preliminary report of our theoretical and field studies on the dynamics, community structure, and life-history evolution of barnacles, a group that often dominates the high rocky intertidal zone throughout the world. We conclude that the "ability of fishery science and management to deal with changes in the marine ecosystem" is presently quite limited. But we also conclude that a consensus among marine scientists about the need for theoretical research, coupled with integrated field studies of oceanographic and benthic processes, will lead to exceptionally rapid progress.

## SPECIAL FEATURES OF COASTAL MARINE ECOSYSTEMS

Perhaps all the basic differences between marine and terrestrial populations stem from the fact that organisms do not permanently occupy air while they do permanently occupy the water column. The water column is a habitat while the air is not.

Study sites in benthic marine ecology generally are open systems quite different in concept from the systems studied in terrestrial ecology. A terrestrial population is tacitly viewed as occupying an area containing most of the adults that produce the future recruits. Migration to such a population is viewed as diffusing across its boundaries. In principle, the perimeter of a terrestrial system can be chosen large enough to make the migration rate relative to the reproductive rate within the system arbitrarily small. In contrast, migration to a benthic marine system is not restricted to the boundaries but is distributed, perhaps more or less uniformly, throughout the area of the system. A metaphor for how to define a benthic marine population system is to imagine an area of substrate immersed in a larval bath, a picture quite different from that imagined for terrestrial studies.

A conspicuous feature of benthic marine ecology is the unanimity with which marine ecologists view space on the substrate as a limiting resource for most populations of sessile organisms. In terrestrial systems, especially for animal populations, it has usually been difficult and controversial to determine if there are any limiting resources and to identify them if there are. This fact suggests that models for the dynamics of many marine populations may include explicit reference to the mechanisms for the occupancy and release of space without incurring a risk of being overly specific.

The larvae in the water column are not, of course, spontaneously generated, and at some scale even marine populations are closed systems. On this regional scale the appropriate metaphor may be to view a population as a collection of space-limited local systems, each coupled to the others by their contributions of larvae to a common larval pool. The larvae in this pool are then redistributed among the local systems according to each's available vacant space. This view leads to a hierarchical population model, one where local models embodying the processes studied in benthic marine ecology are coupled by processes in the domain of coastal oceanography.

Many foraging habits of marine invertebrates have no parallel in terrestrial organisms. On land there are no filter feeders combing the air for particulate matter, spores, and flying birds and insects, spiderwebs notwithstanding. Furthermore, the filter-feeding habit in marine systems would seem very energetically efficient. If so, this efficiency may help to explain why it seems a higher fraction of the biomass in a benthic marine community is made of animal tissue compared to a terrestrial community.

## THE DEMOGRAPHY OF AN OPEN SPACE-LIMITED POPULATION

We introduce a model for the demography of organisms that occupy space on a surface of total area, A, immersed in a water bath containing larvae. The processes that occur are a) settlement onto vacant space, b) growth of organisms that have settled, and c) the mortality of organisms. Both settlement and growth consume vacant space while mortality releases it. The simultaneous kinetics of these three processes has been analyzed by Roughgarden et al. (10).

Let the time between consecutive censuses be one week, and the width of an age class also be one week. The number of new larvae to have settled during the week is

$$n_{0,t+1} = sF_t,$$

where $F_t$ is the amount of free space in the system at time t and s is the number of settlers that land per unit free space that survive to the census day. (Free space is synonymous with vacant space; it is space not occupied by an organism.) The number of animals of older age classes is

$$n_{x+1,t+1} = P_x(F_t)n_{x,t} \quad x \text{ in } [0,w-1],$$

where $P_x(F)$ is the probability that an animal of age x at time t survives to become age x+1 at time t+1; this generally depends on the amount of free space available, F. Finally, the amount of free space is found from a relation that the total area is conserved,

$$A \equiv F_t + \text{sumof}(\, a_x n_{x,t} \,),$$

where sumof() is the sum from x = 0 to w, and $a_x$ is the area occupied by an individual of age x. Once provided with an initial condition, this model predicts the dynamics of a population through time; first, all the n's at t = 1 are found from the n's given for t = 0, then the n's at t = 2 are found from the n's at t = 1, and so forth.

To use the model, the assumption that settlement is proportional to the amount of free space must be empirically justified, and the relation between survivorship and free space must be specified. We have modeled that relation with formulas representing $P_x$ as essentially independent of F until the free space is nearly exhausted, at which point the survivorship quickly drops. One cause of increased mortality when free space is exhausted involves a phenomenon called "hummocking," where crowded animals bulge out from the rock surface. Entire hummocks are susceptible to removal by waves (1). Also, our data show that crowded barnacles are more attractive to predation by starfish than barnacles present in sparse cover. We offer empirical evidence that the assumptions of the model are approximately correct for Balanus glandula in some circumstances, although we know that they are not universally correct for sessile invertebrates, or even for all barnacles.

The settlement parameter, s, varies among locations within the intertidal zone, reflecting variation in time of exposure of the substrate to the part of the water column where the larvae occur, the physical and chemical character of the substrate's surface, and the effectiveness of factors (such as limpets) that cause mortality within the first week after settlement. Also, important extensions of the model are possible taking into account seasonal and stochastic variation in the settlement parameter, and that growth, as well as mortality, depends on the amount of free space.

Here are the main theoretical results from the formulation presented above:

1. As the settlement parameter, s, becomes low, the abundance of animals becomes directly proportional to s. Also, animals of all sizes and ages are intermingled with free space. There is a stable age distribution and a dynamic steady state level of free space resulting from

a balance between recruitment and mortality. In practice, this situation will not appear as a steady state in the sense of an abundance that remains perfectly unchanging through time as recruitment exactly balances mortality. To the contrary, since abundance in the limit of low settlement is directly proportional to the settlement parameter, s, stochastic fluctuations in s will have a large impact on abundance, giving the community an air of unpredictability, as lamented by Underwood et al. (12).

2. Regardless of the settlement parameter, if mortality is sufficiently "faster" than growth, then again there is a steady state in which animals of all ages and sizes are intermingled with free space. The abundance of animals is, however, not directly proportional to s unless s is low enough for result 1 above to apply.

3. If the growth of settled animals is "faster" than mortality <u>and</u> the settlement parameter, s, is sufficiently high, then there is no stable steady state level of free space, nor any stable age distribution. Instead, at a local spot there is an oscillation in the amount of cover. At any particular spot, say 100 sq cm in area, there is a cycle starting with vacant space. This is then quickly filled with recruits. The recruits grow with little mortality and soon all the vacant space is exhausted. Now the animals' survival drops, and most or all of the animals die simultaneously, leaving the spot nearly vacant again. In this situation animals of all ages and sizes are not intermingled with free space. Instead, the landscape is such that animals tend to occur in patches with other animals of the same size. In addition there are patches of vacant space or "gaps" (6). A patch of animals of the same size represents a place where there formerly was a gap; it was then quickly filled by recruits so that it now represents a cohort. The landscape appears as a mosaic of cohorts punctuated by discrete gaps of vacant space.

The technical criterion for whether mortality is "faster" than growth involves the slope near the origin of a function called the cohort area function. Consider the survivorship curve, $l_x$, defined by the $P_x$'s in the limit of ample free space,

$$l_x = P_0 P_1 \cdots P_{x-1} \qquad x \text{ in } [1,w]$$

$$l_0 = 1$$

The $l_x$ is the probability that a settled organism lives to age x, provided there is ample free space. The cohort area function is $a_x l_x$. It represents the area occupied by a cohort as a function of the cohort's age. If the cohort area function monotonically decreases, we say that mortality is faster than growth because the area occupied by the cohort is continually shrinking. (This includes the case of no growth.) If the cohort area function increases to a maximum and then decreases, then we say that growth is faster than mortality because initially the cohort expands in area before eventually disappearing from the system. Result 2 above is that there is always a stable steady state, regardless of s, if $a_x l_x$ is monotonically decreasing in x. Result 3 is that if $a_x l_x$ increases sufficiently rapidly to a maximum before decreasing, then there is a critical value of s, say $s_0$, such that if $s < s_0$ there is a stable steady state, while if $s > s_0$ there is a limit cycle.

## APPLICATION TO BALANUS GLANDULA IN MONTEREY, CALIFORNIA

We introduced the model above as an aid to understanding the community ecology in the high intertidal zone at Hopkins Marine Station, on the south shore of Monterey Bay in California. The study site consists of granitic rocks whose total area is about 2500 sq m. By inspection we could see that the abundance of Balanus glandula varied across the site, with nearly 100% cover of barnacles on rocks with seaward exposure, and decreasing to very sparse cover on rocks adjacent to shore, which lie behind the exposed rocks.

In one initial hypothesis for this gradation in barnacle abundance, the rocks adjacent to shore were supposed to be somehow unsuitable for barnacle growth and survival; perhaps as a result of heat stress during spring low tides when the moderating effect of fog is often absent. However, we also observed that barnacles on the sparsely covered rocks adjacent to shore were approximately as reproductive and could attain the same large size as barnacles in other areas. Thus, we could see no evidence that heat stress should be considered as a leading hypothesis.

In another initial hypothesis, predation was supposed to be responsible for the sparse cover of barnacles on near-shore rocks, much as suggested by Connell (2) for the cause of differences in barnacle cover between spots that could be accessed by predators and inaccessible spots, at Friday Harbor, Washington. Actually, the near-shore rocks at Hopkins are less accessible to starfish than are more exposed rocks, nor are there copious empty tests (from thaid, flatworm, or avian predation) or basal plates (from starfish predation) on near-shore rocks, but nonetheless, the variation in the relative magnitude of predation from all sources across the site might be an important factor. In any case, according to the model, the importance of predation in affecting abundance can only be assessed relative to the settlement rate and the growth rate. Only if the settlement is very high, so as to saturate free space when it becomes available, can predation be viewed in isolation from these other factors.

Briefly, we have found that variation in the settlement rate is the primary cause of variation in the abundance of Balanus glandula across the study site at Hopkins. The settlement rate (measured as cyprids settling per week per sq cm of vacant space during the settlement period of Spring, 1983) is ten to one hundred times higher at seaward rocks than at near-shore rocks. (We conjecture that this gradient in settlement rate across the site results from a reduction of larval concentration in the waters that contact the near-shore rocks. We hypothesize that the cyprid larvae settle onto the first substrate they encounter, and by the time the water contacts the near-shore rocks there are few larvae left.)

This great difference in settlement rate in turn affects the landscape of community structure in the intertidal in ways suggested by the model. Our data, together with Hines' (5), show that the weekly survivorship under uncrowded conditions is essentially independent of age, leading to an $l_x$ curve that is almost exactly an exponential decay. However, the weekly survivorship is very sensitive to the exhaustion of free space, as Fig. 1 shows. The cause of the mortality at low free space levels in our study is predation by the starfish, Pisaster ochraceus, which apparently preys preferentially on barnacles once they have achieved dense cover.

Our data on basal area, together with Hines' (5), can be combined with the $l_x$ data to yield the cohort area function, $a_x l_x$, presented in Fig. 2. Clearly, the function is not monotonic; cohorts initially expand before eventually leaving the system.

Finally, Fig. 3 shows that the net settlement rate into a site is approximately a linear increasing function of the fraction of free space in that site. Our data also show that free space is, by far, the preferred substrate. Settlement occurs on the sides of the tests of already established individuals only if there is less than 5% free space available.

With these data the model predicts, for low settlement, that the population approaches a stable age distribution and a steady state level of free

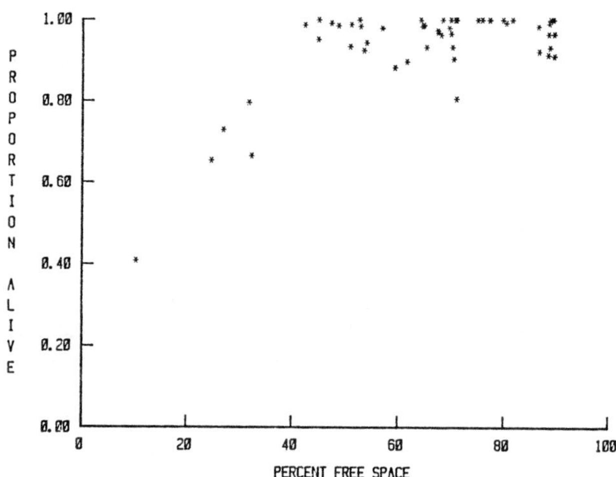


FIG. 1 - Probability of survival through one week for Balanus glandula as a function of the percentage of free space available in small quadrats (72mm x 48mm).

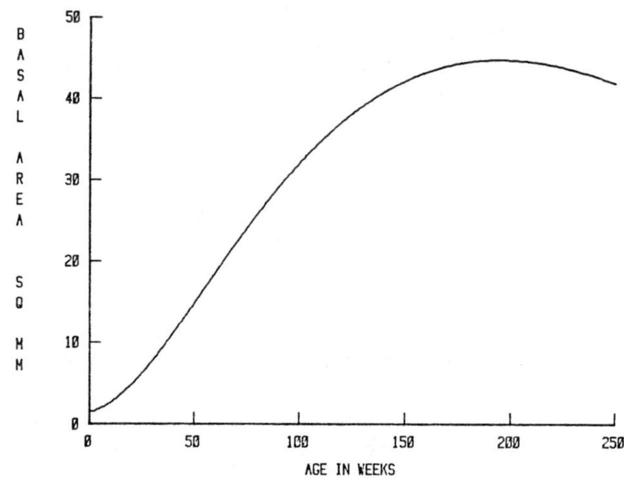


FIG. 2 - The cohort area function for Balanus glandula.

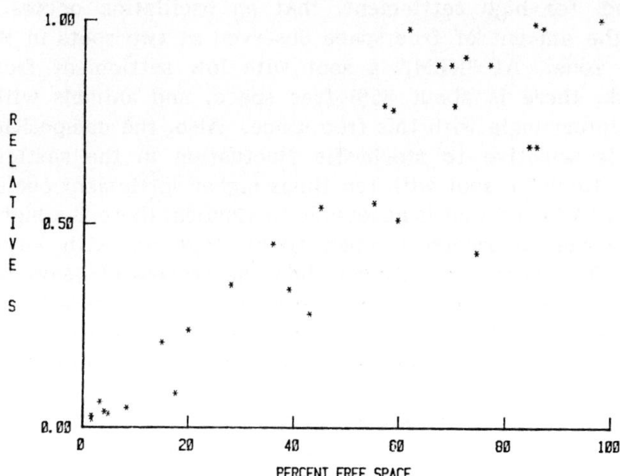


FIG. 3 - Settlement in a sector (1/12 of a quadrat) relative to the settlement in the entire quadrat as a function of the percent free space in the sector.

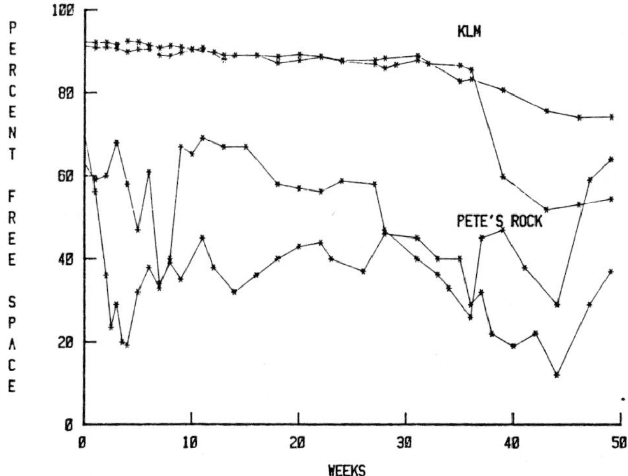


FIG. 4 - Dynamics of free space at two locations during 50 weeks (September 1982 to September 1983) in the high intertidal zone at Hopkins Marine Station, Monterey Bay, California.

space, and, for high settlement, that an oscillation occurs. Figure 4 presents the amount of free space observed at two spots in the Hopkins intertidal zone. At "KLM," a spot with low settlement from a near-shore rock, there is about 85% free space, and animals with all sizes and ages intermingle with this free space. Also, the composition of KLM quadrats is sensitive to stochastic fluctuation in the settlement rate. At "Pete's Rock," a spot with ten times higher settlement per unit vacant space than KLM and that is accessible to starfish, there is a higher average level of cover (averaged through time), together with large "swings" in cover. Our photographs clearly show the process of heavy recruitment onto the space left after Pisaster predation, rapid growth, followed by the recurrence of starfish predation after the recruits have grown to occupy nearly all the space originally available. The cycle then repeats.

## OTHER SCENES OF HARD SUBSTRATE

A third type of community found in the high intertidal zone at Hopkins may be termed a "lattice." At spots on "Bird Rock," a place of very high settlement, Balanus glandula exists in 100% cover and individuals are packed into a lattice, as in a honeycomb. This system seems nearly static. Many of the individuals seem very old, with columnar tests 2 to 3 cm long and a basal diameter of 0.5 to 1 cm. We conjecture that this situation forms where the settlement to vacant space is very high, and where mortality does not substantially increase when free space becomes exhausted (e.g., the spot is inaccessible to starfish and the wave action generally not severe enough to remove hummocks). After settlement the animals grow against one another, packing ever more tightly, until all the free space is exhausted. Then further change grinds to a halt until a rare event, perhaps of a catastrophic nature, should befall the site. Still another possibility is that senescence leads to the spontaneous formation of gaps in this otherwise continuous cover.

Clearly many scenes may be set on hard substrate. In summary, four scenes seem relevant to the population dynamics of barnacles. a) The "high-F steady state community" results from a low settlement parameter. Because settlement is limiting, stochastic and spatial fluctuation in the settlement rate greatly affect abundance. b) The "low-F steady state community" results from high mortality that is independent of F. Here stochastic fluctuation in settlement has less effect than in the high-F community. c) The "patch-mosaic community" results from high settlement, low F-independent mortality coupled with high mortality when free space becomes exhausted. d) The "lattice community" perhaps results from high settlement and low mortality; here, as space is exhausted, growth stops, leaving a static lattice that persists until senescence at a much later time, or until a rare catastrophic event kills the organisms.

In this classification settlement plays a role as important as predation in determining community structure. The fundamental importance of the settlement rate is often overlooked. Connell's classic study (3) of zonation between barnacles in Scotland relies on high settlement relative to mortality as a tacit assumption. If the settlement were not very high, then the barnacles would not have come into extensive contact with one another. Without extensive contact Balanus cannot cause adult Chthamalus to remain only in a higher zone in the intertidal. Indeed, along much of the central California coast, where the settlement rate is lower than that measured in Scotland, Balanus and Chthamalus do not zone but have completely overlapping distributions in the intertidal. The clue that the settlement is not high enough for zonation to result is that there is ample free space where the two species distributions overlap. Similarly, the celebrated "intermediate disturbance principle" of marine ecology has a tacit assumption of high settlement. According to this principle, the highest species diversity in a community occurs at an intermediate level of disturbance. But, if the settlement is too low for extensive contact to develop among the space-using organisms, then there is no opportunity for a competitive hierarchy to develop. In these circumstances diversity and disturbance are probably inversely related.

Thus it emerges that the settlement rate is fundamentally important in marine ecology. What, then, determines the settlement rate? In a proximal sense, we hypothesize that the settlement rate reflects the concentration of larvae in the water column. So the question then becomes, what determines the concentration of larvae in the water column? For organisms with long-lived pelagic larval dispersal, this question can only be answered by passing to a regional context, by changing scales.

## THE REGIONAL DYNAMICS OF SPACE-LIMITED LOCAL POPULATIONS

Suppose there are H local systems all bathed by the same water mass. A simplified picture of the dynamics in each of the local systems is provided by considering only the processes of recruitment onto free space and mortality within the local systems, ignoring growth. This simplified picture is consistent with the more complete local model considered earlier, provided the local systems are not in oscillation. This simplification allows the mortality rate, basal area, and fecundity in each local site to be averaged over the age distribution there. The dynamics of local system, h, can then be described by

$$dn_h/dt = (c_h L)F_h - u_h n_h \qquad h \text{ in } [1,H]$$

$$A_h \equiv F_h + a_h n_h,$$

where the settlement parameter is now replaced by the quantity, $(c_h L)$, the free space in local system h is $F_h$, and the average instantaneous mortality rate in local system h is $u_h$. The total area in local system h, $A_h$, is conserved, and $a_h$ is the average basal area of an animal in local system h. The settlement rate onto vacant space is presumed to be proportional to the concentration of larvae in the water mass, L. The constant of proportionality, $c_h$, may vary among local systems to take account of differing accessibility or attractiveness of the local systems to the larvae. The larvae, in turn, are made by the adults from all the local systems and are contributed to the common larval pool. The dynamics of the larval pool can then be described by

$$dL/dt = \text{sumof}(\ m_h n_h - (c_h L)F_h\ ) - vL,$$

where sumof() means the sum over h from 1 to H, $m_h$ is the average fecundity per animal in local system h, and v is the instantaneous mortality rate for larvae in the water column. The terms in the equation for larval dynamics represent production of new larvae from the local systems, loss of larvae through settlement onto vacant space in the local systems, and mortality of larvae in the larval pool itself.

Our analysis of this model relies on a classification of the local systems into "sources" and "sinks" for larvae. The sign of the quantity,

$$r_h = c_h(m_h/u_h - 1),$$

determines how a local system is classified. If $r_h$ is positive, then the expected production of larvae from a piece of free space in local system h exceeds the death rate of larvae that settle onto that space; this condition represents a "source." If $r_h$ is negative, the area represents a "sink." In general a region must be considered as presenting both source and sink areas to the larvae.

The main result from this regional model is that all possible steady state abundances of larvae in the larval pool, $\hat{L}$, are roots of the equation

$$R(L) \equiv (1/v)\text{sumof}(\, r_h \hat{F}_h(L)\,) = 1,$$

where $\hat{F}_h(L)$ is the steady state free space in local system h, and is given by

$$\hat{F}_h(L) = A_h/[1 + (c_h a_h/u_h)L].$$

The basic relation, that R(L) equals 1 at the steady state, indicates that the production of larvae, summed over all the local systems in which they settle, equals the death rate of larvae in the water column. Furthermore, the necessary and sufficient condition for a positive root to be a locally stable steady state is that the slope of R(L), dR(L)/dL, be negative at that root.

In the special case where all the local systems are sources, there is a unique steady state configuration, and it is globally stable regardless of any other values of the parameters. If some local systems are sources and others are sinks, then generally several distinct steady states are possible. However, the steady state associated with the highest steady state larval abundance is always locally stable (provided it is positive).

Some steady states can be attained only along trajectories such that the free space in sinks becomes low. When the free space in a sink is low, it ceases to function as a drain on the larval pool, thereby allowing the remaining mortality in the larval pool to come into balance with larval production from source locations. A population that can exist in a region under these circumstances cannot be established with a small propagule; the colonizing propagule must exceed a threshold size for it to "take." It is perhaps relevant that the Australian barnacle, Elminius modestus, became established in the British Isles only during a period of extremely active shipping during World War II (4).

## INTERSPECIFIC COMPETITION

Competitive exclusion is impossible in a local system as a result of interpecific competition between two forms whose larvae settle in vacant space. As long as there are larvae of both species in the water column, and vacant space is available, then both species are present in the local system. The extinction of a species, including the possibility of competitive exclusion, is inherently a question of regional dynamics.

We have extended the model of regional dynamics introduced above to include several species that compete for space in the local systems without interacting as larvae while in the water column. Our work indicates that the species diversity in a local system is a reflection of specialization (or partitioning) on a regional scale. The main results are:

1. The number of coexisting species is less than or equal to the number of distinct types of local systems. For example, two species competing for space in one local system, or in many identical local systems, cannot coexist.

2. For any pair of species in a community of coexisting space competitors, if one species has a higher r/v in a certain local system, then the other species has a higher r/v in some other local system.

These results are reminiscent of traditional competition theory ((9), Ch. 24), but with a focus on between-habitat partitioning instead of within-habitat partitioning. However, the regional dynamics of competition do not share a close similarity with the Lotka-Volterra competition equations of traditional competition theory. By developing the special case of two species competing in two local systems, qualitative results can be exhibited that are impossible in two-species Lotka-Volterra systems. If both local systems are sources to both species, then indeed only the familiar cases of two-species Lotka-Volterra equations are found, namely, that a particular species excludes the other regardless of initial condition, that both species coexist, or that either species can exclude the other depending on the initial condition. But if one of the local systems is a sink to at least one of the species, then the following additional possibilities arise: a) the presence of one of the species is necessary for coexistence with the other (the former species is a competitive keystone species; if it is removed the other goes extinct); b) the presence of one of the species is needed for the second to invade, and this invader proceeds to exclude the first and to remain alone in the region; and c) the presence of one of the species is needed for the second to invade, and this invader in turn excludes the first species and becomes extinct itself, leaving no species in the region.

Thus the dynamics of species competing for space within the local systems of a region may be qualitatively different from that encountered in traditional competition theory. This finding strengthens our suggestion that new theoretical research is needed to understand the ecology of populations with sessile adults and a long-lived pelagic larval phase.

## LIFE-HISTORY EVOLUTION

The model of regional population dynamics can also be used as a point of departure for developing a theory of life-history evolution. Traditional life-history theory is focussed on the maternity and survivorship curves of demography, m(x) and l(x) ((9), Ch. 19). The traditional theory can address evolutionary trends involving trade-offs between longevity and fecundity as exemplified by, for example, latitudinal trends in the clutch size of birds. But such theory is silent on the life-history features of populations with sessile adults and pelagic larvae. It is silent on the evolutionary significance of traits like basal area (for this consumes free space), larval selectivity, and the relative duration of larval and sessile phases, nor does it address the evolution of specialized abilities to survive and to reproduce in one type of habitat accompanied by the loss of abilities in other habitat types. The regional model serves as a prototype for how a new life-history theory, one more appropriate to marine life histories, can be developed.

The theoretical task is to determine what constitutes a favorable mutation in a population whose structure is represented by the regional model. This is accomplished by extending the model to include genetic variation at one locus with two alleles. In the genetical model, the condition for the increase of a rare mutant allele can be derived. This condition then tells one what life histories will be favored as new mutations accumulate. It also provides a criterion for an evolutionarily stable strategy (ESS) (8), that is, a life history that cannot be invaded by a mutation to any other possible life history.

We have determined that the condition for increase of a new allele in a population represented by the regional model is

$$R_{Aa}(\hat{L}_{aa}) > 1,$$

where the population is assumed to consist initially of "aa" individuals, and the new mutation is labelled "A."

The implications of this formula for the evolution of barnacle life histories are beginning to be explored. The first interesting result concerns the evolutionary significance of basal area. One may conjecture that selection will favor individuals that have a large basal area, for such individuals will preempt the limiting resource of space. According to the evolutionary model, however, the basal area is selectively neutral. Other results pertain to the evolution of larval specificity during settlement (as modeled by the $c_h$ parameters) and to the evolution of habitat specialization.

**CONCLUSION**

Our studies show that marine populations with sessile adults and pelagic larvae have many features without any parallel in terrestrial ecology. Yet there is nothing uniquely intractable about marine populations. Their study requires a theoretical framework that is tailor-made for the special characteristics of the marine habitat and the kinds of organisms that live there. It is clearly inviting error to transport models that originate in terrestrial situations to a marine context. But, a tailor-made marine theory does not preclude noticing similarities between results from marine and terrestrially oriented models, and the mathematical approaches applied originally to analyze terrestrial models are generally useful in marine models as well.

Our studies demonstrate that models that are tailor-made for the marine habitat can be formulated, solved, and put to good use in benthic study sites. The great unknown lies at the interface between such benthic ecological studies and the processes of coastal oceanography. Our studies suggest that the composition of the water column has a controlling effect on community structure in the intertidal zone. Conversely, at least at certain times, a large fraction of the offshore plankton community consists of larvae produced by coastal benthic creatures. Certainly both biological oceanographers and benthic ecologists can greatly benefit from integrated studies.

What, then, is the "ability of fishery science and management to deal with changes in the marine ecosystem"? To the extent that the "marine ecosystem" includes invertebrates living on benthic substrate, the ability of fishery science is quite limited. The stock-recruitment theory usually identified with fishery science seems inappropriate to sessile invertebrates because it does not revolve around space as a resource and does not deal with open systems.

May (7) has pointed out that the stock-recruitment curve used for prediction of stocks in fisheries science is a phenomenological model whose parameters are fitted to each species and situation. The use of the stock-recruitment curve for prediction of stocks is analogous to the use of autoregressive models from time-series analysis for prediction in other subjects. Hence, stock-recruitment models offer little help in understanding the mechanisms of a population's dynamics. Moreover, a stock-recruitment formulation may not provide an appropriate representation of a population's dynamics, even for the purposes of prediction, when the spatial and genetic structure of the population are involved in its dynamics and response to harvesting. To provide a deeper understanding of the causes of population change in the marine environment, and perhaps also to provide more reliable long-range prediction than current methods offer, specific mechanism-oriented models like that developed here for barnacles should be developed for fish populations. Shepherd and Cushing's (11) attempt to connect the stock-recruitment curve to biological mechanisms of density-dependence is an important step in this direction. Moreover, the theory we have reviewed here for barnacles offers some features that might be incorporated in fishery models for demersal fishes whose larvae are pelagic. The abundance of coral reef fish directly reflects the number of fish larvae that settle out of the plankton (13), in a manner that is possibly analogous to the low settlement limit in our model for barnacles. Similarly, perhaps it might be desirable to view demersal fish populations in the neighborhood of patch reefs, or other required habitat, as open populations in the same spirit as our view of barnacles.

**Acknowledgements.** We thank R. Bailey, C. Baxter, S. Brown, S. Levin, R. May, and R. Paine for their advice on the manuscript. We gratefully acknowledge support from the Department of Energy (Contract EV10108).

**REFERENCES**

(1) Barnes, H., and Powell, H.T. 1950. The development, general morphology and subsequent elimination of barnacle populations (Balanus crenatus and B. balanoides) after a heavy settlement. J. Anim. Ecol. 19: 175-179.

(2) Connell, J.H. 1961. The effect of competition, predation by Thais lapillus and other factors on natural populations of the barnacle Balanus balanoides. Ecol. Monog. 31: 61-104.

(3) Connell, J.H. 1961. The influence of interspecific competition and other factors on the distribution of the barnacle Chthamalus stellatus. Ecology 42: 710-723.

(4) Crisp, D.J. 1958. The spread of Elminius modestus Darwin in North-west Europe. J. Mar. Biol. Assoc. 37: 483-520.

(5) Hines, A.H. 1979. The comparative reproductive ecology of three species of intertidal barnacles. In Reproductive Ecology of Marine Invertebrates, ed. S.E. Stancyk, pp. 213-234. Columbia: University

of South Carolina Press.

(6) Levin, S.A., and Paine, R.T. 1974. Disturbance, patch formation, and community structure. Proc. Nat. Acad. Sci. USA 71: 2744-2747.

(7) May, R. 1980. Mathematical models in whaling and fisheries management. Am. Math. Soc. Lect. Math. Life Sci. 13: 1-64.

(8) Maynard Smith, J. 1974. The theory of games and the evolution of animal conflicts. J. Theor. Biol. 47: 209-221.

(9) Roughgarden, J. 1979. Theory of Population Genetics and Evolutionary Ecology. New York: Macmillan Publishing Co.

(10) Roughgarden, J.; Iwasa, Y.; and Baxter, C. 1984. Theory of population processes for marine organisms. I. Demography of an open population with space-limited recruitment. Ecology, in press.

(11) Shepherd, J., and Cushing, D. 1980. A mechanism for density dependent survival of larval fish as the basis of a stock-recruitment relationship. J. Conseil 39: 160-168.

(12) Underwood, A.J.; Denley, E.J.; and Moran, M.J. 1983. Experimental analyses of the structure and dynamics of mid-shore rocky intertidal communities in New South Wales. Oecologia 56: 202-219.

(13) Victor, B.C. 1983. Recruitment and population dynamics of a coral reef fish. Science 219: 419-420.

Standing, left to right:
Erik Ursin, John Gulland, Serge Garcia, Brian Rothschild, Eike Rachor, Helmut Maske

Seated, left to right:
Bob Paine, Trevor Platt, John Lawton, George Sugihara, Bernt Zeitzschel

Exploitation of Marine Communities, ed. R.M. May, pp. 131-153. Dahlem Konferenzen 1984. Berlin, Heidelberg, New York, Tokyo: Springer-Verlag.

# Ecosystems Dynamics
## Group Report

G. Sugihara, Rapporteur
S. Garcia
J.A. Gulland
J.H. Lawton
H. Maske
R.T. Paine
T. Platt
E. Rachor
B.J. Rothschild
E.A. Ursin
B.F.K. Zeitzschel

### INTRODUCTION

The deliberations of this group represent a dynamic balance between the practical concerns of the fishery manager and the scientific goals of the academic ecologist. Although the title of this report is ecosystems dynamics, our discussions focused almost entirely on the multispecies problem and ways in which insights from ecological systems theory and food web research might feed into fishery management. A major theme in our talks was how best to characterize and simplify complex systems in order to highlight change and understand structure in marine communities.

Our report begins with a general discussion of the need for multispecies and ecosystems perspectives in fishery management and goes on to suggest guidelines for how this might be approached. In particular we report on what is known about reducing the perceived complexity of marine systems by various methods of aggregation and how this relates to model choice and to the importance of a careful selection of spatial and temporal scales in the data. We then discuss questions related explicitly to system dynamics, such as how to treat stability and change in marine communities and how simple knowledge of ecosystem structure (topology) may generate deeper insights into system dynamics. This section is followed by a discussion of both what can be learned by experimental manipulation of marine systems and to what extent fishing can be evaluated as an experimental manipulation. Finally, we suggest a hierarchical approach for understanding ecosystem behavior that follows a continuum of complex to simple nested models on fast to slow time scales.

## IMPORTANCE OF THE MULTISPECIES VIEW

Exploited fish populations are embedded within a complex web of interactions involving species from many different taxa existing together in a variable environment. How important is this broader context for understanding and maintaining marine fisheries?

Despite having a deep tradition in academic ecology, the ecosystems perspective and, indeed, the multispecies view have played only a minor role in marine fisheries science. Classical fisheries models, directed mainly toward managing temperate and boreal stocks, have typically focused on individual species, treating them as independent management units. These single-species models rarely account for abiotic influences such as light, temperature, and turbulence, and explicitly exclude the complications of interspecific linkages. Some early exceptions to this general trend include the pioneering work of Larkin (15, 16) and May et al. (21) on optimal exploitation in Volterra competition and predator-prey systems, and the dynamic simulation model for the North Sea by Andersen and Ursin (1), among others. Not withstanding its potential shortcomings, the single-species approach has met with remarkable success, especially in managing certain long-lived species, and is the basis for most of our present-day management decisions. It was perhaps this lack of clear necessity that kept the ecosystems perspective from advancing in a field whose pragmatic concern is fishery management.

With the rising status of the Third World and rapid growth of its tropical fisheries, however, the limitations of the classical single-species approach are becoming more apparent. Most tropical icthyofauna contain well over 700 species (14), not to mention an abundance of other taxa. The fishery might target on up to a hundred different species, using a diversity of gear, methodologies, and catch limits. Consequently, these systems are producing some of the strongest evidence pointing to the inadequacy of treating species as independent management units ((8, 27, 32), and Larkin et al., this volume). Where single-species approaches produce poor results, models which explicitly contain interactions or which aggregate species to circumvent this need are often more successful (32). Theoretical studies have demonstrated the potential that interspecies effects have for substantially altering stock-yield forecasts (21).

Even in cases where the single-species view has been sufficient, a multispecies approach may be better. As reviewed by May et al. (21) and by Gulland and Garcia and Paine (both this volume), there is growing albeit spotty evidence from temperate and subarctic systems suggesting the importance of interspecific linkages. These range from ad hoc observations on pelagic systems regarding species succession and replacements after heavy fishing, to controlled experiments documenting the interaction web in the rocky intertidal zone (26). Historical inertia toward the single-species view combined with the logistical difficulty of the multispecies problem have no doubt contributed to the present lack of careful documentation of such interactions in pelagic systems. For temperate freshwater systems, where there is a substantial tradition in this problem, hard evidence for interaction in fish populations abounds (e.g., (10, 13, 17, 44, 48-51)).

Without forgetting the success that the single-species approach has had, one should be mindful of the potential long-term risks involved. These risks are amplified by the existence within most marine systems of more than one independent fishery, each targeting on a different species mix and size of fish. Heterogeneous cropping of populations or of age distributions within a group of dynamically related species has the potential to produce nasty results for a fishery, ranging from faulty yield-effort or stock-production forecasts to catastrophic declines in stock abundance. More will be said of this in the section on CHANGE AND PERSISTENCE IN MARINE COMMUNITIES. On a slightly different tack, a narrow focus on individual species runs the risk of failing to detect subtle changes in the associated biota (not necessarily icthyofauna) which might have strong effects on the fishery. For example, a perturbation which depletes kelp beds, a required spawning ground for Pacific herring, could greatly depress future stocks for this fishery. In a related way, knowledge of the ecosystem could in principle point to an inexpensive yet effective indicator of system health. The failure of the Peruvian guano birds to recover from the 1965 El Niño, for example, might have been used as an early indication that the system was becoming less capable of recovering from disturbance, and thus might signal an early warning of the collapse that occurred at the time of the 1972/3 El Niño. It should be mentioned, however, that the meaning of this signal became clear only in retrospect and that there was even a consensus in some circles that depressed bird populations (implying lower predation) should be taken as an encouraging sign for the fishery. A similar air-water link has been described for a freshwater system in Panama where an associated subweb of birds provided a conspicuous indicator of changes occurring within the cichlid populations of Lake Gatún (51). As will be discussed more fully in the section on EXPERIMENTAL MANIPULATION, such data from indicator species requires cautious interpretation and a firm understanding of the biology involved. Nevertheless, they can provide valuable additional pieces in a complex jigsaw of information that might be used by the fishery manager.

Two broad ways in which the ecosystems perspective might benefit fishery science, therefore, are in providing better models for fish population dynamics and stock estimates and in pointing to key linkages and indicators of system health. While the latter case involves describing qualitative relationships among species, the former case, modeling system dynamics, requires quantitative information about species interactions. The following section will address some major concerns involved in constructing such models.

## MANAGING COMPLEXITY

A typical fishery in temperate waters may harvest a dozen or twenty different species, each species consisting of several local groups, perhaps distinct in genetic makeup. A detailed study of each one of these groups (not to mention the associated nontarget taxa) would involve an unacceptable work load. The problem is worse in tropical waters where there may be a hundred harvested species, but where the number of scientists, their support facilities, and the background of previous research may be much less.

Incorporating such complexity into a modeling effort may also be undesirable for technical reasons. A point is reached where further complexity in a model contributes more heavily to the variance of the output, rather than adding more realism (Fig. 1). This variance derives from the experimental and natural variability in the parameter estimates and from the convolution and amplification of this variance within the model itself (22). Moreover, just as in fitting a higher-order polynomial, it may be possible with only a little "tuning" of the parameters to get a complex model to fit a finite sequence of historical data. This gives the misleading impression that the model succeeds in depicting reality, when in fact the model, as with the polynomial, may be completely off in predicting what happens next.

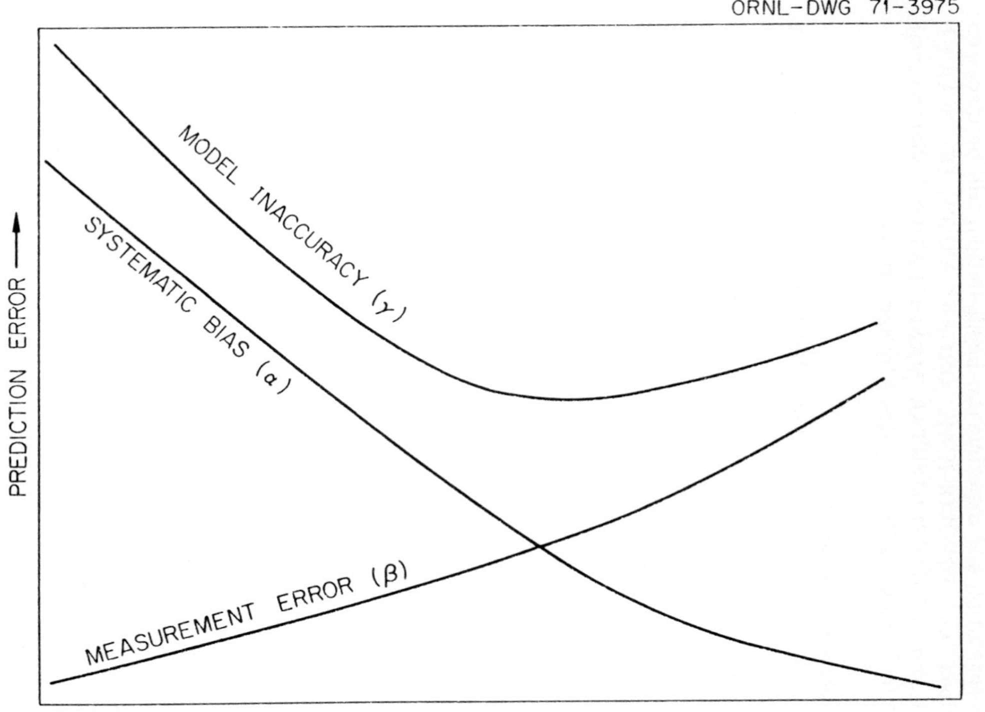


FIG. 1 - Components of error on model predictions as a function of model complexity. Systematic bias or alpha error will usually decrease as models account for more of the complexities of real systems. Measurement or beta error tends to increase as more parameters are added to the model. The resultant model inaccuracy ( $\gamma = \alpha + \beta$ ), therefore, increases for very complex models. The relationships suggest that there is an optimal level of complexity for a model (from (22)).

The need to simplify complex fisheries systems can be met in two interrelated ways: a) by limiting the extent of the system (i.e., by focusing on a small subsystem or taking a single-species stance), or b) by aggregation. The first approach, which may involve a combination of field data interpretation and experimental manipulation, will be discussed in the sections on TOPOLOGICAL ORGANIZATION OF MARINE ECOSYSTEMS and on EXPERIMENTAL MANIPULATION. Our present discussion will focus on simplifying systems by aggregation.

**Reducing Complexity by Aggregation**

A number of intuitive suggestions have been offered for aggregating complex systems. Samuelson (34) has outlined two principal methods for simplifying economic systems which apply equally to ecological systems: first, aggregation by common function, such as by guild, etc., and second, aggregation by interaction strength into strongly interacting subsystems. Some specific recommendations made in the fisheries context include aggregation by: a) species co-occurrence in the fishing gear, b) similarity in growth and/or mortality rates on other model parameters, c) body size, d) value, e) location, and f) total biomass, among others. In most cases such lumpings are done with little appreciation of the fundamental issues involved and of the errors which may be committed. Our aim here will be to review briefly these issues and to make some recommendations which will be of use in studying multispecies fisheries.

**Fundamentals.** The general form of the aggregation problem is outlined in the structure below.

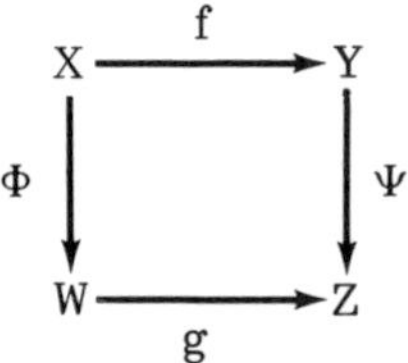


Here $f(x) = y$ is the microsystem (disaggregated), $g(w) = z$ is the macrosystem (aggregated), with $\phi$ and $\Psi$ being the active and passive aggregation functions, respectively. For example, X might represent a vector of species taken individually, W being collections of species into guilds, with Y and Z representing their biomasses at various times. A fundamental question in aggregation theory involves the study of conditions under which

$$\Psi [f(x)] = g[\Phi (x)] \quad (1)$$

for all elements and relations in the microsystem. When, for example, does a species-by-species analysis always give the same forecast of total biomass as an aggregated analysis? Or more generally, under what conditions does routing the analysis through the microsystem always give the same results as routing it through the macrosystem? As we shall see, such conditions for <u>total consistency</u> are extremely restrictive.

A more practical problem is to relax the requirement for total consistency and study the error generated when there is only <u>partial consistency</u>, i.e., where Eq. 1 holds for only certain values of X. Two categories of partial consistency include: a) <u>constrained consistency</u>, where consistency (Eq. 1) is obtained only for certain values of the independent variables (e.g., equilibrium conditions), and b) <u>characteristic consistency</u>, where the macrosystem and microsystem are consistent only for certain characteristic values associated with the range, such as the dominant eigenvalue or the logical output of a decision framework. Constrained consistency has been the more tractable of these problems, although both are important.

As a specific example, consider an exponentially growing population of organisms consisting of $\geq 2$ local subpopulations, each with a different deterministic growth rate. One might naively ask if it is possible to obtain the same growth trajectory for the total population by treating it as separate subpopulations (microsystem) as by treating it as a whole (macrosystem)? Suppose we begin with the less ambitious aim of constrained consistency so that initial conditions are consistent, and we take $\Phi$ and $\Psi$ to be summations within and across the individual subpopulations. f and g are both exponential growth equations, but g operates with an average growth rate. As seen in Fig. 2, unless all growth rates are equal, the growth trajectories for the summed subpopulations will exceed those in the aggregate model. In general, regardless of the choice of exponent, it is not possible to replace the sum of n exponentials with one without introducing error. Several variations on this theme have appeared in the ecological literature (3, 5, 9, 24).

Where f and g have the same functional form (i.e., generalized Volterra), the crude essence of reducing a system from n dimensions to n-k dimensions lies in the errors involved in approximating n exponential terms with n-k terms. For a microsystem of Volterra equations, it is easy to show that zero-error in the macrosystem generally requires time-

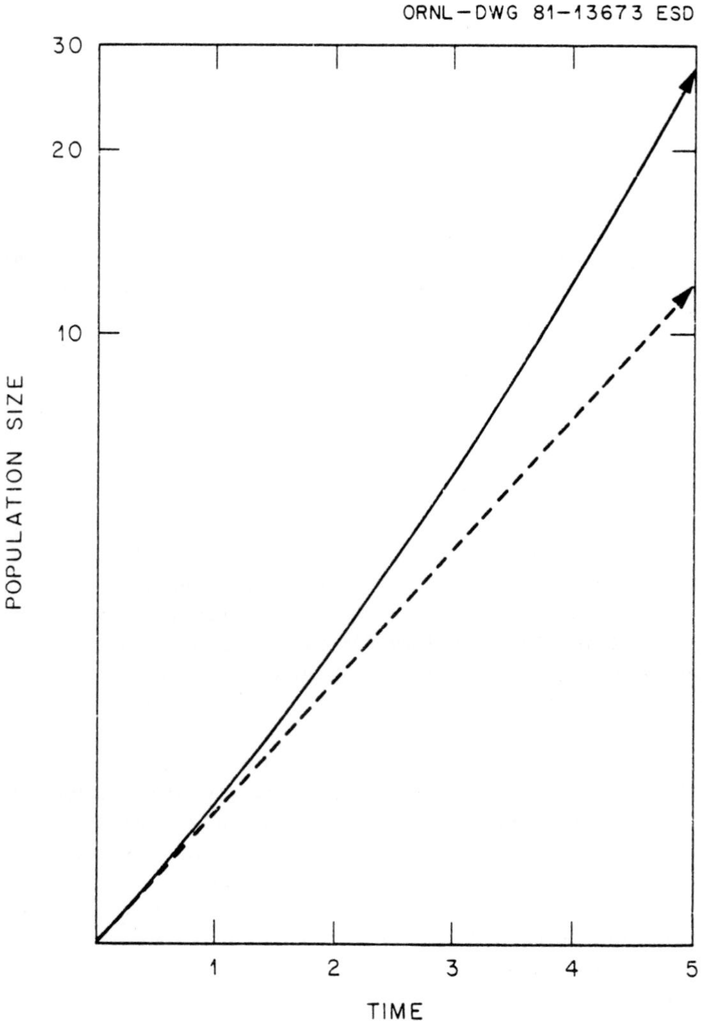


FIG. 2 - Time dynamics of a population growing at an average growth rate (- - - -) compared to the summed population size (——) of the components used to calculate the average (from (9)).

varying coefficients (not constants) in the aggregate. That is, f and g must have different functional form. Indeed, except for a few trivial cases, if a Volterra system works at one level of aggregation, no other level, higher or lower, will be described exactly by a Volterra system with constant coefficients (Sugihara, in preparation). The important message here is that such models tend to be specific to any level of aggregation, i.e., nonlinear models are scale-specific. Errors tend to occur, for example, when population processes are represented by the physiological responses of individuals (24). More will be said of this in the section on Aggregation in the Schaefer Model: Scaling.

Similarly, one must be aware that the data and parameter measurements should be scaled to fit the level of aggregation required by the model. If the model objects are highly aggregated, then the parameters should be measured on a scale that reflects this. Thus it is possible that coarse allometric estimates of growth/mortality parameters may be more appropriate for certain (large-scale) models than are data obtained directly from individuals.

**A Review of Some Specific Recommendations on Aggregation**

In most unconscious applications of aggregation theory, f and g have the same functional form, and the active and passive aggregation functions are summations or summations of partitions (e.g., sums of guild members). In such cases, errors will almost certainly occur unless f and g are linear (e.g., $f(ax) = af(x)$). The linear case has been studied extensively in input-output analysis and economics but has limited utility for multispecies fisheries.

Aggregation error has been studied for a variety of nonlinear ecological models (3, 9, 22, 24, 25). In all of these studies, the functional form of the macrosystem and microsystem is assumed to be the same with constrained consistency at equilibrium, i.e., $\Psi[f(x)] = g[\Phi(x)]$ for $x \in \{\text{equilibrium}\}$. The problem addressed by each of these investigations is to examine the error in the sum of all components (e.g., error in total biomass) as the system returns to equilibrium after a disturbance. Among a variety of interesting results that this work has produced is the recommendation that mass-balance systems should be lumped according to components that have similar loss rates (i.e., respiration or mortality), regardless of the functional form of the inputs or interaction terms. These loss rates may vary by a multiple of three in some cases and still produce less than 10% error in the aggregate. However, when the aggregation involves combinations of elements from different trophic levels, unacceptable errors are almost always the rule.

In a multispecies model containing age structure, this could translate into the recommendation that lumpings be made by age class rather than by species groups. The major differences in fish mortality occur between age classes rather than between species; larval mortality is estimated to be three orders of magnitude higher than juvenile-adult mortality (33).

The most general criterion for lumping dynamical systems is the Lange-Hicks condition. Simply stated, variables that move together may be combined error-free into a single variable which is an appropriately weighted average of the original variables (3, 37). Thus species that remain in constant proportion to one another over some time interval may be treated as one. In studying how a system returns to equilibrium after a fishing disturbance, for example, a reasonable first approximation might be to lump all species that have been disturbed more or less proportionately (i.e., those exposed to similar patterns of fishing mortality). These aggregations might be defined in set theoretic fashion by the multiplicity of fishing gear to which each species is exposed. This type of aggregation would tend to minimize the largest errors which occur shortly after the disturbance (Cale et al., in preparation). A different aggregation scheme based on the specific structure of species interactions might be required in the longer term as the system approaches equilibrium. Likewise, depending on whether control is from above in the form of a common predator, or from below, as with a common resource, the Lange-Hicks criterion suggests that it might be reasonable in some situations to aggregate according to shared predator or shared resource.

How one selects spatial and temporal scales in the data is also affected by this criterion. Given aggregations of species as defined above, the extent of the data-gathering effort should be designed so that members of the aggregates continue to "move together" in different locations. Thus, two locations may be aggregated with no error if the species in each aggregate a) occur in roughly the same proportions, or b) are exposed to similar patterns of fishing mortality. The time span over which this is true, in turn, defines the aggregations temporally. It should be emphasized that the notion of components "moving together" in an aggregate depends as much on the model as it does on the data.

**Aggregation in the Shaefer Model: Scaling**
The ultimate information that the fishery manager requires is knowledge of the relationship between the amount of fishing and the total catch. A number of studies have shown that this relationship depends on whether the analysis is done on a species-by-species basis or in various aggregations.

An especially nice study of this problem involved an analysis of the deep-sea handline fishery in Hawaii (32). A cluster analysis of the thirteen common species suggested the existence of three distinct species groups which apparently segregate on the basis of depth distribution. Application of the Shaefer model on a species-by-species basis produced very poor results. However, when catch statistics were pooled according to cluster group, the correlations between yield and effort were much improved. Furthermore, when the cluster groups were combined into a single aggregate, it was found that slightly less of the variation in total catch could be explained (not statistically significant) than when the intermediate level of aggregation was used.

This example speaks to the scale-specificity of models alluded to earlier. The single-species Shaefer model used here does not account for interspecific interactions, yet one would expect such interactions on the species-by-species level. At the cluster group level, no significant interaction was found between clusters, and thus the model was successful. Although the authors were unclear on this point, it is evident that the fit at the cluster group level was improved by the negative interactions within each cluster reducing the variance in the total biomass relations (the variance of the sum for each group was less than the sum of the variances for the individual species). That is, the cluster group variance was diminished by the negative covariances between species resulting from competition or substitutability in the fishing gear. These effects, competition and substitutability, are difficult to distinguish using catch data, but both lead to a reduced variance in the aggregated Shaefer model.

On the surface, aggregation by cluster group appears to violate the Lange-Hicks criterion. Species in each group are potential competitors and covary negatively, i.e., do not move together. However, as was mentioned earlier, the concept of components moving together is model-specific. In this classical fishing model, where the system is assumed to be in equilibrium, the idea of components moving together applies to how the equilibria move as fishing effort changes. Thus species which may be competitors but which are perfect substitutes as far as the fishery is concerned may be lumped for this analysis because their equilibria move together.

Notice the separation of time scales implied by this example. Aggregations which track slow-moving equilibria are not the same as ones used to track fast-moving competitive adjustments. The importance of this fact for fishery management is that, if one is interested in a Shaefer yield-effort forecast involving slow-moving equilibria, one should not focus attention on micro-level fast processes. Rather, because of the scale-dependence of this model, attention should be given to understanding and using data for higher-level aggregates in terms of either species clusters or expanded time or space frames for single species.

### Risks of Aggregation and Cautions

A risk to be aware of in lumping organisms is the failure to account for some crucial life-history detail which might push the organisms out of the aggregate at some future time. One must also be cautious of so-called common sense aggregations (e.g., to species) which miss subtle but important differences, e.g., growth discrepancies between male and female plaice in the North Sea or the possible genetic differences which might have explained the disappearance of northern groups of California sardine.

One of the clearest shortcomings of pooling organisms, however, is in the failure to detect changes in composition within a relatively constant total. In New England waters and in the North Sea, for example, there has been a rapid succession of dominants but relatively stable total catch biomass. Masking change may not matter for a single fishery but could be vital when one fishery in the system is targeting on a group of fish whose proportion in the catch declines substantially. Clearly, how to aggregate to highlight important changes in a system must be done with caution and on a case-by-case basis.

## CHANGE AND PERSISTENCE IN MARINE COMMUNITIES

How do we recognize changes of state for marine ecosystems? Sudden changes of structure are among the most startling phenomena of living systems. Well-known stock collapses include the Pacific sardine, the Peruvian anchovy, and the Namibian pilchard. Although these cases are characterized as catastrophic, it is not clear whether they represent a collapse from steady state, a return to steady state, or a recurring low-frequency, high amplitude cycle. Controversial data for the Pacific sardine and the Atlantic herring (38) suggest the existence of regular and dramatic fluctuations in stock levels occurring every 50 to 100 years, from levels of relatively constant high abundance to stable low abundance. These patterns appear against a background of large year-to-year fluctuations in recruitment which may be at least partially driven by environmental factors. The question of how to recognize changes in state appears to boil down to how to distinguish between natural variation and the changes induced or maintained by human activities such as fishing. Hence, a frequent turnover in dominant species, changes in species abundance distributions, and wide year-to-year fluctuations in recruitment do not necessarily signal an anthropogenic impact, rather the issues involved appear to be more subtle.

Steele and Henderson (39) offer a novel approach to this problem for the pelagic fish stocks mentioned above, which undergo a rapid fluctuation in abundance roughly every 50 years. They have been able to duplicate this observed behavior with a simple population model driven by natural environmental fluctuations and having multiple equilibria. Furthermore, it is shown that increased fishing effort will significantly increase the frequency of these marked changes in abundance. Therefore, the authors suggest that the rapid succession in dominant species observed every few years in the North Sea might be an effect of increased fishing pressure accelerating the natural 50-to-100-year period. If this is true, the assumption made by fishery management of a natural persistence in stocks will require rethinking.

The potential for lively dynamical behavior and questionable single-state stability of stocks is amplified in the multispecies context. Theoretical studies (e.g., (11, 18-20), and Beddington, this volume) have shown increasing ecosystem complexity to be accompanied by local instabilities (but see (6)), multiple stable domains, limit cycle behavior, and chaos. This work brings into question the conventional wisdom that stability should increase with system complexity and suggests rather that high diversity such as is found in the tropics might be a consequence of environmental stability. Among empirical ecologists there is no clear consensus as to which way the argument should go, and evidence has accumulated on either side, using a multiplicity of definitions for stability (cf. (30)). Within fisheries circles, where the issue is less ripe, the same confusion exists. Do the rapid changes observed in tropical fisheries with high rates of exploitation reflect system fragility, or do they reflect robustness in a fast-tracking response due to the relatively rapid increase in fishing effort? No consensus was achieved on this question within our discussions.

A related and equally provocative issue concerns the importance of alternative stable states in marine ecosystems (cf. Beddington et al., this volume). Multiple stable states are a ubiquitous feature of model systems and might be expected to occur in complex real-world fisheries (particularly with competition between age classes). Although there are a number of potential examples of multiple equilibria (Beddington and Steele, both this volume), there is some difficulty in separating true alternative states from ones maintained by fishing. This is compounded by the fact that shifts between these alternative equilibria need not be sudden, and that real-world spatial heterogeneity, often ignored in such models, may act to dampen these effects. Nonetheless, the theoretical potential for multiple states clearly exists and the fundamental practical consequences of this phenomenon must be recognized.

## TOPOLOGICAL ORGANIZATION OF MARINE ECOSYSTEMS

### What Topology Has to Offer

Nutrient cycles, energy flow, and population interactions represent alternative interpretations of the same basic linkage patterns in ecosystems. Recent studies of the formal topology of these linkage patterns for food webs have uncovered some interesting regularities in natural systems (2, 4, 29, 40, 41, 43), which appear to hold across a wide range of environments including marine, freshwater, and terrestrial habitats. Such uniformity suggests deep, general principles in ecosystem organization that have yet to be fully appreciated.

An immediate benefit of these studies is in defining the specific mathematical subset to which real ecological systems belong. Such information can be a considerable aid in building reasonable analytic models. In the same spirit, it might be profitable to superimpose crude dynamics on a specific, empirical food web in order to find regions of the web where structure makes it vulnerable. In general, real webs have been found to possess a topological property which enhances their dynamical stability.

DeAngelis (7) has shown that resiliency in a food chain, from the point of view of both energy flow and nutrient cycling, is directly related to system throughput. This suggests that webs with shorter, directed cycles should tend to be more stable than ones containing long, meandering chains. This is in rough agreement with the empirical evidence analyzed by Ulanowicz (46), where the length and number of directed cycles in webs from the Crystal River system (a Florida salt marsh stream) were found to be inversely related to disturbance from heat pollution.

Crude data on food web linkages have also been applied to the question of compartmentalization in systems, i.e., localization of strong interactions by surfaces of weak interaction. A thought-provoking review of several empirical webs has led Pimm and Lawton (31) to the conclusion that real systems are not strictly compartmented. Here, compartmentation is defined as a complete lack of connection between parallel trophic chains. On the other hand, an analysis of the topology of niche overlaps leads to the conclusion that potential competitive interactions tend to be localized into numerous small guilds or cliques. In both of these studies, the existence and not the strength of the interaction is taken into account. Thus the links between trophic chains observed by Pimm and Lawton may yet be weak in comparison to the links within each parallel chain, thereby reconciling these apparent differences.

The lack of knowledge as to the quantitative strength of linkages is an obvious shortcoming of these studies. How does one decide whether or not an interaction is important enough to be included? The fact that Cohen (4) and Sugihara (40, 41, 43) were able to find regularities in structure despite these worries suggests that these particular properties are robust. For the question of compartmentation, however, interaction strength cannot be ignored. Nonetheless, as a first step in understanding large systems, even crude data as to system structure can be very useful.

In the fisheries context, future emphasis might be profitably given to the study of food webs in nursery areas rather than of webs involving mature fish, as it is suspected that the post-larval and juvenile stages are important in regulating stock abundance. It should be emphasized that, although we are speaking specifically of food webs, one can expand these analyses to a more generalized notion of an ecosystem web. Such a web would explicitly include non-trophic aspects of ecosystems such as spawning substrates or transitions across age classes (a species could be represented as a subgraph of age categories), as well as unconventional trophic aspects such as webs of bacterial decomposers, etc. Indeed, as suggested in the previous section, it may be reasonable to abandon "species" entirely as objects in the web and focus rather on size or age classes. A study of ecosystem webs has been proposed by Oak Ridge National Laboratory and the University of Tennessee.

**Are Marine Systems Compartmented?**

As mentioned in the section on MANAGING COMPLEXITY, two related ways of reducing the complexity of a problem are by aggregating the variables involved and by limiting its extent. In choosing the latter course, one must be mindful about limiting the analysis to an arbitrarily abstracted subsystem when the larger context is really required. This appears to be the case for tropical marine fisheries where independent single-species analyses are often inadequate. Rather, as discussed in the section on MANAGING COMPLEXITY, the limits to the system should be determined by both the model used and the aggregation required for its success. For fisheries models and ecosystems models this usually translates to identifying the limits (if they exist) of the strongly interacting subsystems. Identifying such subsystems could provide information as to how far and fast a local disturbance might echo through a biotic chain. If marine ecosystems in fact are compartmented into small, strongly interacting subsystems (separated by surfaces of weak interactions), their study will be greatly simplified.

What evidence is there that marine ecosystems are compartmented? Paine (26) has documented the interaction web for intertidal communities in the Pacific Northwest and has found that while there are scores of potential linkages among species, there are only a relatively small number of strong or important ones, and these define particular, tightly interacting subsystems. The cluster groups described by Ralston and Palovina (32) for the Hawaiian deep-sea handline fishery suggest that the interactions in this system might be partitioned by depth. It was suggested by Beverton et al. (this volume) that the success of the single-species approach in the North Sea be taken to indicate that this system was a loosely knit affair. To what extent marine systems are compartmented is still an open question that might profitably be addressed by the manipulative approach described below.

## EXPERIMENTAL MANIPULATION

Numerous insights into highly interactive subsystems have been derived from controlled (experimental) manipulation of small-scale assemblages, either artificially or by natural agents (e.g., (26)). Can these results be applied approximately to pelagic fisheries? Should the effort be made to extend an experimental approach to essentially intractable ecosystems? The answer to the first question is yes: species in their communities must interact, and the smaller-scale models discuss such relevant features as critical species, cascades of effects, or whether the interactions are strong and potentially controlling or weak and inconsequential. Such controlled manipulation may also eventually shed light on a) the relationship between community stability (or resilience) and macroscopic features or organization, or b) whether control is from the trophic level above or below. Unfortunately, although fisheries' exploitation is a manipulation, it is not a controlled one, and resemblance to true experiments ends there. Nonetheless, fishing data on habitat/depth preferences, the temporal variability of stock sizes, and species co-occurrence in the gear may be profitably analyzed as a first step in pointing to potential subsystems. Equally, sensitivities of ecosystems to extreme natural events could provide easily obtainable rough-cut information as to system structure. Although experiments may be both difficult and expensive, we believe that they should be attempted whenever practicable because of their capacity to resolve variables.

A unique situation presently exists off the Spanish coast where fisheries' managers with the cooperation of local fishermen are clearing large hake, chimaera, and sharks from a portion of the continental slope with the hope of creating a shrimp fishery. The possibility of a comparison of data on this system before and after removal of these fishes will make this perhaps the first large-scale, marine fishing/removal experiment deliberately done under quasi-controlled conditions.

## A HIERARCHICAL APPROACH

We conclude this report with a rather speculative section pointing to some general guidelines for future research.

As reviewed by Steele (this volume), key processes in marine ecosystems take place over a wide variety of spatial and temporal scales. In addition to various scale constraints issuing from processes in the physical environment, one must account for the roughly 10,000-fold span in generation times in marine organisms covering bacteria to whales.

Simultaneously embracing processes and dynamics at all levels would be a formidable and unproductive task. Given the logistics of making measurements and the technical problems attending such complexity in models (see section on MANAGING COMPLEXITY), a more reasonable approach would be to develop a hierarchical approach consisting of nested models of intermediate to low complexity. This has been the costly conclusion coming out of the massive and unsuccessful IBP studies of the last decade (23, 24, 47).

In a related vein, Simon (36, 42) has argued that living systems should be composed of stable parts of intermediate complexity and that such systems should, in general, evolve much faster than ones constructed in a single step. Indeed, the ontogenetic reasons for the existence of such structure could be equally applied as a rationalization for choosing to do a hierarchical analysis of complex systems. Rather than attempting to analyze internal complexity as a whole, it would be better to take advantage of the fact that nature is inherently hierarchical and to study it in terms of a layered complexity that reflects this organization (42). The key problem here is to find the surfaces which define these layers, e.g., where levels do not interact as variables in a dynamic sense.

One possible way to define surfaces of weak interaction (in a dynamic sense) is to separate processes which operate on different time scales. Thus lower-level dynamics might be sufficiently fast to equilibrate relative to some given scale of reference, while higher-level slow dynamics would be represented as constants. Hierarchical complexity thereby decomposes into averages or moving equilibria coming from below and constants constraining the system from above (cf. (20)).

A possible first step toward such an analysis might be made using allometric information (28) to define these layers. Schwinghammer (35), for example, found biomass minima in the size spectra for benthic shallow-water communities occurring at between 500 to 100 μm and at 8 μm, a size that separates surficial microbes from mobile interstitial fauna. One can speculate, therefore, on the existence of three distinct macroscopic time compartments for this system, a critically important problem worth looking into further.

## Mathematics of Hierarchies

Some relevant mathematical work which might be brought to bear here should be briefly mentioned.

Conditions under which one can dissect the behavior of a dynamical system into fast and slow components were described by Tychonoff (45). Roughly speaking, if a system consists of a compact set of m fast components $v_j$, and n-m slow ones $u_j$, such that

$$\epsilon \dot{v}_j = f_j (t, u, v_j) \quad , j = 1, \dots , m, \tag{2}$$

then as $\epsilon \rightarrow o$, the system approaches the degenerate solution $u_i = g(t, u, v)$ of dimension n-m, where v is the m-dimensional manifold defined by the equilibrium solutions of v for all u. That is, the system's slow components will tend to be confined to some neighborhood of the manifold defined by the equilibria of its fast equations. The proof of this result can be shown to depend on Tychonoff's more famous theorem that the product of compact sets is again compact. Unfortunately, no general way of measuring the errors of this approximation in terms of $\epsilon$ has been proposed.

Goodwin (12) has proposed a heuristic method for analyzing hierarchies in terms of successive averaging operations, or what he calls "inverse limit systems." Here a sequence of averaging devices for each level in a hierarchy are combined in a convolution integral. Thus, for example, if $\omega_i$ and $\omega_j$ are sequential averaging operators, such that each operates in a time period $s_i$ and $s_j$, respectively, then the output $\langle X(t) \rangle$ is

$$\langle X(t) \rangle = \int_0^\infty \omega_j(s_j) dsj \int_0^\infty \omega_i(s_i)\, X(t_j - s_j - s_i)\, dsi. \tag{3}$$

Here again, there is no discussion of the errors involved in using this composition rule or of the averaging operators involved.

On a slightly different tack, Simon and Ando (37) have analyzed dynamical behavior in a hierarchical decomposition of linear systems of the form

$$X(t) = X_o P^t, \tag{4}$$

where P is a real square matrix, $P = P^* + \epsilon\ C$. $P^*$ is completely decomposable into independent submatrices and C is an arbitrary matrix of the same dimension as $P^*$. P, therefore, is only partly decomposable, consisting of semi-independent block submatrices. The authors study the system as $\epsilon \rightarrow o$, where fast-time dynamics are contained within each subblock and long-run dynamics emerge from an aggregate system having one variable for each subsystem with interactions between subblocks. Although their analysis is for a constant matrix P, it is not difficult to rework it for the case when P changes slightly, by some epsilon, in each time step.

**Acknowledgements.** Research sponsored by the Office of Health and Environmental Research, U.S. Department of Energy, under Contract No. DE-AC05-840R21400 and NSF BSR-8315185 with Martin Marietta Energy Systems, Inc. (Publication No. ----, Environmental Sciences Division, ORNL).

## REFERENCES

(1) Andersen, K.P., and Ursin, E. 1977. A multispecies extension to the Beverton and Holt theory of fishing, which accounts of phosphorus circulation and primary production. Medd. Dan. Fisk. Havunders.

(N.S.) 7: 319-435.

(2) Briand, F. 1983. Biogeographic patterns in food web organization. In Current Trends in Food Web Theory: Report on a Food Web Workshop, eds. D.L. DeAngelis, W.M. Post, and G. Sugihara. ORNL/TM-5983. Oak Ridge, TN: Oak Ridge National Laboratory.

(3) Cale, W.G.; O'Neill, R.V.; and Gardner, R.H. 1983. Aggregation error in nonlinear ecological models. J. Theor. Biol. 100: 539-550.

(4) Cohen, J.E. 1978. Food Webs and Niche Space. Princeton, NJ: Princeton University Press.

(5) Cohen, J.E. 1979. Long run growth rates of discrete multiplicative processes in Markovian environments. J. Math. Anal. Applic. 69: 243-251.

(6) DeAngelis, D.L. 1975. Stability and connectance in food web models. Ecology 56: 238-243.

(7) DeAngelis, D.L. 1980. Energy flow, nutrient cycling and ecosystem resilience. Ecology 61: 764-771.

(8) FAO. 1978. Some scientific problems of multispecies fisheries. Report of the Expert Consultation on Management of Multispecies Fisheries, Rome, 20-23 September 1977. FAO Fish. Tech. Paper 181.

(9) Gardner, R.H.; Cale, W.G.; and O'Neill, R.V. 1982. Robust analysis of aggregation error. Ecology 63 : 1771-1779.

(10) Gatz, A.J.; Sale, M.J.; and Loar, J.M. 1984. Habitat use by rainbow trout and brown trout: influence of available habitat and interspecific competition. Ecology, in press.

(11) Gilpin, M.E., and Case, T.J. 1976. Multiple domains of attraction in competition communities. Nature 261: 40-42.

(12) Goodwin, B.C. 1976. Analytical Physiology of Cells and Developing Organisms. New York: Academic Press.

(13) Gorman, O.T., and Karr, J.R. 1978. Habitat structure and stream fish communities. Ecology 59: 507-515.

(14) Gosline, W.A., and Brock, V.E. 1960. Handbook of Hawaiian Fishes. Honolulu: University of Hawaii Press.

(15) Larkin, P.A. 1963. Interspecific competition and exploitation. J. Fish. Res. Board Can. 20: 647-678.

(16) Larkin, P.A. 1966. Exploitation in a type of predator-prey relationship. J. Fish. Res. Board Can. 23: 349-356.

(17) Loar, J.M., ed. 1984. Assessing Stream Flows for the Protection of Fishery Resources: A Field Evaluation of Existing Methods in Southern Appalachian Trout Streams. ORNL/TM-9323. Oak Ridge, TN: Oak Ridge National Laboratory.

(18) May, R.M. 1974. Stability and Complexity in Model Ecosystems. Princeton, NJ: Princeton University Press.

(19) May, R.M. 1977. Thresholds and breakpoints in ecosystems with a multiplicity of stable states. Nature 269: 431-477.

(20) May, R.M. 1979. The structure and dynamics of ecological communities. In Population Dynamics, eds. R.M. Anderson, B.D. Turner, and L.R. Taylor. Oxford: Blackwell Scientific.

(21) May, R.M.; Beddington, J.R.; Clark, C.W.; Holt, S.J.; and Laws, R.M. 1979. Management of multispecies fisheries. Science 205: 267-277.

(22) O'Neill, R.V. 1973. Error analysis of ecological models. Report Presented at the Third National Symposium on Radioecology in 1971. Deciduous Forest Biome Memo Report 71-15.

(23) O'Neill, R.V. 1975. Modeling in the eastern deciduous forest biome. In Systems Analysis and Simulation in Ecology, ed. B.C. Patten, pp. 49-94. New York: Academic Press.

(24) O'Neill, R.V. 1979. Transmutations across hierarchical levels. In Systems Analysis of Ecosystems, eds. G.S. Innis and R.V. O'Neill, pp. 59-78. Fairlands, MD: International Cooperative Publishing House.

(25) O'Neill, R.V., and Rust, B.W. 1979. Aggregation error in ecological models. Ecol. Mod. 7: 91-105.

(26) Paine, R.T. 1980. Food webs: linkage, interaction strength and community infrastructure. J. Anim. Ecol. 49: 667-685.

D. 1979. Theory and management of tropical multispecies A review, with emphasis on the Southeast Asia demersal ;. ICLARM Stud. Rev. 1.

(28) Platt, T. Structure of marine ecosystems: its allometric basis. (Manuscript).

(29) Pimm, S.L. 1982. Food Webs. London: Chapman and Hall.

(30) Pimm, S.L. 1984. The complexity and stability of ecosystems. Nature 307: 321-326.

(31) Pimm, S.L., and Lawton, J.H. 1980. Are food webs compartmented? J. Anim. Ecol. 49: 879-898.

(32) Ralston, S., and Palovina, J.J. 1982. A multispecies analysis of the commercial deep-sea handline fishery in Hawaii. Fish. Bull. 80: 435-448.

(33) Rothschild, B.J. 1981. More food from the sea? Bioscience 31(3): 216-222.

(34) Samuelson, P.A. 1948. Foundations of Economic Analysis.

(35) Schwinghammer, P. 1981. Characteristic size distributions of integral benthic communities. Can. J. Fish. Aquat. Sci. 38: 1255-1263.

(36) Simon, H.A. 1973. The organization of complex systems. In Hierarchy Theory: The Challenge of Complex Systems, ed. H.H. Pattee. New York: Braziller.

(37) Simon, H.A., and Ando, A. 1961. Aggregation of variables in dynamic systems. Econometrica 29: 111-138.

(38) Soutar, A., and Isaacs, J.D. 1974. U.S. Fish. Wildl. Serv. Fish. Bull. 72: 257.

(39) Steele, J.H., and Henderson, E.W. 1984. Modelling long-term fluctuations in fish stocks. Science 224: 985-987.

(40) Sugihara, G. 1983. Holes in niche space: A derived assembly rule and its relation to intervality. In Current Trends in Food Web Theory - Report on a Food Web Workshop, eds. D.L. DeAngelis, W.M. Post, and G. Sugihara. ORNL/TM-5983. Oak Ridge, TN: Oak Ridge National Laboratory.

(41) Sugihara, G. 1983. Niche Heirarchy: Structure, Organization, and Assembly in Natural Communities. Ph.D. Dissertation, Princeton University, Princeton, NJ.

(42) Sugihara, G. 1983. Peeling apart nature. Nature 304: 94.

(43) Sugihara, G. 1984. Graph theory, homology, and food webs. In Population Biology, ed. S.A. Levin. American Mathematical Society Proceedings, vol. 30.

(44) Swingle, H.S. 1950. Relationships and dynamics of balanced and unbalanced fish populations. Agric. Exp. Stat. A Lab. Polytech. Inst. 274: 1-74.

(45) Tychonoff, A.N. 1950. On systems of differential equations containing parameters. Mat. Sb. 27: 147-156 (in Russian).

(46) Ulanowicz, R.E. 1983. Identifying the structure of cycling in ecosystems. In Current Trends in Food Web Theory: Report on a Food Web Workshop, eds. D.L. DeAngelis, W.M. Post, and G. Sugihara. ORNL/TM-5983. Oak Ridge, TN: Oak Ridge National Laboratory.

(47) Watt, K.E.F. 1975. Critique and comparison of biome ecosystem modelling. In Systems Analysis and Simulation in Ecology, ed. B.C. Patten, vol. III, pp. 139-152.

(48) Werner, E.E., and Hall, D.J. 1976. Niche shifts in sunfish: experimental evidence and significance. Science 191: 404-406.

(49) Werner, E.E., and Hall, D.J. 1979. Foraging efficiency and habitat diversity in competing sunfish. Ecology 60: 256-264.

(50) Werner, E.E.; Hall, D.J.; Laughlin, D.R.; Wagner, D.J.; Wilsmann, L.A.; and Funk, F.C. 1977. Habitat partitioning in a freshwater fish community. J. Fish. Res. Board Can. 34: 360-370.

(51) Zaret, T.M., and Paine, R.T. 1973. Species introduction in a tropical lake. Science 182: 449-455.

Exploitation of Marine Communities, ed. R.M. May, pp. 155-190. Dahlem Konferenzen 1984. Berlin, Heidelberg, New York, Tokyo: Springer-Verlag.

# Observed Patterns in Multispecies Fisheries

J.A. Gulland and S. Garcia
Dept. of Fisheries, FAO
Rome, Italy

**Abstract.** Virtually all fisheries exploit more than one species of fish, but the range of species exploited and the ability and willingness of fishermen to switch attention from one species to another vary. The fisheries off northwest Africa are reviewed as an example of a multispecies, multifishery system. Examples are given of systems in other regions. The impact of having to deal with a multispecies, rather than a single-species situation, on the different stages of fishery management is discussed.

## INTRODUCTION

The basic biological or economic theories actually used to advise fishery managers deal mostly with the simple case of a single fishery exploiting a single monospecific and stable stock of fish. In practice, the same stock of fish may fluctuate and be harvested intentionally or accidentally by a wide range of vessels, from canoes to factory vessels, and a given vessel usually catches during the year, or during a single fishing operation, a number of different species. The biological interactions between species (competition, predator-prey), including the stability of the community structure and the operational interactions between different groups of vessels targeting a particular species, need to be assessed and taken into account, as do the effects on shore, e.g., the interactions between the markets for different species.

The question of interaction and stability in multispecies fisheries has therefore become recognized as a matter deserving of increased practical and theoretical attention, and a number of studies have addressed this problem (14, 30, 40, 41), but these have done little more than open the door on a complex maze of interrelated problems. An important initial difficulty is that there are many types of multispecies fishery, and of multispecies problems, and many different scales at which they can be examined. Simple cases, such as the krill-whale system with one fishery on a predator and another on its prey, are in the minority in the sea. It is probably premature to attempt a definitive taxonomy of multispecies systems, but this paper will attempt to identify some of the main types.

Patterns and problems characteristic of multispecies fisheries will be shown in some detail through the example of West Africa where most of the classical events and some exceptional ones have occurred. This will be followed by a selected review of examples of different types of problems in various parts of the world. The final section deals with the types of question that need to be answered; for convenience this has been arranged following the stages of management identified by the ACMRR Working Party on the Scientific Basis of Determining Management Measures (15).

## CHARACTERISTIC MULTISPECIES PATTERNS: THE EXAMPLE OF WEST AFRICA

The fisheries off West Africa can be considered as a good example of tropical multispecies fisheries. They have all the necessary classical characteristics of these types of resources, and above all, they have undergone in the last ten years a series of dramatic (apparent and actual) changes in species composition and distribution that can be used to illustrate some of the patterns encountered in the exploited multispecies fisheries. The environments range from equatorial lagoons with high rainfall to tropical upwelling systems. The fisheries range from subsistence artisanal ones with canoes to highly industrialized ones with modern long-range factory ships.

### The Resource Is Multispecific

Catches are everywhere characterized by very high numbers of species, the highest diversity being observed in the biogeographical transition zone between the warm temperate and the strictly equatorial areas between Cape Verga and Cape Bojador.

In the Senegal-Mauritania shelf demersal resource as many as 174 species can be identified (10), and any single trawl haul can contain 20-50 species in the coastal area of Cape Timiris. The pelagic resources comprise less species, and most of the pelagic biomass over the shelf of Mauritania is made up of eight species of carangids, clupeids, squids, and cuttlefish.

The official statistics tend to group catches into commercial categories consisting often of several species, but still 48 species are identified in the general statistics of Senegal (24, 34). These species can fortunately be grouped in a limited number of species assemblages (10, 37) on the basis of co-occurrence on biotypes characterized by depth and bottom type. It is interesting to note that according to Longhurst, similar assemblages containing similar species can be found in many other intertropical areas of the world.

The high species diversity is associated with an important space/time heterogeneity in the resource (Fig. 1), according to bottom type, depth, or position of the thermocline. The species composition tends to vary in a given place seasonally in relation to bathymetric movements of

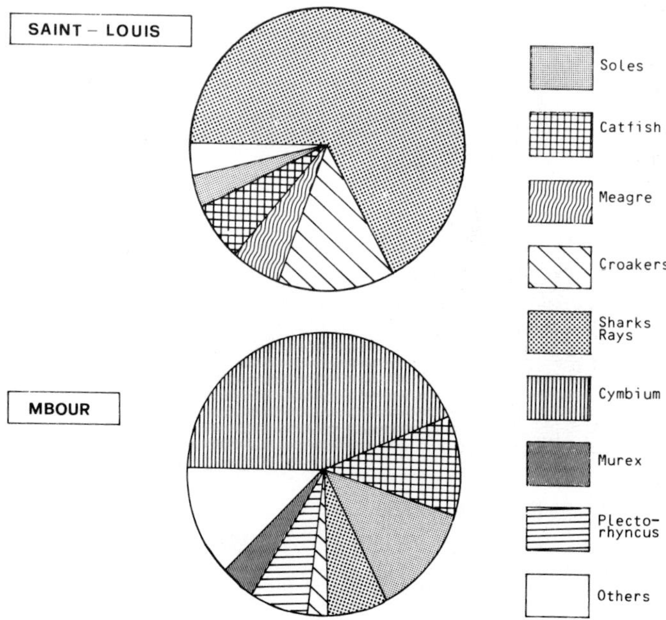


FIG. 1 - Differences in annual species composition of catches of bottom set gill nets in two different villages of Senegal (52).

fish with changes in river discharge (monsoon migrations) or upwelling strength. Important seasonal feeding and spawning migrations occur around the tropics (between 10 and 22° north, for instance), changing also drastically the composition of the resource available to the artisanal fishery, off Senegal, for instance (Fig. 2) (7, 26).

The combination of all these sources of heterogeneity affects the fisheries and leads to high variability in the catches which are often increased by changes in fishing strategy.

**The Catch Composition Depends on the Fishing Strategy**
Species composition of catches differs greatly according to the targets sought. It may vary greatly between two fleets exploiting the same area (Table 1). It may also vary with time for the same fishery. The trawl fishery of Senegal is a good example of this as seen in Fig. 3 which

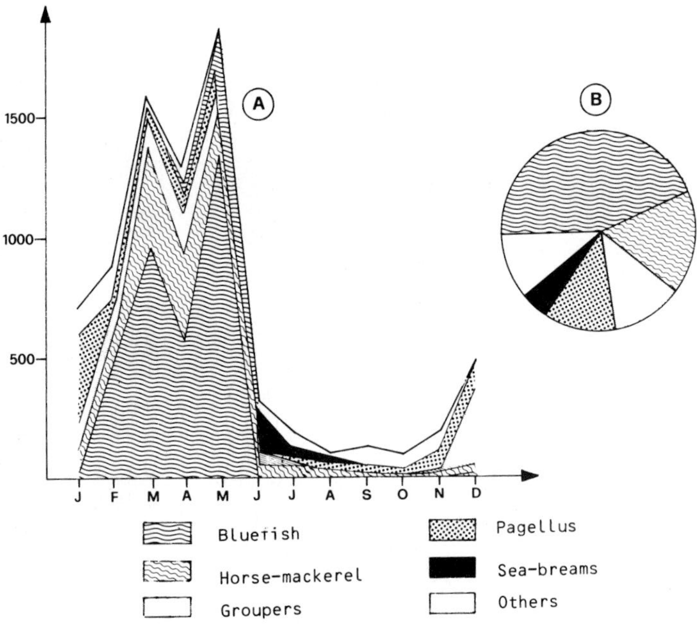


FIG. 2 - Seasonal changes in species composition of landings (A) and annual species composition (B) in the handline/canoe fishery of Kayar, Senegal (52).

TABLE 1 - Comparison of species composition of landings of two long-range trawl fleets on the Guinea Bissau shelf.

| Species | Japanese fleet | Russian fleet |
|---|---|---|
| Sciaenids | 3% | 61% |
| Sparids | 50% | 16% |
| Carangids and clupeids | 22% | 9% |
| Lutjanids | 8% | 0.02% |
| Balistes | 0.3% | 0% |

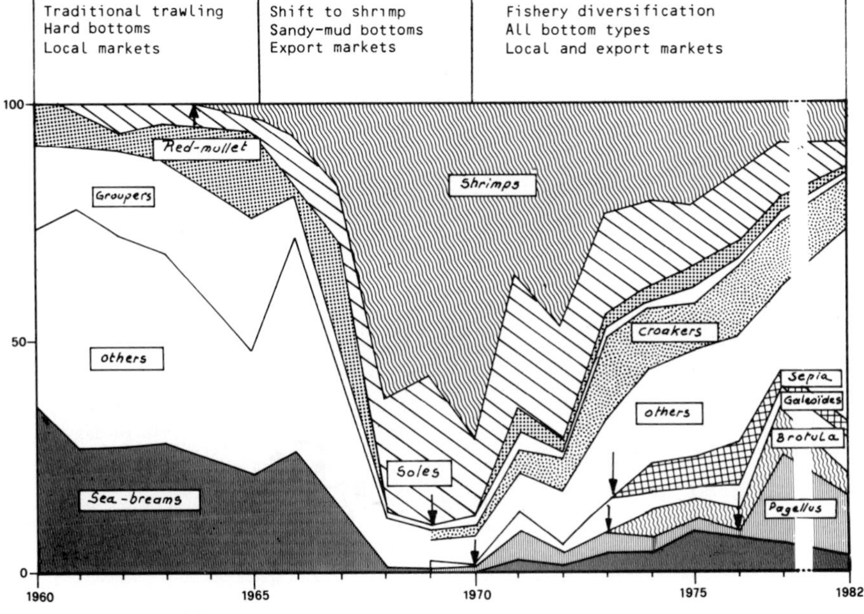


FIG. 3 - Changes in percentage species composition of landings in the Senegalese trawl fishery. The arrows indicate the appearance of new targets in the fishery (27).

depicts the drastic modifications in species composition of landings resulting from changes in the export market opportunities and discovery of new resources. It illustrates nicely a process which probably always occurs as a fishery develops and progressively "colonizes" the whole spectrum of available resources. If not properly documented, these changes in landings can be confused with actual changes in abundance.

**Discards Are a Characteristic Feature**

They are a direct consequence of species diversity contained with unequal market value. They are widespread in all tropical areas and are particularly important off northwest Africa. In Senegal, for instance, Monnoyer (43) reported that discards in the trawl fishery ranged between 65% (for the fishery targeted on red mullet) to 75% (on shrimps) with the discards when fishing on sparids and cephalopods varying from 60-70%.

The total quantity discarded is roughly estimated as 90,000 t, compared with landings of 40,000 t. Even in the "poor" Mediterranean area, it is noted that 44 to 72% of the catch is discarded (1).

Discarding practices generally change with time and, if not properly documented, may be confused with trends in abundance. An example is given in Fig. 4.

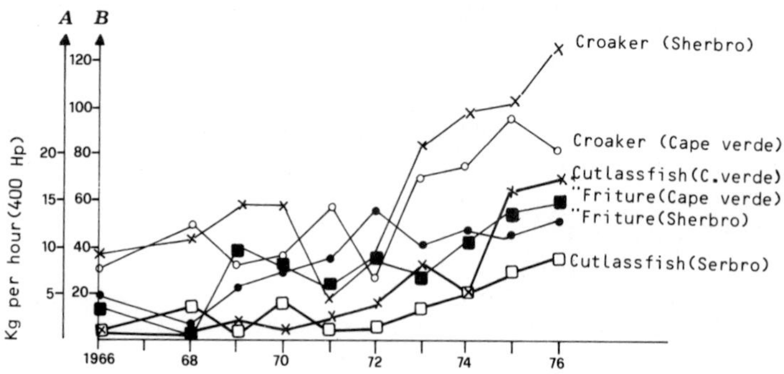


FIG. 4 - Changes in CPUE (catch per hour of a 400 h.p. trawler) of characteristic species landed by Ivory Coast trawlers off Sierra Leone (Sherbro division) and Senegal (Cape Verde Coastal division) related to change in discarding practices (6).

## The Species Composition of the Resource Changes

One basic question about multispecies resources is the stability of their composition. While undocumented changes in fishing strategy or discarding practices may be confused with real changes in species composition or abundance, the West African resources offer a few striking examples of real and large-scale changes.

## The Collapse of Sardinella aurita

The pelagic community of the Ivory Coast, Ghana, and Togo dominated by sardinellas, horsemackerels, and mackerels has suffered from a collapse of its major component, the Sardinella aurita, in 1973 after a 3-4-fold relative increase in catch in 1972 and recovered, apparently, 3-6 years later (Fig. 5B). The phenomenon has been analyzed many times (4, 20, 25). It seems true, from the results presented by Binet and despite the limited number of data points, that before the collapse the catch used to vary with upwelling strength and river outflow (Fig. 5A, Curve A), these two environmental variables being themselves (loosely) inversely related. When river outflow decreased sharply during the 1970s and upwelling strength increased, it seemed that plankton abundance (and food availability) decreased (4) and that stock availability inshore increased, thus potentially affecting larval survival while increasing fishing mortality.

The high 1972 catches due to drastic increase of catchability of sardines to artisanal purse seining, because of exceptionally low river flow, led to collapse, probably aggravated by low larval survival due to poor feeding conditions and low fecundity of the collapsed stock which consisted almost entirely of the 0+ age group. Since the collapse, some faint relationship between catches and environment may exist, but at a much lower level of catches (Fig. 5A, Curve B). The separation of the two sets of data in Fig. 5 is probably a bit artificial, and it is more likely that there exists a continuum in the data set from Curves B to A related to the changes of underlying biomass under the effect of both recruitment and fishing mortality changes during the recovery phase.

It must finally be noted that in this case biomass is affected by environmental changes through changes in recruitment as well as in fishing mortality (despite a relatively constant effort) rendering impossible the separation of the two effects with the data available. A schematic interpretation of Fig. 5A is given in Fig. 5C. This figure may imply that the environment was the only driving factor. However, as catchability is modified by the environment quite drastically in this particular case, the data could just as well be interpreted with the classical production model theory as in Fig. 5D following MacCall (38).

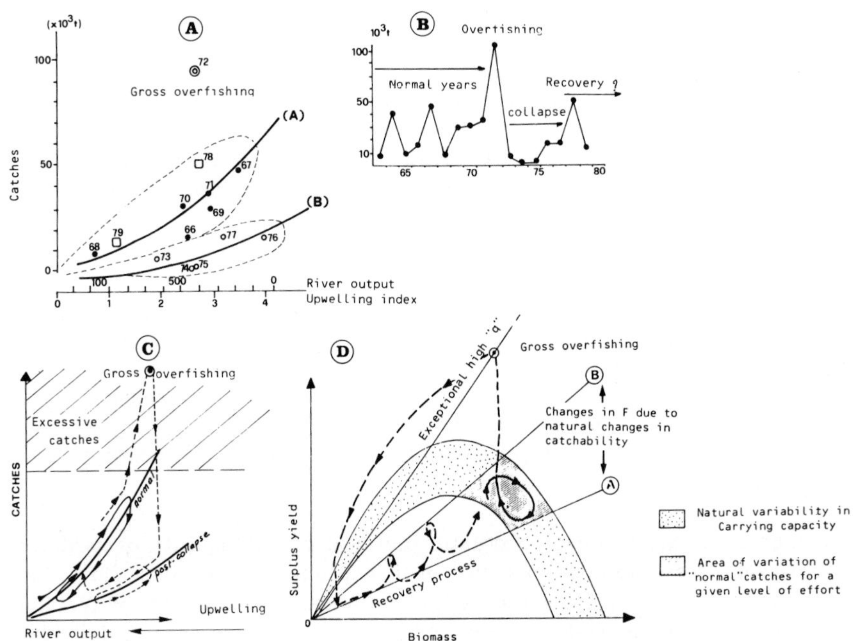


FIG. 5 - The collapse of the Sardinella aurita stock off Ivory Coast/Ghana/Togo.

5A - Relationship between environmental variables and catches before and after the major 1972 overfishing. Curves are drawn by eye. Modified from (4). Black circles = normal, pre-collapse situation; open circles = post-collapse; open squares = recovery.

5B - Catch time series.

5C - "Environmental" interpretation of the phenomenon shown in Fig. 5A and 5B, showing time series of catches under moderate fishing (full line), and overfishing (broken line).

5D - The combination of environmentally driven changes of biological productivity and catchability under constant fishing effort: an example of environmentally induced overfishing.

TABLE 2 - Changes in catch composition of the Soviet pelagic trawl fishery off Guinea Bissau (22).

| Species | 1978-79 (%) | 1982 (%) |
|---|---|---|
| Sardinella | 55 | 0.05 |
| Trachurus | 38 | 29 |
| Decapterus | 7 | - |
| Caranx | 0.6 | - |
| Mackerel | 0.2 | - |
| Balistes | 0 | 64 |
| Others | 31 | 20 |

**The "Invasion" of Balistes carolinensis**

This species was present in the hard bottom shelf/sparid fish community in insignificant quantities during the 1960s in the Gulf of Guinea (rarely more than 10 kg/h) while 20 t/hour could be caught by trawling in Guinea-Bissagos at the beginning of the 1980s (Table 2). According to Caverivière (6), the biomass increased from 1971-72* off the Ivory Coast-Ghana and from 1974-75 off Guinea/Guinea-Bissau, leading to two very important stocks now existing (Fig. 6) whose biomass is estimated to be around $5 \times 10^5$ t in the Ivory Coast-Ghana and $0.4 - 1.3 \times 10^5$ t off the Guineas, and represents in some sectors 80% of the pelagic biomass (53). From 1978 the "Guinean" stock of Balistes started to invade seasonally in summer the Senegalese waters north of Cape Verde and up to southern Mauritania.

The possible reasons for this most spectacular phenomenon are not known. It occurred in Ivory Coast-Ghana before the overfishing of Sardinella aurita and was apparently associated with a decrease in biomass of the traditional species caught by the Ivorian trawl fishery (6). The data show in fact that the data points for catch and effort in this fishery for 1966-80 are well below the production model fitted to the data of 1959-65. We have calculated the yearly anomalies as percent of the expected CPUE and plotted the time series together with river overflow, salinity anomaly, and plankton abundance (Fig. 7). There is evidence in the figure that the trends are comparable, and that the decrease in temperature, river output, food availability, and the increase in upwelling and salinity may have reduced the carrying capacity of the area for the

* Because of the average ages in the catches (about two years) it must, however, be assumed that the recruitment started to increase around 1969-70.

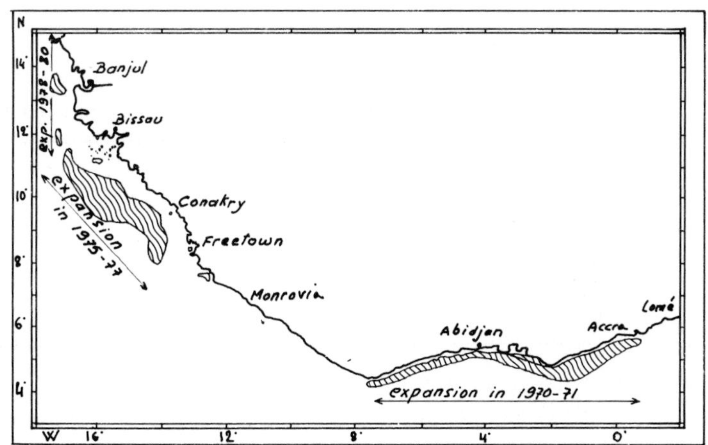


FIG. 6 - Distribution of triggerfish, Balistes carolinensis, from Dakar to Lome in June 1981 (53). Data about expansion periods have been added.

Cpue anomaly in percent(Balistes excl.)
Plankton abundance
River output

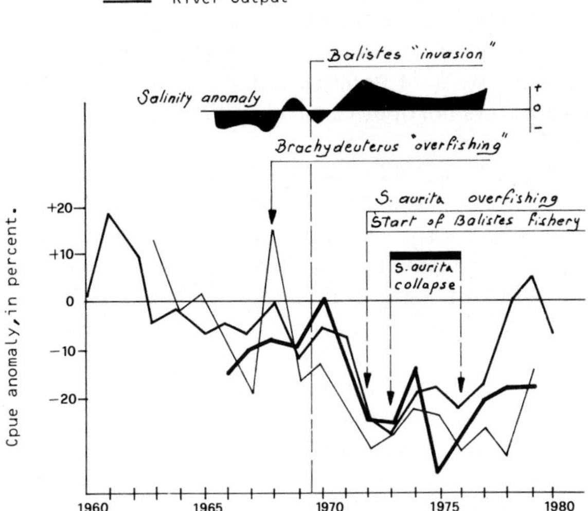


FIG. 7 - Trends in CPUE anomalies in the trawl fishery, river output, plankton abundance, salinity anomalies, and fishery events in the Ivory Coast in relation to the changes in species composition of the resource (Sardinella collapse, Balistes eruption). Environmental data from (4); fishery data original, extracted from (6).

warm water, low salinity sciaenid superthermoclinal community traditionally exploited by the trawl fishery while increasing it for the cool water, high salinity infrathermoclinal sparid community to which the Balistes belongs (37). It can be noted that Fig. 7 seems to show some return to "normal" conditions for the environment and the trawl fishery productivity, while Balistes is still abundant as far as it is known. This may indicate either a lack of relationship or some hysteresis or nonreversibility in the phenomenons involved.

Finally, it is interesting to note that an increase in abundance of the flying gurnard (Cephalacantus volitans) belonging to the same sparid community has apparently occurred at the same time as for Balistes. This is indicated for the Ghanaian area (48) and is seen also to be the case in Guinea Bissau.

**The "Replacement" of Sparids by Cephalopods**

Between Cape Blanc and Cape Garnett off the coast of the Sahara lies one of the most famous fishing grounds for demersal fish of northwest Africa. The fish communities have been studied since the 1940s and a good review exists (39). The fishery fauna has been traditionally largely dominated by gray (in the littoral area) and red (on the shelf) sparids of the genera Pagellus, Pagrus, Dentex, Diplodus, Sparus, etc. However, cephalopods and especially cuttlefish (Sepia) and squids have always been a permanent feature of these species assemblages in this region. Octopus, however, was rarely mentioned in most earlier works (28) and is mentioned as a significant element only in 1962 ((39), p. 63). In various scientific expeditions the relative abundance of cephalopods among the rest of the commercial resources raised from about nothing in 1961-62 to about 30% in 1968 and 90% in 1971 (46). According to Garcia-Cabrera (28), however, the proportion was already 80% in 1967 (17 t of fish for 66 t of cephalopods). Even if the scientific evidence is still weak, the Canarian fishermen who exploited the area for centuries definitely consider the buildup of the cephalopod stock in the early 1960s as a fact. The available recent information on the subject has been summarized in Fig. 8A to 8D.

Figure 8A illustrates the relative changes in catches of sparids and cephalopods since 1964-65 and shows the dates when the stocks have been declared overfished. Figure 8B shows trends in recruitment indexes of octopus and catch rates of big and medium-sized cuttlefish. This figure indicates the drastic increase of recruitments in 1965-66 for both species. Figure 8C shows the decrease of abundance of big and medium-

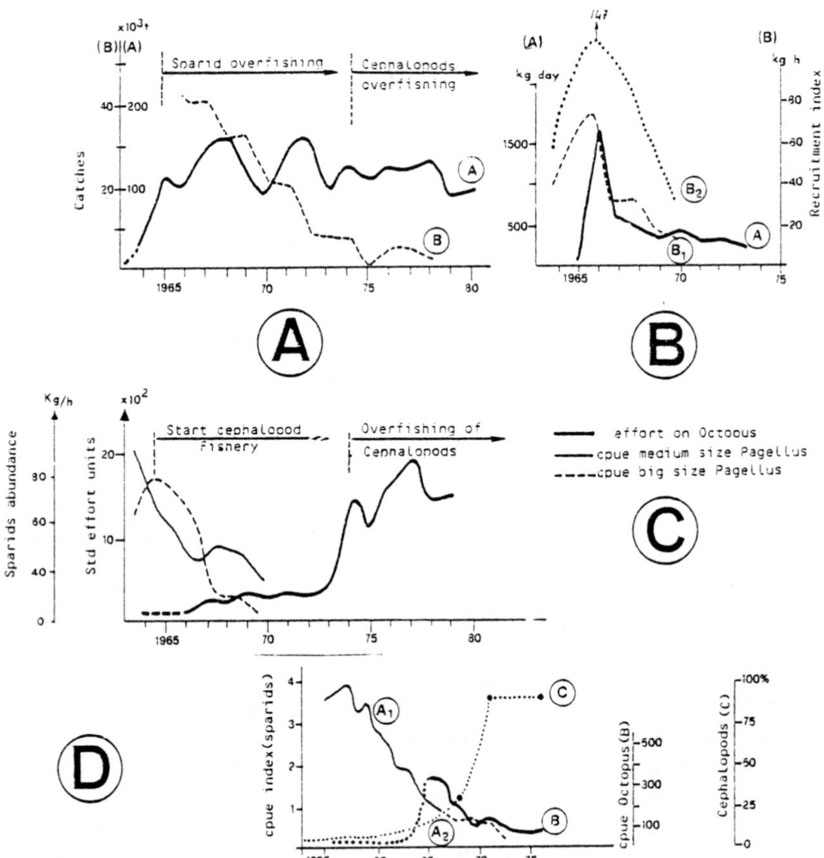


FIG. 8 - Some indexes for monitoring the evolution of the sparid/cephalopod community in the Sahara (Coastal) division.

8A - Evolution of catches of sparids (A) and cephalopods (B). Data source (A) (9, 21), and (B) (16, 17).

8B - Recruitment indexes for octopus (A) in Cape Blanc (21) and abundance of big size (....) and medium size (——) cuttlefish as measured by CPUE (B) (32).

8C - Development of fishing effort on octopus as an indicator of the whole cephalopod fishery (19) and abundance of big and medium-sized Pagellus (33).

8D - Variations in abundance of sparids ($A_1$, $A_2$) and octopus (B). $A_1$ modified from FAO (12), $A_2$ from FAO (13): see text B from (2). C proportion of cephalopods in commercial catches during trawl surveys (46).

sized sparids and the simultaneous increase in fishing effort for octopus (it is believed that the effort increased as well on all cephalopods). As all these figures start from 1965 while the fishery for sparids started after the Second World War at much higher levels of abundance, we have grouped on the same Fig. 8D changes in sparid abundance in northwest Africa in 1955-66 (a period when catch statistics were only grossly geographically allocated) with changes of abundance in the Sahara (littoral) statistical division (1965-73). Both series have been expressed relative to the average CPUE for 1965 and 1966. This figure is only intended to show that sparids have been apparently driven to close to extinction level and that cephalopods increased when sparid abundance was about 30% of their 1955 level.

The figure also shows that, in fact, both sparid and cephalopod stocks have been driven down drastically by fishing. An important fact, not shown in the figures, is that the proportion of small-sized sparids increased drastically in the 1970s, possibly as the fishery moved further inshore searching for spawning concentrations of octopus. According to Garcia-Cabrera (28), huge amounts were reduced into fish meal and later on were discarded when the fishery specialized on cephalopods. These data show that 94% of the fish catch was discarded in 1967 and this observation is largely valid today.

It seems, therefore, clear that the Saharan sparid community has been drastically affected by fishing which reduced the biomass and average individual size of sparids inducing a transfer of effort to cephalopods at the same time as their biomass, especially for octopus, increased. Later, heavy fishing reduced also the biomass and average size of cephalopods. Garcia-Cabrera (28) indicates that the thinning down of sparids suppressed an important source of larval mortality of octopus in the short pelagic phase and that in addition the huge quantities of fish discarded were a good source of food supply for cephalopods. It should probably also be noted that a long-term upward trend in upwelling has been identified in the region between 1967 and 1980 (3) and that we do not know how the continuous cooling down of the region has affected also the larval survival of the sparid community by Pagellus bellotti and other sparids of "Guinean" affinity whose concentrations off the Sahara are at the extreme northern limit of the distribution of these species in West Africa. Some possible mechanisms for such replacements are discussed by Caddy (5).

## The Sardine Expansion

Off West Africa, between 20-36°N, there are three important fishing areas for Sardina pilchardus: in the North (36-33°N), the Center (32°30 to 27°N), and the South (26° to 21°N). It is assumed that they correspond to separate stocks, but some seasonal partial mixing might well occur. The biomass in the central and southern concentrations has widely fluctuated in the last twenty years (3), drastically modifying the catch composition (Table 3).

In the central area, long-term changes in availability to Moroccan traditional fleets and in recruitment have been related either to extreme droughts and/or fluctuation in upwelling strength (Fig. 9). In the southern area the spectacular expansion of Sardina pilchardus catches from $37 \times 10^3$ to $1 \times 10^6$ t in seven years was related to the strengthening of the upwelling in the early 1970s (11, 31). In this area the sardine has been progressively disappearing again since 1973 (35), and it can be seen in Fig. 10 that this seems to correspond with the decrease of the upwelling. These changes have, of course, affected the whole pelagic fish community, and Fig. 10 gives a representation of changes occurring in the catch composition and showing the apparent "replacement" of the tropical species Scomber and Sardinella by the warm temperate Sardina, while the effect on horsemackerel is less clear. The "replacement" of Sardina by Scomber in the central area has also been described and is related to changes in coastal water masses (3).

TABLE 3 - Changes in species composition of Polish fishery in the northern sector of CECAF (22). (The same changes can be observed in Bulgarian catches.)

| Species | 1968-72 (%) | 1975-77(%) |
|---|---|---|
| Sardine | 2 | 87 |
| Horsemackerel | 30 | 5 |
| Mackerel | 20 | 5 |
| Hairtails | 5 | 0 |
| Bluefish | 7 | 0 |
| Sparids | 4 | 0 |
| Sardinella | 6 | 3 |

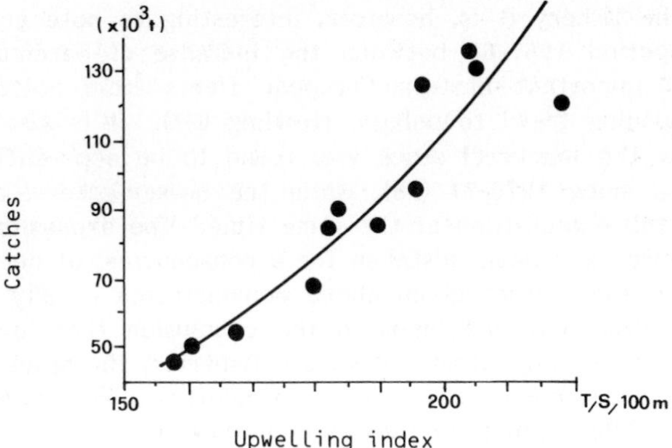


FIG. 9 - Relationship between annual catches and upwelling in the central area (3).

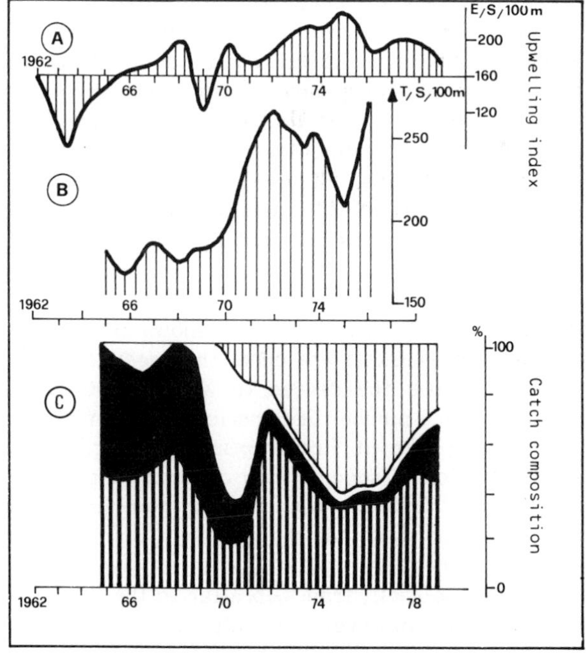


FIG. 10 - A) Variations in upwelling strength in the central area (3). B) Variations in upwelling strength in the southern area (50). C) Changes in species composition of catches (data sources (13, 17)).

Regarding the fishery it is, however, interesting to note a coincidence during the period 1967-68 between the increase of Sardina stock and the shift of important Eastern European fleets from bottom trawling with high opening trawl to pelagic trawling (22). It is also interesting to note that the mackerel stock was found to be apparently exploited beyond $f_{MSY}$ since 1970-71 (18), while the horsemackerel appeared to be close to full exploitation at the same time. The expansion of sardine could therefore have been mistaken for a consequence of heavy or overfishing, while the picture given above demonstrates clearly enough the effect of environment and leads to the conclusion that, owing to the difficulty of measuring effort on pelagic fisheries, the stock assessment of mackerel and horsemackerel and particularly their apparent "fishing down" in the 1970s ought to be reassessed in view of the long-term changes in biomass of sardine and availability of mackerels. (This possibility was in fact mentioned in the papers quoted above.)

**The Various Fisheries Interact (Competition, By-catch)**

As the number of species increase, so do the number of potential interactions between neighboring fisheries. These interactions complicate the assessment as well as management by rendering difficult the monitoring and regulation of fishing mortality. In Senegal, for instance, strong interactions exist within specialized sectors of the trawl fishery as well as between different fisheries. Figure 11 shows the distribution of the relative weight (in percent) of the species sought in a great number of well identified trips aimed at a particular target*. The wide range of percentages observed precludes any accurate a posteriori definition of the target from an examination of catch composition, and Captain's interviews are necessary. Experience also shows that as diversification of targets increases and markets are open, the specialization of the boats decreased. Table 4 shows the degree to which most species occurring off Senegal are caught in a number of fisheries and illustrates the difficulty of regulating effort or size at first capture for a given species and a fortiori for a combination of species with different population parameters under these conditions. It is in order to ease this last difficulty and to facilitate enforcement as well that the CECAF Committee recommended the adoption of a compromise and unique mesh size of 60 mm for all demersal fisheries on the shelf, from Gibraltar to the Congo River. These sorts of compromises are, however, difficult to work out when the species

* A small percentage of trips aimed at a mix of targets was discarded.

TABLE 4 - Interactions between various fishing gears in the Senegalese fisheries (x = present; xx = important; xxx = very important).

| SPECIES | GEARS | | | | | | | | | |
|---|---|---|---|---|---|---|---|---|---|---|
| | | | Artisanal fishing | | | | | Industrial fishing | | |
| | Purse-seine | Ring net | Handline (trad. & modern) | Shrimp fishery | Drift net | Bottom set net | Beach seine | Tuna Purse-seine | Coastal Purse-seine | Trawl |
| PELAGIC SPECIES | | | | | | | | | | |
| Thunnus albacares | | | | | | | | xxx | | |
| Thunnus obesus — big tunas | | | | | | | | xxx | | |
| Katsuwonus pelamys | | | | | | | | xxx | | |
| Euthynnus alletteratus | x | | | | | | x | | | |
| Sphyraena spp. | x | | | | | | x | | x | x |
| Caranx spp. | xxx | | x | | x | | xxx | | xxx | x |
| Trachurus spp. | | | xxx | | | | | | xxx | xxx |
| Sardinella maderensis | x | xxx | | | xxx | | xxx | bait | xxx | x |
| Sardinella aurita | xxx | xxx | | | x | | xxx | bait | x | x |
| Ethmalosa fimbriata | xxx | xxx | | | x | | | | x | |
| Brachydeuterus auritus | x | | x | | x | | xxx | | | xx |
| Pomadasys spp. | x | | | | | | xxx | | xxx | xxx |
| Loligo | | | xxx | | | | | | | x |
| BOTTOM SPECIES | | | | | | | | | | |
| Pagellus bellottii | | | xxx | | | | x | | | xxx |
| Pseudupeneus prayensis | | | | | | | | | | xxx |
| Sepia officinalis | | | xxx jig-line | | | | x | | | xxx |
| Epinephelus aeneus | | | xxx | | | | | | | xxx |
| Palinurus spp. | | | | | | xxx | | | | x |
| Pseudotolithus spp. | | | | | | x | x | | | xxx |
| Arius spp. | x | xxx | x | | | xxx | | | | xxx |
| Cymbium spp., Murex spp. | | | | | | xxx | xxx | | | x |
| Penaeus notialis | | | | x | | | | | | xxx |

Data Sources: (24, 34). Only a limited and typical sample of the 64 species or groups of species identified in Senegalese statistics.

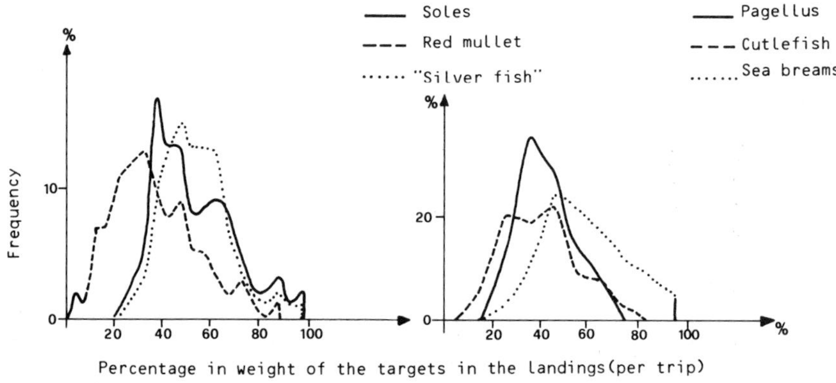


FIG. 11 - Distribution of the relative weight (in percent) represented by the species sought in landings from individual trips with a priori clearly identified targets. Trawl fishery of Senegal.

caught together differ drastically in value as well as in biological characteristics (e.g., hake and deep sea shrimps).

The by-catch mortality is an important problem in multispecies fisheries. The sparid stock off the Sahara is probably held down by at least growth overfishing, despite the nonexistence of an aimed fishery for sparids. The effect of the horsemackerel pelagic fisheries on the stocks of hakes are certainly important due to the high mortality they can generate on juvenile hakes. The direct effect of increased "by-catches" is a change in total mortality as well as in age-specific fishing mortality vectors with the result that the use of production models for management by quotas becomes at least questionable.

## TYPES OF MULTISPECIES FISHERIES

### The Exploitation of a Simple Predator–prey System

Apart from relatively minor fisheries (maximum 400,000 tons in any one season) for finfish (notothenids) and icefish (Champsocephalus) in restricted areas (mostly round South Georgia and Kerguelen), the Southern Ocean presents the best, and possibly the only, example of the simple system of the exploitation of either the prey (krill) or of one or more competing prey species (whales, seals, etc.) (40). Although much the same countries are concerned in harvesting whales and krill, the vessels used are distinct, and the problems are those of biological interaction. Important questions are whether, and to what extent, the recovery of whale stocks could be affected by an increased krill harvest or whether the depletion of the stocks of larger whales has caused an irreversible change in balance between these whales and their competitors (seals, penguins, and possibly minke whales) (36, 42).

**The Exploitation of Unstable Resources**

The upwelling system from central Chile to northern Peru is enormously productive. This production supports correspondingly enormous fisheries, and at its peak around 1970 the Peruvian anchoveta fishery was by far the largest single-species fishery in the world. Since then it has collapsed, and 1984 catches are likely to be near to zero, but there has been an almost equally dramatic rise in the catches of sardine (Table 5). While other species, e.g., hake, contribute to the fisheries in Chile and Peru, especially for the inshore fleet, the main commercial fisheries in both countries are dominated by small shoaling pelagic species (anchoveta, sardine, and horsemackerel). Though the catches by the local fisheries are nearly all taken by purse seiners and turned into fish meal, the fisheries are not wholly indifferent to the species composition. The larger and more active sardines and horsemackerel cannot easily be caught with the same gear used for anchovy, while horsemackerel is usually found in deeper waters, out of reach of purse seiners. These species also can be used for direct human consumption (canning or freezing) and have a potentially higher unit value.

While there has been a definite switch in target species, independent observations (e.g., surveys of fish eggs and larvae) show that there has been a real increase in sardine abundance. This has occurred at about the same time as the collapse of the anchoveta, which is presumed to have been due to a combination of heavy fishing and unusual environmental effects (especially the El Niños in 1972, 1976, and 1982/3). The big biological question is the extent to which the sardine increase was caused by the anchoveta collapse. This then raises the operational questions of whether the abundance of anchovy can be adjusted by controls on the fishing for sardine, and vice versa, and if so, what would be the optimum balance of sardine and anchovy.

Similar situations associated with heavy fishing and environmental changes arise in the other major upwelling areas of the world - off northwest and southwest Africa, off California (where it has been shown that there

TABLE 5 - Catches of major species in the Southeast Pacific ($10^5$ tons).

| Species | 1955 | 1960 | 1965 | 1970 | 1971 | 1972 | 1973 | 1974 | 1975 |
|---|---|---|---|---|---|---|---|---|---|
| Anchoveta | 1 | 35 | 77 | 131 | 112 | 48 | 20 | 40 | 33 |
| Sardine | + | + | 1 | 1 | 2 | 1 | 2 | 5 | 3 |
| Jack Mackerel | + | + | + | 1 | 2 | 1 | 2 | 3 | 3 |
| Other Species | 3 | 4 | 4 | 5 | 4 | 5 | 6 | 5 | 4 |
| Total | 4 | 39 | 82 | 137 | 120 | 56 | 30 | 53 | 44 |

| | 1976 | 1977 | 1978 | 1979 | 1980 | 1981 | 1982 | 1983 |
|---|---|---|---|---|---|---|---|---|
| Anchoveta | 43 | 8 | 14 | 14 | 8 | 16 | 18 | 3(a) |
| Sardine | 5 | 15 | 20 | 33 | 33 | 28 | 33 | 35(a) |
| Jack Mackerel | 4 | 8 | 10 | 13 | 12 | 17 | 22 | |
| Other Species | 4 | 6 | 7 | 11 | 8 | 9 | 8 | |
| Total | 58 | 39 | 56 | 69 | 62 | 69 | 78 | |

Notes: + quantity less than 0.5; (a) preliminary estimates.

have been long-term changes in the balance between sardine and anchovy quite independent of fishing (51)), as well as in the fisheries for small pelagic fish around Japan (30, 44, 54). It is interesting to note that similar changes (from sardine to anchovy) have been observed in the last few years all over the western Mediterranean either in the presence of heavy fishing (Spain) or not (Algeria, Morocco).

**Progressively Diversifying Fisheries**

Changes in species in the catches do not always reflect equivalent changes in the sea. Until the early 1960s, harvesting of the rich resources of the Bering Sea was confined to a few high-valued species (fur seal, salmon, and halibut) and the total weight caught, and the impact on the ecosystem as a whole was small. After 1960, large-scale trawling was carried out, mainly by Japan, targeting in succession on flounders, Pacific Ocean perch, and Alaska pollock. While all these species are caught by trawl, the way the gear is rigged and the areas fished are different, so that the rise in catches of the second and third species is in fact no evidence that these species increased. Indeed, the first "multispecies" question is whether there is any "multispecies" problem at all, other than determining the true effort on one or other species. It is possible that, given proper "effort" data, the dynamics of each species could be described by a simple single-species model, with the decline of the flounder and Pacific Ocean perch catches being due more to the fishing out of accumulated stocks of very long-lived fish, i.e., the so-called "pulse-fishing" than to any real decline in the productivity of the populations.

Similar situations, where the main problem is knowing the species on which the fishery is targeted at a particular stage of its evolution, have occurred on a number of occasions as industrial fishing has spread into new areas, e.g., the changes between sparids and cephalopods off northwest Africa, between plaice, haddock, and cod in the English trawl fishery off Iceland and northern Russia in the first half of this century, and even the changes from blue whale to fin, sei, and minke whales in the Antarctic. In some cases it is clear that the rise in catches of an alternative species is due purely to a switch in attention (e.g., Senegal), in relation to changes in gear technology (wide opening trawl) or market opportunities. In the Antarctic the minke whales are the only species of whales suspected of increasing because of the depletion of the larger species. In other cases, there may have been a real increase of the new target biomass, e.g., in the northwest African cephalopods, Balistes, and sardines.

**Multi-Single-Target Fisheries**

The North Sea contains 50 or more species of fish and invertebrates of commercial interest, of which about a dozen can support major fisheries. While some vessels, especially trawlers, may catch several species in a single haul, and there is some switching of attention by individual vessels at different seasons of the year, the simplified picture of the North Sea fisheries as consisting of a set of individual fleets, each targeting all year round on one species, or a narrow group of related species, is fairly realistic. Thus it is possible to distinguish the Danish trawl fishery for small fish (sandeels, Norway pout, and, in the past at least, small herring) for reduction to meal and the Dutch beamtrawl fishery for sole and other species; the Scottish seine-net fishery for haddock, whiting, and cod; various national fisheries for herring for human consumption, and so on. On a different scale, the first diversification period of the Senegalese trawl fishery was undertaken as a multi-single-target fishery as the various types of the heterogeneous fleet tended to target consistently to particular groups of targets: traditional Mediterranean type of wooden trawlers on red mullet and Japanese type of steel stern trawlers on sparids and cephalopods. This specialization, however, may disappear with time.

For most of the major species it is possible to find one, and often several, fisheries (i.e., a fleet, or identifiable group of vessels of roughly similar type coming from a particular country or fishing port) whose activities have remained consistently targeted on that species. The catch and effort statistics of this fleet or fleets therefore provide a reasonable measure of the abundance of the species, and thus within the limitations of the models, each major species can be analyzed with the standard single-species techniques. This has been to a large extent the standard practice in ICES, particularly in respect to its numerous stock assessment working parties.

Multi-single-target fisheries might be considered as a necessary intermediate stage of evolution of fisheries in their development process from the stage of artisanal, highly diversified fishing strategy of the littoral zone to the full exploitation of the high seas. It can be foreseen that for global economic reasons each element of the fishery will not stay targeted on one species but, at least in the long term, will switch targets in response to trends in relative abundance, changes in market preference, and in response to management measures. As the new opportunities open, as markets become used to the different varieties of targets coming from the open sea, as the fish processing techniques evolve, and as effort limitations are imposed on traditional species, the boats will develop flexibility, the multipurpose boat being one way to reduce the cost of uncertainty raised by heavy fishing and subsequent stock fluctuations. The fishery might then evolve into what we may call a single-multi-target fishery.

**Single-Multi-Target Fisheries**

This is the case when a wide range of targets is exploited by a limited number of types of multipurpose boats shifting seasonally or from one trip to another, or from one group of targets to another in order to stabilize the revenues despite variations in the resource.

We described it above as the likely ultimate stage of development of fisheries, and paradoxically it can also be the departure stage of a fishery because artisanal fishing falls very often in this category.

A lot of medium-range demersal fisheries on the shelf fall in this category. The present state of the Senegalese trawl fishery is close to it as well as most of the traditional trawling on the Mediterranean shelf and especially in the Adriatic Sea. There a limited number of multipurpose boat types using a range of selective (dredge) to nonselective (Italian trawl) gears exploit at one time or another most of the available species for a market able to accept with minor adjustments whatever species mix is offered (very often mixtures of species and sizes are sold in batch, unsorted). There the scientists and the fishermen tend to consider the high space-time variability of the resource (few fish are older than one year) as noise. While no drastic change in species composition has apparently been observed, the overall fishery has a stable production, and the essential concern seems to be in terms of total value of the catch, species interactions being of relatively minor immediate practical interest. However, as trawlers are directing purposely more and more effort to pelagic species by using high opening trawls (in the Gulf of Lions, for instance), conflicts between the pelagic and demersal fisheries can already be perceived.

A similar situation arises in a number of tropical trawl fisheries where it is not clear to what extent the fisherman can alter his fishing practices to change emphasis on one species or another. Since the rapid expansion of the Thai trawl fisheries, whose catches grew from near zero in 1960 to over a million tons a decade later, there have been big decreases in overall biomass (45, 49, 55), as well as changes in species composition. Since these results come from surveys by research vessels, they probably reflect real changes in abundance, but they are fairly closely matched by changes in commercial landings.

The changes in species composition raise interesting scientific questions, since some, e.g., the severe decline of most long-lived species, fit single-species theory, but others, e.g., the increase in absolute abundance of squid, must involve some interaction. However, these questions may be irrelevant for immediate practical purposes, since the fisherman is interested in the total value of his catch, irrespective of the species composition. It seems that for purposes of deciding on current management policy, it may be adequate to use the simplest type of production model, relating the total catch (or CPUE) of all species (preferably expressed as value) to the total amount of fishing.

When analyzing and comparing the effect of fishing on stock composition, the rate of increase of effort should be considered. The biological resource may have some capacity to adapt to increased effort without catastrophic changes, provided the increase is slow enough. In the same way, the fishing strategies and the market can be smoothly adjusted to some slow rate of change of the resource base but not to catastrophic ones. Considering the example of the Adriatic fishery, one may wonder whether a single-multi-target fishery operating for a very diversified market, at high effort levels, is not a valid economic strategy and also a possible biological one, if the problem is considered in terms of global energy balance.

## PROBLEMS AND QUESTIONS

### Definition of Objectives

This is no place to repeat the arguments about MSY, MEY, etc. Most practitioners of the art or science of management would probably follow John Pope's objectives of MSW (Minimum Sustainable Whinge, i.e., that policy which, over a period, leads to least complaints reaching the Director of Fisheries). Most of the theoretical arguments, though based on single-species analyses, translate to the multispecies situation without difficulty. The new element is the degree to which the balance between species should be an objective in itself. Assuming, for example, the greatest total return from the North Sea (in weight value, net economic return, or whatever) would be taken by a pattern of fishing that virtually eliminated cod, to what extent should this mean that that pattern of fishing is unacceptable? Since there is no group, as yet, dedicated to the conservation and protection of cod, this question would probably resolve itself into a second question of the degree to which those fishermen currently fishing for cod can switch, with or without compensation, to other species. The problem of species balance is particularly alive where marine mammals are concerned, and the best formal attempt to deal with it is in Article 2 of the new Convention for the Conservation of Antarctic Marine Living Resources*, which inter alia states the following principles of conservation:

(a) prevention of decrease in the size of any harvested population to levels below those which ensure its stable recruitment. For this purpose its size should not be allowed to fall below a level close to that which ensures the greatest net annual increment;

(b) maintenance of the ecological relationships between harvested, dependent and related populations of Antarctic marine living resources and the restoration of depleted populations to the levels defined in subparagraph (a) above; and

(c) prevention of changes or minimization of the risk of changes in the marine ecosystem which are not potentially reversible over two or three decades.

This statement of objectives was undoubtedly a good solution to the immediate obstacles to establishing the Commission, but it will be

---

* Text of Convention available from the Commission for Conservation of Antarctic Marine Living Resources, 25 Old Wharf, Hobart, Tasmania 7000, Australia.

interesting to see how well it will turn out to be a usable guide to management procedures in practice.

In the policy field, the test will be whether the definition of objectives is clear enough to determine specific actions when there is a clash of interests between conservation and harvesting interests. In the scientific field the test will be whether the implied models (of some level of population ensuring greatest recruitment, or of some constant and maintainable set of ecological relationship) come any closer to the reality of the dynamic multispecies systems of the Southern Ocean than did the model underlying the MSY objective to the dynamics of a single species.

**Determination of Boundaries**

By recognizing a multispecies problem, the manager has implicitly looked beyond the limited boundaries of a single-species situation. The question is how far should the boundaries of his consideration be extended so as to make it more realistic without making it so wide as to be impossible to handle. In the sea, by considering the movements of currents, and of prey and predators (perhaps at several steps removed), it is possible to connect any one species in a particular place to any other species in any other location. How small an area is it reasonable to look at? Is it in fact possible to set definite geographical boundaries except in a few enclosed areas (e.g., the Baltic, the Adriatic), or must subjective, but operationally acceptable boundaries be chosen for each occasion? Is it possible to use other divisions, e.g., is it realistic in a particular region to consider managing pelagic and demersal systems separately? (Probably not, since in several areas - Mediterranean, Gulf of Thailand - trawlers are turning to typically pelagic species - sardines, Rastrelliger spp. - as valuable secondary targets.)

Similar questions arise on the shore side. The arguments about MSY, MEY, etc., have highlighted the fact that the manager must look beyond the biological yield to the producers and consumers of fish. To what extent are there important interactions between species on the market so that management policies on, say, sardine off northwest Africa should have to take into account the events in fisheries for other stocks supplying the same markets (fishmeal, or cheap canned fish) in any part of the world, and not just off Africa?

**Data Needs**

The data needed for the biological assessment of single-species fisheries are relatively well-defined and do not need further discussion, though some items, e.g., on discards, may have increased importance in a multi-species context. There is less explicit agreement on what specific data are needed for economic and social analyses, but that some such data are necessary has become generally accepted. For multispecies fisheries it is sometimes supposed that these data are needed for each important species, plus information on the degree of biological and other interactions, e.g., who eats what; how the demands for different species interact.

When many species occur, the volume of the resultant demand for data is frightening and certainly beyond the capacity of a small fishery department in a typical developing country to supply. There is a need to identify certain key data which will allow analysis of advice about multispecies fisheries.

It is also important to collect data as soon as possible. Experience in single-species fisheries has shown that one piece of good information collected when the fishery was just beginning can be worth volumes of data from a well-developed fishery. Unfortunately, we do not have the models to tell us which are the key data, especially the key data for priority attention in a new fishery. Identification of key data, which may require identification, at least in general terms of the main outline of the models to be used, is therefore a matter of some importance.

Since virtually any kind of data is useful to someone, the real question in practice is how limited resources for data collecting can be best deployed. How should they be divided between data used for stock assessment, economic analyses, or other purposes (bearing in mind some data are useful for several purposes)? Considering just the collection of data for stock assessment, should resources be concentrated on the quantities - catch/effort/age/size - that have been proved useful for single-species work, or should there be a greater concentration on more qualitative observations (food, predators, distribution)? Should they be spread across all species (evenly, or in proportion to their commercial importance), or is it better to concentrate on a few key or typical species? Are there types of data that do not occur in single-species studies, e.g., could the intentions of fishermen when they leave port be valuable information in interpreting catch and effort data in terms of relative species abundance?

**Interpretation and Analysis**

The two approaches to analyzing multispecies biological systems are a) to look at each species separately and add, within the analysis of each species, appropriate terms for the interaction with other species (bottom-up approach), or b) to look at the whole system in terms of energy flow, total catch, etc., and separating, e.g., the production of secondary carnivores into species only to the extent that is possible and desirable (top-down approach).

The first approach seems to work comparatively well in the Southern Ocean, with very few species and considerable interest in what happens to individual species, especially the great whales. It runs into problems of increasing data demands and large numbers of degrees of freedom in fitting parameters as soon as the numbers of species increase (47). These problems become extremely difficult even in areas such as the North Sea or the Bering Sea where the number of species are relatively few and the data supply good. They offer little practical help in many tropical areas with many species and little data.

The integrated approach has proved useful in some places. The analysis of total catch, and catch per unit effort of all species combined in the Gulf of Thailand or Ivory Coast (23), has been sufficient to show that there are far too many trawlers, and there is little call for more detailed analysis and advice until action has been taken to reduce the fleet size (in fact, in the Ivory Coast the marine resource program is now limited to a small monitoring program). The total production-energy flow approach applied, say, to the whole North Sea or the Peruvian upwelling system has given results that are compatible with, and supportive of, the species-by-species analysis. However, the practical value of total biomass, Gulf of Thailand approach, is limited to those situations where the fishery is homogeneous and is not sensitive to changes in species composition. Energy flow analyses tend to have wide confidence limits, e.g., in the efficiency of transfer between trophic levels. Thus, while they can distinguish between areas where the production is hardly touched, and areas where the fisheries are taking much of what is available (which can be useful), they are less able to distinguish between situations where the resources are fully or overexploited, and those where some moderate expansion, perhaps even a 50% increase in catch, is possible. It is also difficult to use this approach to determine what action might be appropriate when stocks are known to be heavily fished.

This implies that, accepting that single-species analyses are inadequate, with both current multispecies approaches having only limited usefulness, fishery scientists have problems. This is true, but there is some relief. Single-species analyses are not completely useless. Most present management decisions are based on them, and if they were abandoned today, the world's fisheries would be in a worse condition. Indeed, with a narrow enough focus, single-species analyses can be perfectly adequate. For example, accepting for the purposes of illustration that in the North Sea the year-class strength of cod is determined to some extent by the abundance of herring and whiting, and that the abundance of food consumption of cod affects the yield of whiting and other small fish, the present advice on cod, based on a single-species model, may still, in principle, give the optimum fishing pattern for cod so far as cod fishermen are concerned, and if controls are set in terms of catch quotas and these are adjusted in the light of information on current year-class strengths, they will result in the correct cod quotas from the point of view of cod fishermen. The same is true for herring and other species, where the policies for herring, etc., are, in principle, the optimum for the fishermen harvesting these species. (We ignore for the present discussion that in practice the policies actually set correspond to fishing mortalities which are in most cases well above any long-term optimum, and that many of the corresponding quotas are poorly enforced.)

Two cautions should be considered: the cod analysis may not necessarily give the correct advice even for the cod fishery, and pursuing the optimum for individual species for fisheries may result in a pattern for the fishing as a whole that is very much suboptimum. The first caution gives rise to a clear scientific question - under what circumstances will a single-species analysis give rise to advice that is bad in terms of that species and the fishery on it? To what extent does the answer depend on the nature of the species (large predator, etc.) and the types of action taken (e.g., the optimum fishing mortality on cod may be insensitive to changes in year-class strength and factors that affect year class, while catch quotas are highly sensitive to these factors). This would seem to be a relatively simple question, and it is tempting to suggest that the answer is that with obvious precaution (e.g., the need to adjust year classes), single-species advice is relatively insensitive in species interactions.

The second caution is more important, and looking only at single species can lead to poor management. It seems, for example, that countries that adopt a flexible opportunistic approach to their pelagic fisheries, taking whatever species is present at the time (e.g., Japan, Chile), are more successful than those focussing on particular species (California-sardine, Peru-anchovy). Gulland (29) suggested, on some assumptions about the nature of the interactions, that the value of the North Sea catch could be significantly increased by taking advantage of those interactions, e.g., by deliberately overexploiting some species to let their competitors expand. The first example concerns a possible or partial replacement of one species by another (sardine/anchovy, etc.). The second depends on the fact (which is in part an assumption) that the recruitment of valuable species (i.e., the larger demersal species such as plaice and cod) increases when smaller, less valuable species (Norway pout, herring) are relatively depleted. Also relevant is the fact that the demand for herring is inelastic, and once catches increase above a certain point, they must go to fish meal. There is therefore little economic attraction in a complete rebuilding of the herring stocks.

Two basic biological questions arise that deserve further examination. First, how reasonable is it to expect that if one species declines, it will be replaced (wholly or to some specified extent by another)? Intuitively, replacement seems likely, and there is evidence to suggest that it sometimes happens in practice. However, as Daan (8) points out, the evidence for direct and complete replacement is poor. Replacement is not automatic, and, if it is, the timing and degree of replacement are not predictable. Nevertheless, even partial replacement could well influence management policies. A manager faced with a collapsing anchovy stock would be less inclined to impose drastic measures to preserve the anchovy stock if he thought that if it did completely disappear, there would be a fair chance of getting increased sardine catches.

The second question is of the degree to which recruitment of one stock can be influenced by the abundance of another. To what extent, if at all, was the increased recruitment of gadoids in the North Sea since 1960 linked to the declines of mackerel and herring? This is clearly connected with the important but unsolved single-species problem of the degree to which the adult stock of the typical, highly fecund commercial fish can be reduced before there is a significant fall in the average recruitment.

Answers to the questions in the preceding paragraphs should warn the scientist, and those he is advising, as to when a single-species analysis could be seriously misleading. They do not allow him to avoid doing the individual, and possibly numerous, species analyses. An approach to this is to treat the species in groups, e.g., large predators, detritus feeders, etc. Pauly (45) has done this for the Gulf of Thailand and suggests that the trends in abundance within a group are consistent, whereas there are differences between groups which can be large. This implies that, for example, production modeling could be done for n groups rather than m species (n < m), and that it might be sufficient to do detailed analyses (looking at growth, mortality rates, etc.) for only-one or two species within a group.

**Formulation of Action**

The problems here do not concern so much the multispecies question as the multifishery question, i.e., the degree to which action concerning one fishery may affect, or be affected by, events in another. These effects may arise from biological interactions between species but are equally likely to be due to operational and economic factors. For example, the implementation of a limited entry scheme for one fishery on a particular species group in a region can have an immediate impact on most other fisheries on other species, as surplus effort is displaced from the controlled fishery and, often, as fishermen seek to establish their rights in the other fisheries in case limited entry is applied to those fisheries, too.

The immediate question is to what extent the choice of management technique to control either the sizes of fish (mesh sizes, minimum market sizes, closure of nursery areas, etc.) or of fishing mortality (catch quotas, licence limitation, etc.) in individual fisheries needs to be modified to take account of species interactions between fisheries.

The more fundamental question is whether it would be desirable, in a multifishery region such as the North Sea or Western Sahara shelf, to put less emphasis on detailed fishery-by-fishery or species-by-species controls, and more on some general overall approach. There are at present far too many fishing vessels in the North Sea, and successes in controlling excess effort in, say, the cod or sandeel fisheries are to a large extent only successes in that they shift the problem to the fisheries on herring or haddock. It may be, and this is a question deserving more careful analysis, that reducing the overall size of the fleet would make the management of the individual species much easier, with less time of research scientists, administrators, and enforcement officers tied up in species-by-species detail.

**Implementation and Enforcement**

On first sight it might appear that the legal and administrative activities of implementing and enforcing fishery management programs do not concern scientists and need not be discussed in a scientific paper. This is not completely true. No fishery regulation is easy to enforce, but some are much more difficult to enforce than others. A measure that might appear optimum on biological or economic grounds may prove impossible to enforce, or enforcement may involve the deployment of so many costly patrol ships or aircraft that the net benefits obtained may be much less than those from a theoretically suboptimum, but more easily enforceable, measure. Since the choice of measure has implications on the work done at each of the preceding stages (models and methods of analysis, data collection), the fishery scientist must take some account of enforcement problems. This is particularly true of multispecies fisheries. Differences in the sizes of the various species caught in a trawl fishery may make some techniques, e.g., mesh regulation, impracticable, while others, which may be inappropriate in respect to a single species, e.g., limits on the total number of fishing vessels, may become attractive when all species become heavily fished.

## REFERENCES

(1) Arena, P. 1978. Indagine qualitativa et quantitativa sui materiali di scarto della pesca e sulle possibilità di una loro conveniente utilizzazione. Mimeo. ESPI.

(2) Belvèze, H., and Bravo de Laguna, J. 1980. Les ressources halieutiques de l'Atlantique Centre-Est. Deuxième partie. Les ressources de la côte ouest-africaine entre 24°N et le détroit de Gibraltar. FAO Doc. Tech. Pêches (186.2).

(3) Belvèze, H., and Erzini, K. 1983. The influence of hydro-climatic factors on the availability of the sardine (Sardina pilchardus Walbaum) in the Moroccan Atlantic fishery. FAO Fish. Report 291(2): 285-328.

(4) Binet, D. 1982. Influence des variations climatiques sur la pêcherie des Sardinella aurita ivorio-ghanéennes: relation sécheresse-surpêche. Oceanologia acta 5(4): 443-452.

(5) Caddy, J. 1983. The cephalopods: factors relevant to their population dynamics and to the assessment and management of stocks. FAO Fish. Tech. Paper 231.

(6) Caverivière, A. 1982. Les especes démersales du plateau continental ivoirien. Biologie et exploitation. Ph.D. Thesis presented at the University of Aix-Marseilles II.

(7) Champagnat, C., and Domain, F. 1978. Migration des poissons démersaux le long des côtes ouest-africaines de 10° à 26° de latitude nord. Cah. ORSTOM ser. oceanogr. 16(3-4): 239-261.

(8) Daan, N. 1980. A review of replacement of depleted stocks by other species and the mechanics underlying such replacement. Rapp. P-v. Réun. Cons. Int. Explor. Mer 177: 405-421.

(9) Domain, F. 1976. Mauritanie. Les ressources halieutiques de la côte ouest-africaine entre 16 et 24° lat.N. Rapport, FAO, FI:MAU 73/007/1.

(10) Domain, F. 1980. Contribution à la connaissance de l'écologie des poissons démersaux du plateau continental Sénégalo-mauritanien. Les ressources démersales dans le contexte du golf de Guinée. Ph.D. Thesis presented at the University Pierre et Marie Curie, Paris VII and the Muséum national d'histoire naturelle, Paris.

(11) Domanewsky, L.N., and Borkova, N.A. 1981. État du stock de la sardine, Sardina pilchardus Walb. dans la région de l'Afrique du nord-ouest. COPACE/TECH. 81/31: 19-30.

(12) FAO. 1968. Supplement 1 to the report of the 5th session of the ACMRR. Report of the ACMRR/ICES Working Party on the Fishery Resources of Eastern Central and Southeast Atlantic. FAO Fish. Report 56(Suppl. 1).

(13) FAO. 1975. Report of the 2nd session of the Fisheries Committee for the Eastern Central Atlantic Fisheries (CECAF) Working Party on Resource Evaluation, Rome, 3-6 December 1973. FAO Fish. Report 158.

(14) FAO. 1978. Some scientific problems of multispecies fisheries. Report of the Expert Consultation on Management of Multispecies Fisheries. Rome, 20-23 September 1977. FAO Fish. Tech. Paper 181.

(15) FAO. 1980. ACMRR Working Party on the Scientific Basis of Determining Management Measures, Report of the ACMRR Working Party on the Scientific Basis of Determining Management Measures. Hong Kong, 10-15 December 1979. FAO Fish. Report 236.

(16) FAO/CECAF. 1976. Nominal catches 1964-74. Statistical Bulletin No. 1.

(17) FAO/CECAF. 1979. Nominal catches 1967-77. Statistical Bulletin No. 2.

(18) FAO/COPACE. 1979a. Rapport du groupe de travail ad hoc sur les poissons pélagiques côtiers ouest-africains de la Mauritanie au Liberia (26°N à 5°N). COPACE:PACE Series 78/10.

(19) FAO/COPACE. 1979b. Rapport du groupe de travail spécial sur les stocks de céphalopodes. COPACE:PACE Series 78/11.

(20) FAO/COPACE. 1980. Rapport du groupe de travail ad hoc sur les sardinelles des côtes de Côte d'Ivoire-Ghana-Togo. COPACE:PACE Series 80/21.

(21) FAO/COPACE. 1982. Rapport du groupe de travail spéciale sur les stocks de céphalopodes de la région nord du COPACE. COPACE:PACE Series 82/24.

(22) FAO/COPACE. 1984. Rapport du groupe de travail ad hoc sur les chinchards et les maquereaux de la zone nord du COPACE. COPACE:PACE Series 84, in press.

(23) Fonteneau, A., and Bouillon, P. 1971. Analyse des rendements des chalutiers ivoiriens. Définition d'un effort de pêche. Doc. Scient. Centre Rech. oceanogr. Abidjan ORSTOM 21(1-2): 1-10.

(24) Fontana, A., and Weber, J. 1983. Apperçu de la situation de la pêche maritime sénégalaise (Déc. 1982). Dakar: CRODT/ISRA (Miméo).

(25) FRU/CRO/ORSTOM. 1976. Rapport du groupe de travail sur la sardinelle (S. aurita) de Côte d'Ivoire-Abidjan, 28/6-3/7/1976. Dakar: ORSTOM.

(26) Garcia, S. 1982. Distribution, migration and spawning of the main fishery resources in the northern CECAF area. CECAF/ECAF Series 82/25, and 10 sets of maps.

(27) Garcia, S.; Lhomme, F.; Chabanne, J.; and Franquerville, C. 1979. La pêche démersale au Sénégal: historique et potentiel. COPACE:PACE Series 78/8: 59-88.

(28) Garcia-Cabrera, C. 1968. Pulpo. Biologia y pesca del pulpo (Octopus vulgaris) en aguas del Sahara español. Publ. Tech. Junta Estud. Pesca, Madrid 7: 161-198.

(29) Gulland, J.A. 1981. Long-term potential effects from the management of the fish resources of the North Atlantic. J. Cons. Int. Explor. Mer 40(1): 8-16.

(30) Hayasi, S. 1984. Some explanation for changes in abundance of major neritic pelagic stocks in the northwestern Pacific Ocean. In Proceedings of the Expert Consultation to Examine Changes in Abundance and Species Composition of Neritic Fish Resources, eds.

G.D. Sharp and J. Csirke. FAO Fish. Report 291(1): 37-56.

(31) Holzlohner, S. 1975. On the recent stock development of Sardina pilchardus Walbaum off Spanish Sahara. ICES C.M. 1975/J: 13.

(32) Ikeda, I. 1971. Observations sur les stocks de seiches au large de la côte ouest de l'Afrique. FAO Rapp. Pêches 103: 92-100.

(33) Ikeda, I., and Sato, T. 1971. Renseignements biologiques sur Pagellus bellotti Steindochner au large de la côte nord ouest de l'Afrique avec une evaluation préliminaire des stocks. FAO Rapp. Pêches 103: 100-104.

(34) Institut Sénégalais de recherches agricoles (ISRA). 1982. Statistiques de la pêche maritime sénégalaise en 1982. Achives No. 120. Dakar: CRODT.

(35) Lambouef, M.; Burczynsky, J.; Bencherifi, S.; and Chbani, M. 1981. Campagne de prospection des stocks pélagiques du cap Cantin (Maroc) au cap Timiris (Mauritanie) en juillet 1980 (résultats préliminaires). Trav. Doc. Dev. Pêche. Maroc 28.

(36) Laws, R.M. 1977. Seals and whales of the Southern Ocean. Phil. Trans. Roy. Soc. London (B) 279: 81-96.

(37) Longhurst, A.R. 1969. Species assemblages in tropical demersal fisheries. Actes sycup. oceanogr. Res. halieutiques Atlant. Trop. Unesco, pp. 167-170.

(38) MacCall, A. 1980. Population models for the northern anchovy (Engraulis mordax). Rapp. P.-v. Réun. Cons. Int. Explor. Mer 177: 292-306.

(39) Maurin, C. 1968. Ecologie ichthyologique des fonds chalutables atlantiques (de la baie ibéro-marocaine à la Mauritanie) et de la mediterranée occidentale. Rev. Trav. ISTPM 22(1).

(40) May, R.M.; Beddington, J.R.; Clark, C.W.; Holt, S.J.; and Laws, R.M. 1979. Management of multispecies fisheries. Science 205(4403): 267-277.

(41) Mercer, M., ed. 1982. Multispecies approaches to fisheries management advice. Can. Spec. Publ. Fish. Aquat. Sci. 59.

(42) Mitchell, B., and Sandbrook, R. 1982. The Management of the Southern Ocean. London: International Institute for Environment and Development.

(43) Monnoyer, P. 1979. Contribution à l'étude des rejets à la mer de la faune ichthyologique capturée par les chalutiers commerciaux dans la zone 34.3.1. Rapport dactylographié du COPACE. Non publié.

(44) Nagasaki, F. 1973. Long-term and short-term fluctuations in the catches of pelagic fisheries around Japan. J. Fish. Res. Board Can. 30(12)Pt. 2: 2361-2367.

(45) Pauly, D. 1979. Theory and management of multispecies stocks. ICLARM Stud. Rev. 1: 1-35.

(46) Pereiro, J.A., and Bravo de Laguna, J. 1980. Dinamica de la población y evaluación de los recursos del pulpo del Atlantico centro oriental. Serie CPACO/PACO 80/18.

(47) Pope, J. 1979. Stock assessment in multispecies fisheries, with special reference to the trawl fishery in the Gulf of Thailand. FAO/ UNDP South China Sea Fisheries Development and Coordination Programme, Manila. Doc. SCS/DEV/79/19.

(48) Pupyschev, V.A. 1982. To the increase in the abundance of Balistes capriscus (Gmel. 1789) and Cephalacanthus volitans in the Gulf of Guinea in VNIRO fishery investigations in the east tropical Atlantic. Proceedings, Lyoghaya I Pishchevayo Promyshlennost, pp. 50-60.

(49) Ritsraga, S. 1976. Results of the studies on the status of demersal fish resources in the Gulf of Thailand from trawling surveys 1963-1972. In Fishery Resources and Their Management in Southeast Asia, ed. K. Tiews, pp. 198-223. Berlin: German Foundation for International Development.

(50) Sedykh, K.A. 1979. Étude de l'upwelling près de la Côte de l'Afrique du Nord ouest per l'Atlant-NIRO. COPACE:PACE Series 78/11: 93-99.

(51) Soutar, A., and Isaacs, J.D. 1974. Abundance of pelagic fish during the 19th and 20th centuries as recorded in anaerobic sediment off the Californias. Fish. Bull. NOAA/NMFS 72: 257-275.

(52) Stequert, B.; Brugge, W.J.; Bergerard, P.; Fréon, P.; and Samba, A. 1979. La pêche artisanale maritime au Sénégal. Études des résultats de la pêche en 1976 et 1977. Aspects biologiques et économiques. Doc. Sci. Centre Rech. océanogr. Dakar-Thiaroye 73.

(53) Strømme, T. 1983. Final report of the R/V DR. FRIDTJOF NANSEN fish resource surveys off West Africa from Agadir to Ghana. May

1981-March 1982, Bergen, Norway.

(54) Tanaka, S. 1984. Variation of pelagic fish stocks in waters around Japan. In Proceedings of the Expert Consultation to Examine Changes in Abundance and Species Composition of Neritic Fish Resources, eds. G.D. Sharp and J. Csirke. FAO Fish. Report 291(2): 17-36.

(55) Tiews, K.; Sucondhamarn, P.; and Israngkura, A. 1967. On the change in the abundance of demersal fish stocks in the Gulf of Thailand from 1963/1964 to 1966 as a consequence of the trawl fishery development. Contrib. Dep. Fish. 8.

Exploitation of Marine Communities, ed. R.M. May, pp. 191-207. Dahlem Konferenzen 1984. Berlin, Heidelberg, New York, Tokyo: Springer-Verlag.

# Some Approaches to Modeling Multispecies Systems

R.T. Paine
Dept. of Zoology, University of Washington
Seattle, WA 98195, USA

**Abstract.** A qualitative taxonomy of models is presented. Fisheries models with a single-species outlook are traditional and uninvolved with the complexities of an interactive, multispecies system. Ecosystem models, because of the required aggregation of many species into units of convenience, lose track of, or become insensitive to, biological relationships. They tend to be descriptive, usually are incapable of forecasting change subsequent to specific disruption, and often run afoul of their own internal intricacies. Intermediate to these extremes are multispecies models, most of which, though idiosyncratic, are focused on the dynamics of specific ecological phenomena. These suggest that aggregation into guilds is unwarranted; that there is no correlation between primary production and community complexity; that, dependent on the coupling involved, removal of one species may work to the benefit or detriment of presumed competitors; and that community complexity (= food web design and even species composition) is imposed by higher trophic level species. Three examples illustrating that small-scale or model relationships can be recognized in open, large natural communities are given. How, but not whether, fisheries managers should respond to the difficulties presented by multispecies interactions remains unanswered.

## INTRODUCTION

The richness of composition of natural communities and their complexity and subtlety of interaction have bedeviled the increasingly pressing necessity of understanding whether such communities display repeatable patterns (are organized) and how they might respond to anthropogenic influences. Here, I assume that all communities are complex, that the complexities are apt to bear biological significance, that these assemblages tend to be organized in the sense that common patterns can be identified, and, finally, that understanding cannot be achieved just by observation/sampling. Thus simplification, either by quantitative or heuristic (conceptual) models preferably coupled with some form of controlled manipulation, is an essential ingredient of understanding.

Models provide a format for orderly, not necessarily realistic, simplification. Single population models are mentioned briefly. I then turn to a more detailed description of both ecosystem and intermediate level models, the latter attempting to relate four or more species or entities of two or more trophic levels. There is an especially rich literature here, both theoretical and experimental, and I believe it to be the starting point for understanding and predicting the effects of multispecies interaction. Finally, I ask whether the results of such simplified models can be identified in systems both vastly richer in species and more extensive in space.

## SINGLE POPULATION AND ECOSYSTEM MODELS

Ecological models range from those for specific single-species populations to highly aggregated box and arrow diagrams. They guide or even mold intuition and experiment; they cannot prove or show that natural relationships exist, or that one or another approach is correct. Some are rooted in mechanism and are capable of prediction. Others are more phenomenological and descriptive. They are not easily classified into general categories but all embodying two or more species must involve an interspecific structural component. A wide range of single-species fisheries models are discussed by Pitcher and Hart (28). Simulation modeling has been clearly reviewed by Wiegert (40); an overview of ecosystem approaches is provided by Hall and Day (12).

### Single-Species Models

The oldest ecological models are those of growth or other dynamical aspects of single populations. Although by implication they are linked to an external support system, say, through constraints on saturation density or stock carrying capacity, or by resource-imposed restrictions on early growth, little consideration is given to interplay with these. The models can be made to incorporate such rich and significant detail as age structure, sex ratio, reproductive value or contribution, and they have served as the basis for investigating the dynamics of commercially exploited stocks. Equally, most of the current crop of strategic or adaptational models are essentially examinations or predictions of single-species behavior. Much of their ecological detail or evolutionary perspective is lost upon expansion to include other species, guilds (functional groups), or even whole trophic levels.

**Ecosystem Models**

At the other extreme lie system models: often complex, always highly aggregated, but usually involving more than two trophic levels and with provisions for permitting inorganic entities (nutrients, weather) to be involved. These are the traditional box and arrow approaches. Their strengths lie in a capacity to unite both biotic and abiotic components. They appear attractive because, unlike their intermediate-scale cousins discussed below, they often include primary productivity as an integral feature. On the other hand, there has been a temptation to emphasize emergent properties at the expense of understanding the dynamics within or between component levels. At its worst, the approach has been libeled as "describe, analyze and conquer," often with the natural, biologically based complexity believed to impede understanding. The question is, can an ecosystems approach, with its focus on macroscopic properties (a whole greater than sum of parts philosophy) or emphasis on extra-specific units, provide the detailed answers required for single-species management? Equally, do such models permit predictions to be generated about multispecies responses in a system under specified regimes of disturbance? I believe not. By their very nature, macroscopic systems' descriptions cannot be of much value to resource managers charged with maintaining the health of specific stocks under exploitation. Four problem areas are identified below: although they may appear to provide a scanty or cavalier basis for dismissal of these large and ambitious models, I believe the reasons provide adequate grounds.

First, a fundamental dilemma facing resource managers and ecosystem ecologists is the fact that most natural selection occurs at an individual level and in a biologically complex environment. There appear to be no shortcuts to the understanding necessary to interpret how populations respond to each other or to natural or imposed stresses that do not involve inspection of the individual and its immediate environment. Adaptationist and coevolutionary insights are invisible in large models.

Second, whole system models aggregate or lump species into common units such as guilds or trophic levels. While this may be reasonable, and perhaps necessary, the process masks effectively all of the more subtle biological structure. Just as all birds or all fish are not ecologically equal, neither will be the individual members of any taxonomically or functionally collective unit. Thus the capacity to recognize and account for the influences of critical or keystone species, species whose activities bear special significance for other community members, will be lost. The fact that ecosystem properties can persist even though populations change should be of little reassurance. It simply implies that within guilds of comparable organisms, functional replacement is possible. It does not deal with the question of resource value measured in either monetary or esthetic terms, or provide clues to why the changes have occurred.

Third, where the model is purposively made more inclusive and complete, validation becomes increasingly unlikely with the resultant gains in realism. To my eye, ambitious models (for example, (25, 29)) seem descriptive at best and must be made to portray events near an idealized steady state. While useful for identifying what needs to be known about relationships within and between the chosen components, such models have a poor record in generating new ideas or facilitating our understanding of where the individual species fits into nature's economy.

Fourth is the problem of common units. In multilevel models, currency as calories or joules or even grams of carbon is appropriate. Use of energy units permits costs assayed as respiratory heat loss to be considered. What is disguised in such representations is that biological materials of unconventional origins or fates - mucus, hair, molted exoskeletons, etc. - are ignored or assumed to be unimportant. More significantly, the assimilation ratio, known to be highly variable depending on the qualitative nature of the energy source, is usually only estimated or obtained indirectly. For marine systems the potentially important role of bacterial production remains controversial (14); the role of dissolved organic carbon (DOC) as an energy supplement has generally been ignored despite wide recognition of its ubiquity and the capacity for uptake by a phyletically broad group of species. Finally, calories or joules do not translate readily into population age structure or even numbers, and the magnitude of an observed material or energy pathway may have little to do with how the system responds under stress (23).

It seems unlikely that ecosystem models based on mass, nutrient, or energy considerations can be profitably employed for the description or prediction of behavior of stressed single-species populations imbedded within their nexus. One major prohibition is the complexity of the required data base. All large-scale models require accurate biological description as a starting point, with details on both the primary and derived relationships. If, for model convenience or tractability, spatial pattern is averaged and ecologically disparate species amalgamated into functional units (i.e., birds, fish, algae, etc.), the resultant loss of information totally isolates such models from most of population biology. Even if scaled down to increase their biological sensitivity, it is generally not in the philosophy of such models to do more than account for the fates of energy or material flowing into the boxes, or to diagram the exchange rates between major units. Steady-state conditions are usually assumed, and it is not the mission of such models to identify the effects of severe perturbation. Thus, although ecosystem models may be useful in other endeavors, their complexity and size, aggregation, and proclivity towards information loss minimizes their utility for understanding problems of exploitation of interactive, multispecies communities.

## MULTISPECIES ECOLOGICAL MODELS

These are models of intermediate complexity which lie somewhere in the void between single-species and ecosystem descriptions. Most are focused on issues of theoretical or applied interest and cover a wide range of subjects only loosely related to each other, but subjects whose further exploration is apt to bear on the issues of how to manage an exploited multispecies system. Most are oblivious to the social, economic, legal, and political factors known to dominate management practices. Many pay attention or attempt to develop the significance of linkage patterns or food web structure. Collectively, they reflect the rather shaky and highly variable common grounds occupied by theoreticians and empiricists jointly interested in multispecies dynamics. Here I consider interrelated features of such models or approaches. Unanimity is apparent in none of them, yet they all carry significance for each other. The phenomena examined are generally recognized as constituting important attributes of marine systems, and many of their conclusions are supported by extensive observational or experimental results.

### Model Formats

The approaches considered below range from highly quantitative to solely qualitative. They all ask questions of how species relate and react to one another in a variety of arrays and configurations. Some consider the role of primary production: for others this basic biological and potentially vital attribute is of no consequence. All share the common liability that spatial patterning is ignored and genotypes are considered static. Figure 1 illustrates both the relatedness and range of such models; only slight rearrangement is required to transform them into portrayals of both standard and more complex food web linkages (15, 23, 26). Model

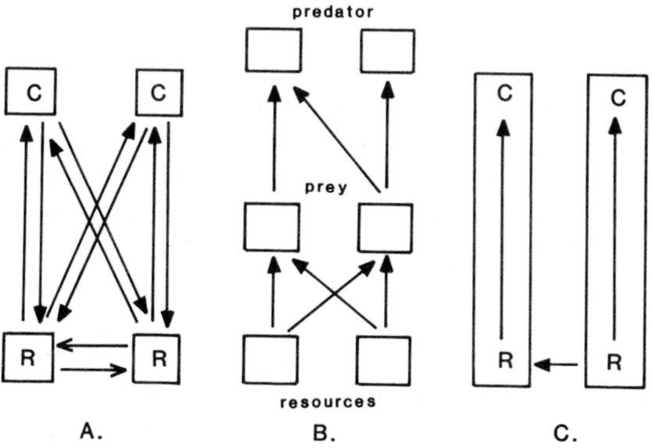


FIG. 1 - Three modeling approaches to simplified assemblages. C = consumer, R = resource. Arrows indicate an interaction and point towards that category or entity likely to benefit or prevail over the short term (see text).

A (37) represents one view of how "indirect mutualisms" might occur within multispecies mixes. It is a two-level model solidly rooted in Lotka-Volterra dynamics with carrying capacities, competition coefficients, and prey-predator relationships all analyzed. Model B (17) adds complexity while retaining most of the quantitative basis, also in a Lotka-Volterra framework. When competitive asymmetries are invoked, fishing or exploitation of one of the predatory species can generate unexpected population declines in its apparent competitor. Model C (23) is based on a strictly qualitative portrayal of an extensively studied natural community. No attempt has been made to quantify interaction terms or search for stability. On the other hand, expected broad patterns of response are based on experimental evidence of why and in which direction the community responds to manipulation.

**Primary Productivity**

There is no doubt that the rate of primary production is a necessary component of community level ecology. The majority of the world's major fisheries are found in upwelling regions, and whole zones are generally considered rich or poor on the basis of these rates. On the other hand, there is little agreement on how productivity might influence general community features. In many pelagic systems, highly productive regions are apt to be characterized by shorter, not longer, food chains (30, 34). Long-term study of the English Channel has identified major shifts in zooplankton and fish species composition, and no dramatic changes in productivity (32). Furthermore, Pimm (26) was unable to identify a relation between measured rates of production spanning four orders of magnitude ($10^1$ to $10^4$ mg C $m^{-2}$ day $^{-1}$) and trophic complexity. Some microcosm studies suggest why the relationship might be vague: Elliott et al. (8) have shown experimentally that the rate of primary production at constant nutrient input levels is highly dependent on the associated trophic structure. When only grazers are present, productivity is low due to effective cropping; when a third level is superimposed, productivity increases again because of grazer limitation. Such reverberations should be at least as apparent in all marine communities, where grazers are estimated to consume most of the plant cells in open water systems (34) and are known from manipulation studies of benthic communities to be able to regulate such features as plant productivity, species composition and richness (23). Outstanding problems exist because most ecological theory is constructed around the assumption that community structure, diversity, and stability are significant properties. It remains to be discovered whether, or how, any of these relate to productivity.

**Food Web Design**

Webs come in an enormous array of complexities, linkage patterns, and behaviors. Is there significance to these variations, do they make much difference, what needs to be known to specify system behavior? Webs provide the ecological scaffolding on which all multispecies relationships are cast. As such, they form a necessary structural component to any modeling effort. A spate of recent books or papers (4, 15, 16, 17, 26) has taken an optimistic stance that the intricacies can be reduced to rational and useful generalities. Here I present a slightly less sanguine view, not in opposition to the above statement but because I believe that basic and vexing problems are yet to be resolved.

Webs are easily drawn (Fig. 1). The fundamental problem is to decide what constitutes a link, if and when the conceptual diagrams are to be compared with or calibrated against field data. Does one count all prey, or only the most important ones? How should importance be assessed: abundance, trophic position, functional role? If the web shows temporal or spatial variation in complexity, should these finer patterns be averaged out of existence? What portion of the linkage patterns are induced, in the sense that substantial modifications (in the extreme case, additions or deletions) are well-known to generate major community changes? Should the web be partitioned to incorporate dietary change as a function of age or size, as seen in Hardy's classic web of herring linkage pattern?

The answers to such questions require both an empirical base, without which precise management will be difficult, and also theory, more essential for generalized understanding and prediction. For instance, I (23) have claimed that the dynamics of only one link need to be understood to model a community-level influence of the starfish Pisaster, yet Pisaster is linked trophically to at least forty other species. Similar claims could be made for sea otters (10) and other generalized, high trophic level consumers. In the aquatic web examined by Kerfoot (13), the addition of a single consumer increases web species composition, complexity, and number of trophic levels. Counterexamples are known. In summary, webs are variable in nature, flexible in a dynamical sense, and often arbitrarily described, with subjective judgments at all levels producing biases of unexplored significance. Web complexity, a fundamental associate of stability analyses, seems unrelated to, and might even be inversely correlated with, levels of primary productivity. On the other hand, secondary and primary production must be directly related. I suggest that the ambiguity of relationship between production and complexity remains the central problem in these ecological models: productivity influences stock abundance, yet web complexity probably determines in a broader sense the capacity to resist or recover from exploitation.

**Guilds**

The concept of the guild, a collection of species utilizing a shared resource in roughly similar ways, has proven enormously valuable in turning the attention of ecologists to relations between component species within multispecies assemblages. It has also provided a rationale for aggregation of ecologically similar species into single boxes where they are treated collectively with their potential individualism averaged. Is this procedure acceptable? I argue below that it is not.

An increasingly general literature exists on the phenomenon of ecological complementarity in which higher, often only distantly related taxa show signs of replacing one another geographically or competing directly with one another. Examples span the range of ant-rodent interactions, those between lizards and birds or birds and fish, or even the level of ecological similarity between marine mussels and primitive chordates. The message is that taxonomic relatedness is a poor criterion, by itself, for aggregating species into functional units. It is even reasonable to contend that whales, seals, and penguins compete (17); comparable trends obtain in the mechanistic study of Menge (18) where competitive asymmetries exist, with the "winner" always being larger, often orders of magnitude so. A related and important problem of within-guild dynamics, assuming that membership has been accurately determined, is the following. An increasing body of experimental evidence implies that in guilds where space is competed for, the probability of multispecies coexistence is reduced in the absence of extrinsic forces (24). Terrestrial plant ecologists have suggested the same. Most of these extrinsic forces are related to exploitation in some form, though physical disturbance or "contemporaneous disequilibrium" will yield the same result. In exploited, space-limited communities, simplification can be expected at lower trophic levels once the regulating influences are sufficiently reduced. If this is a plant level, changes in productivity of unknown magnitude and unpredictable duration can be anticipated. Whether comparable patterns exist within guilds of such mobile organisms as zooplankton and fish remains unknown, although there are numerous suggestions (2, 13) that they would. A diverse quantitative literature has been developed on the ecological conditions conducive for coexistence, with most of the models assuming that extrinsic forces are absent. From a management perspective, it would be enlightening to know how commercially valuable species are ranked by pure competitive prowess, for their steady-state yield is unlikely to be independent of their associates' abilities, and of how the guild membership in general responds to increased ecological opportunity. From a small-scale perspective, it seems irrational to accept that all species within an interacting set behave as equal yet independent entities.

### Interaction Strengths and Compartments

The behavior and stability of food web models are intimately related to the strength of interspecific interaction (15) and the relative complexity of the chosen assemblage. In general, as interaction intensity increases, these randomly assembled matrices become increasingly unstable. Whether this finding applies to the real world depends, among others, on both defining strength and determining whether natural communities can be realistically decomposed into subunits (= compartments, blocks, modules). Not much progress has been made in either area. Some species clearly have the capacity to influence the destinies of their associates more than others. Elephants are a recognized major agent of landscape modification; gazelles are not. A tropical starfish can devastate coral reefs, whereas other corallivores do not. Similar statements can be made about other starfish, sea otters, and even alligators: these species, unlike other members of their respective guilds, tend to be disproportionately important because they either directly or indirectly influence the physical structure of their environment. Although structure is less easily identified in aquatic systems, similar strong effects have been recognized where fish have been introduced or experimentally removed. However, though evidence for individual influence is usually obtained by population manipulation, it is not readily translated into the model's dynamical language. Finally, it is just conceivable in a fisheries context that exploitation of a weakly interacting species would have little influence; removal of a strongly interacting one might generate the system-wide cascades known from benthic communities. The trend towards mixed-species fisheries will only confound these effects, making it still more difficult to apply insights from ecological theory to fisheries management.

Another aspect of May's (15) analyses involved determining the influences of decomposing more complex assemblages into subsystems. If simpler units are more apt to be stable than complex ones, then a compartmentalized assemblage is more apt to persist than one organized randomly. It is not known whether natural systems are compartmentalized, though I (23) and Gilbert (11) believe they exist while Pimm and Lawton (27) and others reject the notion. Unfortunately, recognition of a compartment requires both extensive knowledge of the species involved and usually experimental identification. Compartments are uninteresting if their presence can only be associated with major environmental discontinuities. Furthermore, most natural food webs are described so incompletely or clumsily, with emphasis solely on the major constituents, that minor participants may not be recognized. If compartments are only associated with the less significant species, then they will have little importance to fisheries management. However, if compartments are induced either by substantial changes in primary production (22), or by changes in the stock of predators (2, 5, 13, 22), as seems generally true in freshwater communities, then there may be commercially important consequences.

The studies mentioned above suggest that dynamically generated compartments exist in some communities. Although no fisheries implications have yet been identified, it is not difficult to propose candidates. The presumed indirect relationship of ctenophore predation to herring year-class recruitment (35) might provide an example. If herring and ctenophores compete for copepods, and if these ctenophores are themselves consumed by another ctenophore (Beroe), itself a major prey item of cod and haddock, the stage is appropriately set. Thus when cod are abundant, copepods and therefore herring will be scarce. The current inverse population trends of cod and haddock increasing with herring stocks decreasing in the North Sea are suggestive, although the ctenophore connection remains highly speculative. On the other hand, interactions such as those reported by Vesin et al. (38), in which extensive exploitation of capelin probably led to more food and hence substantial population increases in Arctic cod and squid could be invoked. In either case, the evidence favors a view of interactively linked species. If such, even obscure, links are not sought, they obviously will not be incorporated into fisheries models, much to the detriment of prediction rooted in biological understanding.

### System Control

It is not a trivial matter to consider via model, manipulation, or observation whether assemblage structure is controlled primarily from below by nutrient inputs and changing production rates or phytoplankton size spectra (14), or from the top by consumers. Data exist on both sides of the question. For instance, over the short term, the structure of upwelling systems seems unaffected by exploitation patterns. On the other hand, many benthic studies (23) suggest that if a major consumer population predictably occurs, its removal will bear importance for the associated species. Numerous freshwater studies (i.e., (2, 13)) imply the same: structure, when examined at small spatial scales, is determined from above. Further thought on the subject is apt to be useful, for although it should not lead directly to specific management practices, it will be important for interpreting changes which might develop following exploitation of some stock.

## APPLICATION TO NATURAL COMMUNITIES

Previous paragraphs have developed the theme that theoretical and small-scale experimental studies can help interpret both the presence of and patterns resulting from multispecies interactions. The questions remain: a) are such effects recognizable at larger spatial scales in "open" communities? b) Do they matter in an economic or esthetic sense? Three examples are provided of system perturbation and eventual response at large spatial scales.

### The Lake Washington Plankton Community

This moderately large freshwater lake has been repeatedly stressed, and changes at all trophic levels continue to develop despite nutrient input, measured as dissolved phosphorus, being constant since 1968. There are many suggestions that alteration in species composition initiates cascades of change measured at several trophic levels. For instance, the virtual disappearance of Mysis has been associated with a major increase in its prey, Daphnia. Associated with the resurgence of Daphnia has been an increase in lake transparency ((7), and unpublished data) and a possible decline in annual primary productivity. The current introduction of rainbow trout, an effective planktivore, could initiate the type of change shown by Elliott et al. (8) in their 200L experimental systems. Thus questions of resource management are apt to become more pressing if trout, introduced for the benefit of fishermen, depress Daphnia to the point where water clarity and apparent quality begin to diminish. In summary, shifts are quite detectable at all major trophic levels and support a view of complex interdependence in a mixed-species system.

**Pacific Coast Sea Otters**

The sea otter was once widely distributed from Japan to Baja California before its near elimination by fur hunters. Although recovery is well under way, both island and mainland sites lacking otters exist. The evidence for ecosystem-wide otter effects is based on comparisons of these two "states." Otters effectively reduce the abundance of sea urchins to the point where these important grazers exert little influence on the associated kelp community. Dependent solely on the presence of otters, urchin standing crop varies by a factor of nine (10), and productivity of large algae from unmeasurable to a maximum of 13 g C $m^{-2}$ $day^{-1}$ (6). Documentation of further influences - extensive siltation behind kelp beds in the absence of urchins, and even effects on terrestrial bird populations - are known (9). The changing patterns generated by the removal and/or subsequent reintroduction of a single species of high trophic status native to the community can be understood as a cascade of interaction. Smaller-scale experiments illuminate some of the critical relationships, and the general interpretation garners much support from spatial repetition of the patterns subsequent to otter reintroduction. The commercial/esthetic implications are many and are well illustrated in the multiple user conflict in California over exploitation or preservation of many components of the otter-dominated food web.

**Northern Hemisphere Seabird Population**

The changing status of piscivorous seabird populations, often in competition with mankind for commercially valuable fish stocks, implies that a direct trophic link is involved, and that it might be measurable at wide geographic scales. The effect is not novel. Shaefer (31) noted the link between decreasing stocks of the Peruvian anchoveta and the fate of sundry guano birds. Murphy (21) even suggested that the failure of these birds to recover to previous levels following El Niño events might serve as a sensitive barometer of fish stocks. Comparable coupling and population trends characterize fish-seabird interrelationships in South African waters (3). In California, Anderson et al. (1) have shown that pelicans, though consuming $\ll 1\%$ of the anchovy stock, are ecologically dependent on these small fish and therefore provide independent assessments of anchovy abundance. In the North Atlantic, many seabird populations are showing marked declines in status, although the causes are not always clear. Habitat alteration and pollution working with diminished food resources due to expanded fisheries development seem most likely. For instance, development of a capelin fishery has probably led to widespread starvation of puffins in eastern Canada. In Norway, puffins compete with mankind for sandeel and herring, their preferred prey, and stocks there are on the decline as well (19).

A comparable situation appears to characterize exploitation of the walleye pollock in the Bering Sea. The fish is the most important winter food of murres. As might be expected, murre populations are declining. On the other hand, pollock consume copepods which are the primary resource of auklets, and while few data exist, there is the suggestion that population trends in the planktivorous auklets are opposite those of the larger, piscivorous murres (33). These are the sorts of trophic reverberations anticipated when some primary interaction is significantly altered, with some indirectly influenced species benefitting while others lose. Such inverse population trends need have nothing to do with direct interspecific competition. Rather, the changes are induced through modification of their prey resources, with the birds passively responding. As an aside, it appears that such between-ecosystem couplings exhibit donor-controlled dynamics, with the avian population size influenced by events within the aquatic community but with no or unmeasurably slight reciprocity.

I have chosen three examples (others are (20, 36, 39)) to suggest that interactions understandable at small, usually experimental, spatial scales can be extended to help interpret community dynamics and change detectably at much larger ones. Although examining exploitation effects in ocean basins might produce enough averaging (spatial or temporal) to reveal only the grossest patterns, this cannot be assumed. On the contrary, both direct and indirect effects can readily be detected when sought within large systems, and they may be of considerable economic import if they provide signs of otherwise unobservable community well-being or change.

## CONCLUSIONS

When a harvested species exhibits significant decreases in numbers, at least three explanations can be offered: a) the simplest answer is that of overexploitation and its anticipated consequences. b) Another might be that environmental caprice or variation has caused the change, or that cyclical system behavior with a long time base has failed to be detected by short-term study. c) Last, one might argue that more complex couplings within the community's infrastructure have indirectly led to the decline. Although all possibilities are likely, most fisheries managers have only considered the first two. Here, I have tried to provide reasons and natural examples why the last category should not be summarily dismissed. The most compelling ones are to be found in the extensive literature on interactions within experimentally manipulated intertidal and small freshwater assemblages, though plausible interpretation of pattern can be extended to larger-scale systems. Ignoring the challenge presented to managers and ecologists alike by complex and highly variable natural communities will benefit no one. Nature, though tolerant, is hardly totally forgiving of human excesses. The whole, like the whale, cannot possibly be managed without adequate knowledge of the parts.

**Acknowledgement.** Supported in part by the U.S. National Science Foundation, Grant OCE 80-25578.

## REFERENCES

(1) Anderson, D.W.; Gress, F.; Mais, K.F.; and Kelly, P.R. 1980. Brown pelicans as anchovy stock indicators and their relationships to commercial fishing. CalCOFI Report 21: 54-61.

(2) Brooks, J.L., and Dodson, S.I. 1965. Predation, body size, and composition of plankton. Science 150: 28-35.

(3) Crawford, R.J.M., and Shelton, P.A. 1978. Pelagic fish and seabird interrelationships off the coasts of Southwest and South Africa. Biol. Conserv. 14: 85-109.

(4) De Angelis, D.L.; Post, W.M.; and Sugihara, G., eds. 1982. Current Trends in Food Web Theory. Report on a Food Web Workshop. Oak Ridge Natl. Lab. 5983: 1-137.

(5) Dodson, S.I. 1970. Complementary feeding niches sustained by size-selective predation. Limnol. Oceanogr. 15: 131-137.

(6) Duggins, D.O. 1980. Kelp beds and sea otters: an experimental approach. Ecology 61: 447-453.

(7) Edmondson, W.T., and Litt, A.H. 1982. Daphnia in Lake Washington. Limnol. Oceanogr. 27: 272-293.

(8) Elliott, E.T.; Castanares, L.G.; Perlmutter, D.; and Porter, K.G.

1983. Trophic-level control of production and nutrient dynamics in an experimental planktonic community. Oikos 41: 7-16.

(9) Estes, J.A., and Palmisano, J.F. 1974. Sea otters: their role in structuring nearshore communities. Science 185: 1058-1060.

(10) Estes, J.A.; Smith, N.S.; and Palmisano, J.F. 1978. Sea otter predation and community organization in the western Aleutian Islands, Alaska. Ecology 59: 822-833.

(11) Gilbert, L.E. 1980. Food web organization and the conservation of neotropical diversity. In Conservation Biology, eds. M.E. Soule and B.A. Wilcox, pp. 11-33. Sunderland, MA: Sinaur Assoc., Inc.

(12) Hall, C.A.S., and Day, J.W. 1977. Ecosystem Modeling in Theory and Practice. New York: John Wiley and Sons.

(13) Kerfoot, W.C. 1982. Propagated effects along food chains: vaulting. In Current Trends in Food Web Theory. Report on a Food Web Workshop, eds. D.L. De Angelis, W.M. Post, and G. Sugihara. Oak Ridge Natl. Lab. 5983: 105-109.

(14) Landry, M.R. 1977. A review of important concepts in the trophic organization of pelagic communities. Helgol. Wiss. Meeres. 30: 8-17.

(15) May, R.M. 1974. Stability and complexity in model ecosystems. Second ed. Monogr. Pop. Biol. 6: 1-265.

(16) May, R.M. 1983. The structure of food webs. Nature 301: 566-568.

(17) May, R.M.; Beddington, J.R.; Clark, C.W.; Holt, S.J.; and Laws, R.M. 1979. Management of multispecies fisheries. Science 205: 267-277.

(18) Menge, B.A. 1972. Competition for food between two intertidal starfish species and its effect on body size and feeding. Ecology 53: 635-644.

(19) Mills, S. 1982. On the edge of the precipice. Birds 9: 57-60.

(20) Moreno, C.A.; Sutherland, J.P.; and Jara, H.F. 1984. Man as a predator in the intertidal zone of southern Chile. Oikos 42: 155-160.

(21) Murphy, G.I. 1973. Clupeoid fishes under exploitation with special reference to the Peruvian anchovy. Hawaii Inst. Mar. Biol. Tech.

Report 30: 1-73.

(22) Neill, W.E., and Peacock, A. 1980. Breaking the bottleneck: interactions of invertebrate predators and nutrients in oligotrophic lakes. In Evolution and Ecology of Zooplankton Communities, ed. W.C. Kerfoot, pp. 715-724. University Press of New England.

(23) Paine, R.T. 1980. Food webs: linkage, interaction strength and community infrastructure. J. An. Ecol. 49: 667-685.

(24) Paine, R.T. 1984. Ecological determinism in the competition for space. Ecology 65, in press.

(25) Patten, B.C., and Finn, J.T. 1979. Systems approach to continental shelf ecosystems. In Theoretical Systems Ecology, pp. 183-212. New York: Academic Press.

(26) Pimm, S.L. 1982. Food Webs. London: Chapman and Hall.

(27) Pimm, S.L., and Lawton, J.H. 1980. Are food webs divided into compartments? J. An. Ecol. 49: 879-898.

(28) Pitcher, T.J., and Hart, P.J.B. 1982. Fisheries Ecology. London: Croom Helm.

(29) Robertson, A., and Scavia, D. 1979. The examination of ecosystem properties of Lake Ontario through the use of an ecological model. In Perspectives on Lake Ecosystem Modeling, pp. 281-292. Ann Arbor, MI: Ann Arbor Science.

(30) Ryther, J.H. 1969. Relationship of photosynthesis to fish production in the sea. Science 166: 72-76.

(31) Shaefer, M.B. 1970. Men, birds, and anchovies in the Peru current-dynamic interpretations. Trans. Am. Fish. Soc. 99: 461-467.

(32) Southward, A.J. 1980. The western English Channel - an inconstant ecosystem? Nature 285: 361-366.

(33) Springer, A.M.; Roseneau, D.G.; Murphy, E.C.; and Springer, M.I. 1984. Population and trophic studies of seabirds in the northern Bering and eastern Chukchi Seas, 1982. In Environmental Assessment of the Alaskan Continental Shelf. Final Reports of Principal Investigators. Boulder, CO: BLM/NOAA OCSEAP.

(34) Steele, J.H. 1974. The Structure of Marine Ecosystems. Cambridge: Harvard University Press.

(35) Swanberg, N. 1974. The feeding behavior of Beroe ovata. Mar. Biol. 24: 69-74.

(36) van der Elst, R.P. 1979. A proliferation of small sharks in the shore-based Natal sport fishery. Env. Biol. Fish. 4: 349-362.

(37) Vandermeer, J.H. 1980. Indirect mutualism: variations on a theme by Stephen Levine. Am. Nat. 116: 441-448.

(38) Vesin, J.-P.; Leggett, W.C.; and Able, K.W. 1981. Feeding ecology of capelin (Mallotus villosus) in the estuary and western Gulf of St. Lawrence and its multispecies implications. Can. J. Fish. Aquat. Sci. 38: 257-267.

(39) Wharton, W.G., and Mann, K.H. 1981. Relationship between destructive grazing by the sea urchin, Strongylocentrotus droebachiensis, and the abundance of American Lobster, Homarus americanus, on the Atlantic coast of Nova Scotia. Can. J. Fish. Aquat. Sci. 38: 1339-1349.

(40) Wiegert, R.G. 1975. Simulation models of ecosystems. Ann. Rev. Ecol. Syst. 6: 311-338.

Exploitation of Marine Communities, ed. R.M. May, pp. 209-225. Dahlem Konferenzen 1984. Berlin, Heidelberg, New York, Tokyo: Springer-Verlag.

# The Response of Multispecies Systems to Perturbations

J.R. Beddington
Dept. of Biology, University of York
Heslington, York Y01 5DD, England

**Abstract.** This paper considers evidence from ecological theory and observation relevant to the response of multispecies marine communities to exploitation. The problems of reversibility, multiple stable states, and the differences between complex and simple communities are reviewed.

## INTRODUCTION

The title of this paper might be more appropriate for a series of monographs rather than a short paper, for clearly all ecosytems are continually perturbed, whether by climatic or other environmental factors, or indeed by the various activities of man in agriculture, fisheries, settlement, or industrialization. However, its generality does permit selection of those aspects of the subject relevant to the management of multispecies marine communities.

There are a number of scientific questions which are central to the rational management of marine communities, but all revolve around the question of sustainability.

What levels of mortality imposed by a fishery will permit a sustainable yield? Are there levels below which a fish population will not recover? Can judicious manipulation of the catch composition of the fishery alter the potential of the community to produce yields of a particular type, e.g., high value species? Can a community be depleted to a level where its potential for producing a harvestable resource is reduced?

With the exception of the first question, these questions and others like them are rarely explicitly addressed in the scientific bodies of the various fisheries' organizations. Instead, such bodies concentrate on the estimation of stock abundance and the calculation of allowable catch levels, although often implicit in the advice given by these bodies to management are a set of beliefs about the answers to such questions.

Part of this paper is aimed at considering the sort of evidence from both theory and observation, which bears on these beliefs.

**ECOLOGICAL THEORY**

Although communities are subject to continued environmental change, in many cases the component populations appear to fluctuate around characteristic levels of abundance. There are exceptions, often rather better studied, where cyclic fluctuations in species abundance and dramatic changes in the species composition are purported to occur.

One paradigm to explain these phenomena is to view the community as a dynamic system involving the interaction between the species. These interactions are described by equations defining the rates of change of each species with respect to the abundance of other members of the community. Such a paradigm has been around since the work of Lotka (26) and Volterra (47). The paradigm is useful for explaining a variety of the observed behaviors of ecological communities. Dynamic systems can display: single globally stable equilibria, multiple equilibria with alternative locally stable states, or cyclic behavior in which the attractor is a stable limit cycle. Lewontin's paper to the Brookhaven Symposium in 1969 is probably the clearest exposition of this set of ideas (25). It is also Lewontin who indicated the way in which the typical observations on ecological communities, showing fluctuations in abundance, can be reconciled with the predicted steady behavior of model systems. In effect, the link is made by assuming that the system behavior described by the equations is that produced by an average environment; fluctuations around that average can be envisaged either as producing "wobbles" on the parameters of the equations or as perturbations to the component system. The technical literature on the appropriate methods for formally incorporating environmental fluctuations into population models is extensive and cannot be discussed within the scope of this paper. Nisbet and Gurney (32) provide a review of much of the relevant literature.

A central question of relevance to the exploitation of multispecies communities that is posed by the paradigm concerns the potential for the existence of multiple locally stable states. It is with this question that much of this paper will be concerned. The question is often expressed as one of philosophical determinism (16, 29). Are the ecosystems we observe the result of chance and could other possible states have been observed? Or has the community inexorably approached the state we now observe?

For example, when man first explored the Southern Ocean the system was dominated by abundant mammalian predators of krill, the whales and the seals, and bird predators, mainly penguins. Is this inevitable, or could a rather different system be possible with perhaps cephalopod or fish predators dominating the system?

Such a question is not merely academic. It is important to the exploitation of marine communities for there is an obvious danger that the perturbations caused by a fishery could be irreversible if there are alternative stable states. Indeed, in the text of the Convention for the Conservation of Antarctic Living Marine Resources this idea is enshrined in Article II which states as an objective of the convention that changes which are not potentially reversible over two or three decades should be avoided.

There are two avenues of study in the ecological literature which are relevant to the question of multiple stable states. The one is concerned with the construction of model communities, the other with detailed mechanisms of population control in systems with a few species.

## MODEL COMMUNITIES

### Complexity and Stability

Most of the studies of model ecological communities have been concerned with the problem of complexity and stability. In essence this can be expressed as the question: how does the stability of systems alter with complexity?

A recent review by Pimm (40) usefully summarizes what has been rather a growth area in theoretical ecology. This indicates two main results which are of relevance to multispecies fisheries. The first is that for the various measures of complexity, the more numerous the species, and the more often and the more strongly they interact, the less chance there is of stability. This might be a cause for optimism in that simplifications to marine communities by reductions in species abundance or composition might not be inexorably linked to instability. It is certainly in contrast to the warnings of a variety of authors who have proclaimed of the dangers of simplification (12, 48).

However, a central question of relevance to management remains unstudied. This is concerned, not with all possible systems, but with the set of systems that are locally stable. Of these systems it is reasonable to query, what is the effect of a change in which additional mortality is added to various components of the system? Do complex communities retain their stability more readily than simple ones?

The mathematical apparatus for investigating this problem is straightforward, and it is curious that so far the various theoretical studies have omitted this question. A complementary question is concerned with the effect of the distribution of fishing mortality within the trophic web. For example, are certain patterns of harvesting, such as the joint harvesting of predator and prey, particularly destabilizing?

A subset of this problem has been considered in which, instead of an increase in mortality, the effect of the deletion of species on the stability of model communities was explored (38, 39). In essence these studies indicate that the chances of stability decrease with the number of species and the number and strength of the interactions. From this it might be concluded that the more complex the community, the more dramatic the effect of the removal of a species on the species composition.

There are two exceptions to this conclusion: the first is concerned with a rather special subset of models termed "donor-controlled." In such models the abundance of the prey is unaffected by the abundance of the predator. In systems composed of donor-controlled dynamics the chance of stability is enhanced by the increase in complexity. An open question is then, which, if any, marine communities are donor-controlled? However, there is a wealth of evidence that the removal of top predators has a marked effect on the community, implying that donor control is uncommon.

A second exception is that the species density of all or a group of species may be more resilient to change in complex than in simple systems (2). This would imply that although the abundance of individual species alters, the overall biomass of the system may remain relatively constant. This would seem worthy of some exploration in the context of marine communities.

**Multiple Stable States**

Most analysis of model communities has concentrated on local stability. It therefore begs the question of whether or not global stability exists. An exception is the work of Gilpin and Case (16) who explored the behavior of a system of normalized Lotka-Volterra equations.

Clearly these models are rather specialized, but the results obtained are interesting enough to attract comment. The authors explored by numerical integration the relationship between the potential number of domains of attraction, their size and type as the number of species in the community increased. They obtained an empirical relationship between the number of domains of attraction, D, the species number, m, and the mean interaction strength, a. This showed an exponential increase in the number of domains of attraction with both species number and interaction strength. Gilpin and Case found this equation to be a good approximation in the range m = 1-20, and for relatively small values of "a" used.

Domain of attractions were also ranked according to size, and it was found that for all the species numbers investigated, there was always one state with a large domain of attraction with the size of the next largest being some 20% to 50% of the size of the largest.

Although their paper does not present the distribution of species composition as the number of species was increased, the mean number of species in a domain of attraction was found to reach a rather small asymptote with approximately 2.5 species around an initial species composition of 10. It would thus appear that the stable states with many species present have rather small domains of attraction and are rare. The implications of this result, if it is at all general, are clearly important for fisheries. Complicated communities, once perturbed by fishing, are unlikely to return to the same state even if fishing ceases.

The models used by Gilpin and Case in their investigation are rather simple and contain little of what the biologist might term "real biology." It is therefore reasonable to enquire whether this very simplicity has led to the result and whether more complex models would show less of a predilection for multiple stable states.

Austin and Cooke (3) developed a much more detailed ecological model system and investigated its behavior in response to a variety of perturbations. They noted a variety of behaviors, but by far the most common was aperiodic shifts between alternate states of the system. This study, although involving more complex models, is still abstract and lacks empirical underpinning. Accordingly, in the next section the implications of some detailed biological mechanisms which are well supported by evidence are explored.

## DETAILED MODELS OF POPULATION CONTROL

A few years ago the literature on this topic was reviewed by May (29). May concentrated on simple one- and two-species model systems and demonstrated the ubiquitous character of multiple stable states with examples from models of grazing ecosystems (33), insect pest control (16), and host-parasite systems (28). Since this paper there have been a variety of other studies (1, 7, 24, 27, 30, 37, 43), but two basic mechanisms appear to be central.

The first involves the effect of a death rate (whether by grazing, harvesting, predation, or parasitism) imposed on some "prey" population. The "prey" population is, in the absence of this death rate, expected to be ultimately regulated by density-dependence. The death rate as a function of "prey" density is usually assumed to be of a Holling-type III functional response (19), although type II responses are also possible.

Taken together, these processes produce multiple equilibria in the following way. At very low levels of "prey" abundance, the prey growth rate exceeds the level of predation. At somewhat higher levels of prey abundance, increasing predation first equals and then exceeds the prey growth rate. This creates a locally stable equilibrium at low prey densities. As the prey abundance increases further, its growth rate increases to a point where it first equals and then exceeds the predation. This creates an unstable equilibrium. Finally, beyond its maximum production level, the prey growth rate declines and becomes equal to and then less than the predation level. This creates a second locally stable equilibria. This mechanism is illustrated in Fig. 1.

The second type of mechanism occurs in host-parasite systems and potentially in all sexually breeding populations. In such systems the rate of change of the population declines to zero at low population densities. This can be caused by an absence of suitable mates or from other problems in the transmission of reproductive material, for example, in host-parasite systems with an intermediate vector (1). At high densities the population is controlled by resource limitation.

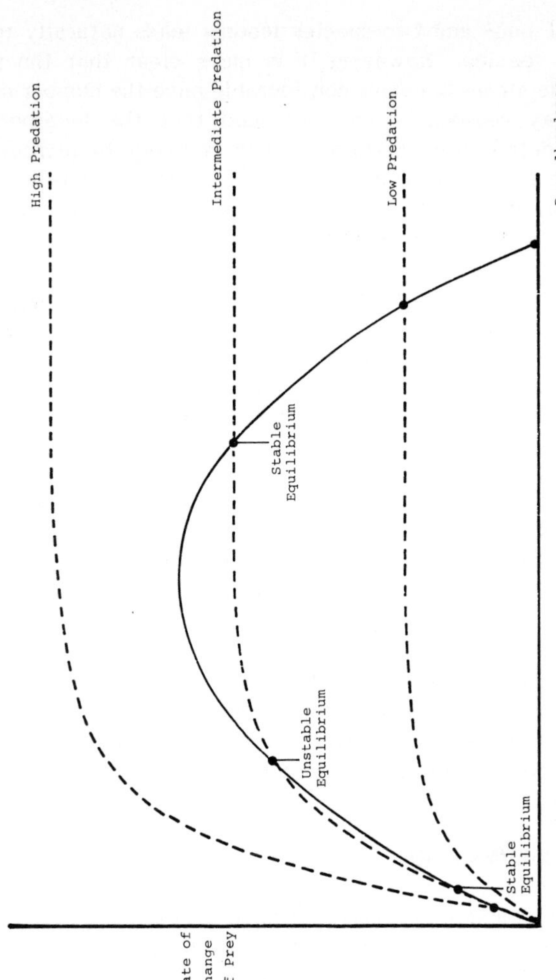


FIG. 1 - The way in which different levels of predation may produce multiple stable states. At high and low levels of predation, only a single stable equilibrium is generated; however, at the intermediate level, two alternate stable equilibria are produced.

The generality of these mechanisms is, of course, open to question. Nevertheless, there appears little doubt that individual mechanisms such as density-dependent growth and functional responses of type II and III are common.

The discussion of one- and two-species models leads naturally to models containing more species. However, it is quite clear that the potential for multiple stable states becomes considerable once the number of species increases. It may reasonably be concluded that the incorporation of more biological detail into abstract models is likely to increase rather than decrease the potential for multiple stable states. From the viewpoint of this paradigm, it is thus implausible that the changes produced in marine systems by fisheries are formally reversible.

At this point it seems appropriate to consider the confrontation of theory with experiment and to ask to what extent there is empirical corroboration of the theoretical ideas. Unfortunately, direct confirmation is difficult as the environment varies continually, and hence the effect of random variation on what might be observed needs to be considered.

Earlier it was indicated that deterministic models could be viewed as providing a picture of the behavior of communities in average environmental conditions. Random fluctuations would be expected to move the system between the domains of attraction of various stable states, the duration of stay in any particular region being determined by the interplay of the fluctuations and the size of the domains of attraction.

In a complicated dynamic landscape with many stable states, the observed behavior of the component populations would be indistinguishable from random variation. Only in relatively simple dynamic landscapes with only a few significant stable points would one expect to observe distinct types of community composition.

## ECOLOGICAL OBSERVATION

### Multiple Stable States

In a recent paper, Connell and Sousa (9) have performed a critical review of what they term "the evidence needed to judge ecological stability or persistence." In their paper they ask "whether real ecosystems (as opposed to model ones) are stable and consider the evidence for the existence of multiple stable states." They argue that: a) "there is no demarcation between assemblages that may exist in an equilibrium state and those that do not," and b) "there was no evidence of multiple stable states in unexploited natural populations or communities" and go on to conclude: "Rather than the physicists classical ideas of stability, the concept of persistence within stochastically defined bounds is, in our opinion, more applicable to real ecological systems."

Regrettably, there is insufficient space in this paper to consider in detail either the evidence they consider leading to their concluding view or the opinion itself. Nevertheless, it perhaps should be said that the distinction between the density-dependent processes which they accept and the population regulation mechanisms involved in most models of a community smacks of scholasticism.

In their paper Connell and Sousa make some important general points concerning the spatial and temporal scale needed for sampling. Their points have relevance to many exploited marine communities. For example, the appropriate spatial scale for sampling fish stocks can vary widely from the local to the transoceanic. Similarly, the different time scales of reaction for small pelagic fish such as anchovies through such predators as the cods or, indeed, at the extreme to the great whales is clearly important in assessing their dynamic behavior.

As indicated earlier, a variety of theoretical studies have demonstrated mechanisms in which multiple stable states can be seen to occur. A number of these studies cite examples to support the existence of alternate states. May (29) states, "a large body of empirical observations show that many communities have a multiplicity of stable states." Connell and Sousa (9), by contrast, in commenting on the various examples cited, indicate that all have shortcomings: either the evidence is insufficient, or the physical environment is different, or there is intervention by man.

It is hard to imagine any system observed over a reasonable period when some environmental variable does not change. Similarly, what constitutes satisfactory evidence in ecology is an open question, although it should be said that the critical comment on the studies of Sutherland and his co-workers (44-46) seems valid. The point made is that although historical events did determine the composition of the marine fouling community of invertebrates studied by Sutherland, both the spatial and temporal scale of sampling were insufficient to determine whether alternate states had been reached, rather than a transition from different initial conditions towards a similar state.

The examples of primary interest are those in which it is accepted that the communities have different states, but Connell and Sousa argue that these have only been preserved by man's intervention. These examples include the fish communities in Lake Windemere (23) and in the Great Lakes (20), both of which were altered by overfishing certain species, and in grazing systems in Australia and New Zealand (33). In all these systems there is some continued human activity, and the argument appears to be that if such activity ceased, the system would return to its previous behavior. This may well be the case, but the implications for management of fisheries are dubious. It would appear to imply that overfishing is reversible, simply by a cessation of activity. It is not clear in other studies of exploited marine communities that this is the case. These studies are considered below.

**Predator Removal or Addition**

Pimm (39) reviewed nineteen studies in which there had been removal of a predator. Most of these studies involved aquatic invertebrates, although one involved a marine mammal and in a number, a fish was the predator removed. Of the nineteen studies, only two showed neither a further species loss nor marked changes in species abundance. Other studies tend to support this result (8, 17, 34). In a similar manner, the addition of a predator or other natural enemy in the biological control of pests has been shown in a number of studies to have dramatically affected the species composition (5, 11).

This first body of evidence has considerable relevance for fisheries management. Top predators are often the initially preferred species when a community is exploited. On the basis of theory and observation, it should be expected that there will be marked changes in species abundance and composition following this removal.

The addition of species to an exploited marine community is a rare event. However, major increases in predator populations, perhaps following a cessation of exploitation, may be expected to alter the community structure.

**Concluding Remarks**

There are three main areas where ecological theory and observation suggest implications for the exploitation of multispecies fisheries.

First, more complex communities may be expected to react in a more dramatic manner to perturbations; that is, dramatic in the sense that the changes in the community produced by fishing may be entirely unexpected with no simple identifiable biological mechanism implicated as a cause.

Second, the removal of a top predator may be expected to have "knock-on" effects in the remaining community. Such effects might involve changes in species composition within the prey community or perhaps increases in preferred prey of the predator at the expense of lower trophic levels. This might be expected to be particularly pronounced in complex communities.

Third, ecological theory points unequivocally to an array of possibilities for multiple stable states and points out that it is unlikely that changes in a community produced by a fishery will be reversible. The evidence from ecological observation is more equivocal.

More complex communities are likely to have more stable states but, by contrast, are unlikely to show such marked changes of species composition, as environmental change will move the community through a spectrum of states. It is an open question whether in complex communities observation would be able to determine one state from another, although it seems unlikely. In simpler systems, with fewer possible stable states, it is more likely that marked changes of species composition and abundance will be observed.

## RESPONSE OF MULTISPECIES MARINE COMMUNITIES TO EXPLOITATION

The potential scope of this section is daunting. The world marine fish catch is several tens of millions of tonnes, and in 1982 there were more than twenty fish stocks where the catch exceeded half a million tonnes (13). Perturbations to marine communities clearly abound, although if seen as experiments, both design and data collection are usually lacking.

In order to impose feasible limits on the length of this section, only three problems will be identified and addressed. They can be posed as a series of complementary questions: a) following cessation of fishery activity, are the changes originally produced by fishing reversible? b) Is there evidence that exploited marine communities can exist in alternate states? c) Is there evidence to indicate differences in the response to fishing between complex and simple communities?

### Are Community Changes Reversible When Fishing Ceases?

Fishermen have been catching fish for a long time and in a variety of ways: since neolithic times, in Europe and North America, with relatively sophisticated methods (41). Unfortunately, though predictably, fishing communities have rarely stopped fishing once started, so that evidence for the reversibility of community changes when fishing ceases is comparatively rare.

One such example is afforded by the effective cessation of fishing activity in the North Sea during the Second World War. Following the war both the total catch of all species and the catch per unit effort of many individual species increased, implying that there had been an increase in stock size following a lowering of fishing mortality (18). Although fishing increased markedly in the decades after the war, the ability of the whole community to return to some pristine composition has not been assessed.

Fishery collapses following stock depletion are more common than global wars and provide, in principle, the opportunity for assessing reversibility. In addition, some fisheries management authorities have been successful in closing fisheries for particular stocks when they had been substantially reduced.

Unfortunately, quantitative assessment of the reversibility of changes in fish populations is bedevilled by their natural fluctuations. Recruitment to a fish stock is highly variable and only loosely linked to the adult stock (4, 15). Accordingly, the assessment of whether a stock has failed to recover requires a rather detailed treatment involving the calculation of the likelihood that it remains at different levels of abundance for particular periods. Such assessment has rarely been made, but on cursory inspection, the evidence appears equivocal, even for the same species. The North Sea herring, following protection in 1977, appears to be recovering at something like the expected rate (21). By contrast, the same species off Georges Bank and the Atlanto-Scandia stock appear to have remained at low levels for longer than would be expected. The ICES symposium on pelagic fish stocks (42) contains a wealth of examples of both types.

Interestingly, some relatively unequivocal indications of reversibility come from marine mammals. The Pacific gray whale (22) and the Southern fur seal (36), following the cessation of exploitation, appear to be recovering or have recovered at something like the expected rate.

**Evidence for Alternate Stable States**

Typical data obtained from exploited fish communities come from the fishery. Accordingly, the information suffers in a number of ways. Catch levels or even catch rates may not reflect the abundance of the species. Changes in migration patterns and the relatively long time scale of fish and marine mammal dynamics put a heavy burden on sampling which needs to be conducted on large spatial and temporal scales if satisfactory data are to be produced.

Unsurprisingly, few data satisfy such stringent criteria. Nevertheless, Daan (10) has reviewed a number of examples of what might be termed alternate states of systems. In particular, the sardine-anchovy systems of the NE Pacific, the NW Pacific, and the SE Atlantic appear to indicate alternate periods of dominance by one or another species. More recent information on these communities would appear to corroborate the earlier data (14).

In addition there appear to have been changes in the species composition in both the North Sea and the Baltic, associated with the removal of herring and mackerel in the former case, and cod in the latter (10).

In all cases such evidence is equivocal as climatic changes of some sort can be associated with any change of state.

One aspect of the marine environment that is often ignored is the ease with which migration can occur through the spatially heterogeneous environment of the ocean. Many of the models discussed in this paper are framed in a spatially homogeneous environment. It is a general tendency that incorporation of spatial heterogeneity into models tends to promote persistence in systems that might otherwise collapse. This poses a whole range of questions concerning the appropriate spatial and temporal scales of sampling touched on earlier, as well as questions about the spatial scale of the interactions that are occurring within marine communities. One aspect of this problem has been posed in an interesting form in some terrestrial systems where national parks have been created. In certain parks, large mammalian predators appear to become "overabundant" and in doing so seem to be capable of considerably changing the ecology of the system. It is interesting to speculate that this might result from a limitation of the migration that would occur in the larger, pristine system. No parallel problems appear within the marine environment (31). Accordingly, it may well be that the lack of evidence for alternate stable states is a result of the migration between different "patches" of the ocean.

**Complex Communities Versus Simple Communities**

Even simple marine fish communities are quite complex, as different life-history stages occupy different trophic levels. Nevertheless, in broad terms tropical communities appear more complex than temperate, and these in turn appear more complex than the communities in some subtropical upwelling areas and in the polar seas.

The response of tropical communities to fishing has received considerable attention of late (35). The most documented community is that fished in the Gulf of Thailand, although similar communities are exploited elsewhere in the South China Sea and off Indonesia. Interestingly, exploitation of this community appears to have resulted in a series of secondary effects with certain species declining at rates that cannot be attributed to either fishing (6) or dramatic climatic change.

By contrast, in other systems most changes are attributable to either fishing or some major oceanographic phenomenon.

Finally, it is worth noting that the evidence for alternate stable states is strongest for the simplest communities, an observation satisfactorily in line with an argument developed above.

## REFERENCES

(1) Anderson, R.M. 1979. The influence of parasitic infection on the dynamics of host population growth. In Population Dynamics, eds. R.M. Anderson, B.D. Turner, and L.R. Taylor. Oxford: Blackwell Scientific.

(2) Armstrong, R.A. 1982. The effects of connectivity on community stability. Am. Nat. 120: 391-402.

(3) Austin, M.P., and Cooke, B.G. 1974. Ecosystem stability: A result from an abstract simulation. J. Theor. Biol. 45: 435-458.

(4) Beddington, J.R., and Cooke, J.G. 1983. The potential yield of fish stocks. Tech. Paper 242. Rome: FAO.

(5) Beddington, J.R.; Free, C.A.; and Lawton, J.H. 1978. Modelling biological control: on the characteristics of successful natural enemies. Nature 225: 513-519.

(6) Beddington, J.R., and May, R.M. The harvesting of interacting species in a natural ecosystem. Sci. Am. 247(5): 66-69.

(7) Botsford, L.W. 1981. The effects of increased individual growth rates on depressed population size. Am. Nat. 117: 33-63.

(8) Connell, J.H. 1975. Some mechanisms producing structure in natural communities. In Ecology and Evolution of Communities, eds. M.L. Cody and J.M. Diamond. Cambridge, MA: Harvard University Press.

(9) Connell, J.H., and Sousa, W.P. 1983. On the evidence needed to judge ecological stability or persistence. Am. Nat. 121: 789-824.

(10) Daan, N. 1980. A review of replacement of depleted stocks by other species and the mechanisms underlying such replacement. Rapp. P.-v. Reun. Cons. Int. Explor. Mer 177: 405-421.

(11) De Bach, P. 1974. Biological Control by Natural Enemies. London: Cambridge University Press.

(12) Elton, C.S. 1958. The Ecology of Invasions by Animals and Plants. London: Chapman and Hall.

(13) Food and Agriculture Organisation of UN. 1984. Yearbook of Fisheries Statistics, Nr. 52. Rome: FAO.

(14) Food and Agriculture Organisation of UN. 1984. Report of Special Conference on Neritic Fish Stocks. Rome: FAO/IOC, in press.

(15) Garrod, D.J., and Colebrook, J.M. 1978. Biological effects of variability in the north Atlantic Ocean. Rapp. P.-v. Reun. Cons. Int. Explor. Mer 173: 128-144.

(16) Gilpin, M.E., and Case, T.J. 1976. Multiple domains of attraction in competition communities. Nature 261: 40-42.

(17) Harper, J.L. 1969. The role of predation in vegetation diversity. Cold Spring Harbor Symp. 22: 48-62.

(18) Hempel, G., ed. 1978. North Sea fish stocks - Recent changes and their causes. Rapp. P.-v. Reun. Cons. Int. Explor. Mer 172.

(19) Holling, C.S. 1959. The components of predation as revealed by a study of small mammal predation on the European sandfly. Can. Ent. 91: 293-320.

(20) Holling, C.S. 1973. Resilience and stability of ecological systems. Ann. Rev. Ecol. Syst. 4: 1-23.

(21) International Council for Exploration of the Sea. 1983. Report

of the Herring Assessment Working Group for the area south of 62 N. CM 1983/Assess. 9.

(22) International Whaling Commission. 1981. Report of the Sub-committee on Other Protected Species and Aboriginal Whaling. Report 31, pp. 133-140.

(23) Le Cren, E.D.; Kipling, C.; and McCormack, J.C. 1972. Windermere: effects of exploitation and eutrophication on the salmonid community. J. Fish. Res. Board Can. 29: 819-832.

(24) Levin, S.A. 1978. Pattern formation in ecological communities. In Spatial Pattern in Plankton Communities, ed. J.H. Steele, Conf. Ser. IV, Marine Science Ser. 3, pp. 433-465. NATO.

(25) Lewontin, R.C. 1969. The meaning of stability. Brookhaven Symp. 22: 13-24.

(26) Lotka, A.J. 1925. Elements of Physical Biology. Baltimore: Williams and Wilkins.

(27) Ludwig, D.; Jones, D.D.; and Holling, C.S. 1978. Quantitative analyses of insect outbreak systems: the spruce budworm and the forest. J. Anim. Ecol. 47: 315-332.

(28) Macdonald, G. 1973. Dynamics of Tropical Disease. London: Oxford University Press.

(29) May, R.M. 1977. Thresholds and breakpoints in ecosystems with a multiplicity of stable states. Nature 269: 431-477.

(30) May, R.M. 1979. The structure and dynamics of ecological communities. In Population Dynamics, eds. R.M. Anderson, B.D. Turner, and L.R. Taylor. Oxford: Blackwell Scientific.

(31) May, R.M., and Beddington, J.R. 1981. Notes on some topics in theoretical ecology, in relation to the management of locally abundant populations of mammals. In Problems in Management of Locally Abundant Wild Mammals, eds. P.A. Jewell, S. Holt, and D. Hart. New York: Academic Press.

(32) Nisbett, R.M., and Gurney, W.S.C. 1982. Modelling Fluctuating Populations. New York: Wiley.

(33) Noy-Meir, I. 1975. Stability of grazing systems: an application of predator-prey graphs. J. Ecol. 63: 459-481.

(34) Paine, R.T. 1980. Food webs: linkage interaction and community infrastructure. J. Anim. Ecol. 49: 667-686.

(35) Pauly, D., and Murphy, G.I. 1982. Theory and Management of Tropical Fisheries. Manila: ICLARM.

(36) Payne, M.R. 1977. Growth of a fur seal population. Phil. Trans. Roy. Soc. Lond. B 279: 67-79.

(37) Peterman, R.M. 1977. A simple mechanism that causes collapsing stability regions in exploited salmonid populations. J. Fish. Res. Board Can. 34: 1130-1142.

(38) Pimm, S.L. 1980. Food web design and the effects of species deletion. Oikos 35: 139-149.

(39) Pimm, S.L. 1982. Food Webs. London and New York: Chapman and Hall.

(40) Pimm, S.L. 1984. The complexity and stability of ecosystems. Nature 307: 321-326.

(41) Rau, C. 1884. Prehistoric fishing in Europe and North America. Smithsonian Contributions to Knowledge 509.

(42) Saville, A., ed. ICES Pelagic Fish Symposium. Rapp. P.-v. Reun. Cons. Int. Explor. Mer 177.

(43) Southwood, T.R.E., and Comins, H.N. 1976. A synoptic population model. J. Anim. Ecol. 45: 949-965.

(44) Sutherland, J.P. 1974. Multiple stable points in natural communities. Am. Nat. 108: 859-873.

(45) Sutherland, J.P. 1981. The fouling community at Beaufort, North Carolina: a study in stability. Am. Nat. 118: 499-519.

(46) Sutherland, J.P., and Karlson, R.H. 1977. Development and stability of the fouling community at Beaufort, North Carolina. Ecol. Monogr. 47: 425-446.

(47) Volterra, V. 1926. Fluctuations in the abundance of species, considered mathematically. Nature 118: 558-560.

(48) Watt, K.E.F. 1973. Principles of Environmental Science. New York: McGraw-Hill.

Standing, left to right:
Garry Brewer, Wolf Arntz, Victor Smetacek, Jean-Paul Troadec, Michael Glantz, Carl Walters

Seated, left to right:
Walter Nellen, Roger Bailey, Alain Laurec, John Beddington, Bob May

(Not Shown: Fritz Thurow)

Exploitation of Marine Communities, ed. R.M. May, pp. 227-244. Dahlem Konferenzen 1984. Berlin, Heidelberg, New York, Tokyo: Springer-Verlag.

# Management under Uncertainty
## Group Report

J.R. Beddington, Rapporteur
W.E. Arntz
R.S. Bailey
G.D. Brewer
M.H. Glantz
A.J.Y. Laurec
R.M. May
W.P. Nellen
V.S. Smetacek
F.R.M. Thurow
J.-P. Troadec
C.J. Walters

**INTRODUCTION**

The response of individuals to uncertainty is itself unpredictable. Some entrepreneurs thrive successfully in the most uncertain circumstances, and the existence of futures markets and insurance companies indicates that others are prepared to pay to avoid carrying the burden of uncertainty.

Working to control the decision setting so as to minimize surprises is one very general way of coping with uncertainty. There are both legitimate and somewhat unsavory mechanisms worth noting as well. As to the latter: consider bribes and protection monies paid to secure predictable environmental conditions. Or reflect on the fact that most patents for alternative energy supply means and sources have been systematically purchased by the major oil production companies.

Initially the report considers the different sources of uncertainty; it then moves on to consider the sorts of management problems associated with this uncertainty. One particular problem faced by the scientific community is how to frame advice to managers. The others are concerned with the way different sources of uncertainty can be handled by management.

## SOURCES AND KINDS OF UNCERTAINTY

The natural variations of climatic and oceanographic variables lead inevitably to uncertainty about the behavior of marine ecosystems. In addition, interactions amongst species, particularly in the prerecruit stage of the life history of marine species, are poorly understood and often are believed to involve unmonitored biological components of the system (Sissenwine, this volume).

Another source of unpredictability arising from biological interactions among species is the mortality produced by pathogens or pollutants. Although it can be argued that pathogens and parasites are important in the density-dependent regulation and geographical distribution of many populations of marine animals (19), dramatic "outbreaks" of a pathogen - and the concomitant uncertainty in prediction - are more likely in the high density situations found in aquaculture or in some forms of shellfish production. By contrast, pollution in the marine environment can affect many coastal stocks. One example where disease is acting to increase uncertainty is the plague affecting the sea urchins (Strongylocentrotus) on the east coast of Canada and the USA. This is likely to make the dynamics much less predictable on a medium-term time scale.

It is useful to distinguish between two sources of uncertainty. The one is produced by those natural fluctuations which affect the dynamics and abundance of fish stocks in an unpredictable way. The other is produced by the limitations that accompany scientific attempts to monitor abundance and productivity of stocks and the causal mechanisms underlying their dynamics.

## NATURAL VARIATIONS

In dealing with natural variation in the environment, it is reasonable to pose a number of questions. First, what are the different levels of variability associated with different geographical regions and ecosystems? Second, is there a relationship between the life-history characteristics of a species, for example, its position in a trophic web, and the amount of uncertainty associated with its dynamics? Third, is there any emergent pattern in the response of marine ecosystems to exploitation? For example, are changes produced by exploitation reversible when fishing is stopped or drastically reduced?

These questions are posed in order to see if there are some guidelines which may be used for assessing the likely behavior of a particular system about which little is known. In a Baysian sense we are attempting to query what the prior probabilities are, for example, of high recruitment variation, stock collapses, and so on.

## GEOGRAPHICAL DIFFERENCES

It is rather difficult to disentangle the effects of life-history characteristics and geographical location as different species occur in different environments. Nevertheless, it is reasonable to pose questions of this sort, as the answers can provide a rough guide to the levels of variability that may be expected from a fish resource.

Few, if any, oceanographic events are sufficiently dramatic to attract the attention of the set of scholars who have examined the properties of the El Niño phenomenon in the Southeast Pacific. The special properties of such upwelling systems indicate that they may be exceptions to geographical rules, which might indicate that there was less variability as one moved from polar to tropical seas. There are, of course, other exceptions: for example, the population explosion of balistes species off tropical West Africa or the apparent stability of the Southern Ocean. In order to examine evidence for such rules, it is necessary that some attention be paid to the temporal and spatial scale of monitoring and to the relevant parameters of the fish stock. For example, viewed on a time scale of decades, even the properties of highly variable upwelling systems may be predictable, while on a time scale of months, the apparently constant abundance of a reef community may be highly variable. In a similar manner, changes in migration patterns may present an illusion of high variability in fish stock abundance if the spatial scale of the sampling is too restricted, although it should be noted that such spatial variations may be reflected in the catch rates of the fishery.

For fish stocks with similar levels of natural mortality, the level of variation in recruitment is what distinguishes the highly variable stock from the less variable. Similarly, it is the difference in mortality rates which demarcates variable stocks from the less variable when their recruitment variation is similar. However, it should be noted that there appears to be a rough relationship between mortality and recruitment variation. High mortality rates are only rarely associated with high levels of recruitment variation (9). When this happens, several management problems are posed (see below).

At the system level, there appear to be a number of candidate hypotheses for explaining the overall level of variability. One such hypothesis is that there is a relationship between the diversity of a community and its variability. Simple systems which contain few species are apparently more variable than the more complex. Again there are exceptions, the salt marsh community being a case in point.

One such hypothesis is generated from analysis of benthic communities where there are two measures of variability: persistence and resilience. Less diverse systems are often not very persistent but highly resilient (6, 12).

A complementary hypothesis is that variability is related to primary productivity: the higher the productivity, the more variable the system. It was claimed in support of this hypothesis that upwelling systems were highly variable, open pelagic systems less so. Similarly, inshore demersal stocks were variable, deep water demersal stocks less so.

In investigating any such broad hypotheses, there is a problem that information on complex tropical communities is sparse and at times misleading. Species identification is sometimes only to the level of families, and hence the averaging implicit in such classifications may mask variability.

## ECOLOGICAL DIFFERENCES

As indicated in the preceding section, there are two measures of variability, that in the stock and that in the recruitment. Variation in recruitment can be readily measured either by the variance in the natural logarithm of recruitment or from the residual variation around some stock-recruitment curve. Recent studies appear to indicate that certain types of species differ markedly in the levels of recruitment variation. Gadiformes appear to have the largest variation with Perciformes and Clupeiformes intermediate and Pleuronectiformes the least variable. However, it must be emphasized that within these families there are numerous exceptions to this generalization (9). Information on molluscs and crustaceans indicates that bivalves may be more variable than certain crustaceans with strong territorial behavior, for example, some lobster species.

In almost all studies, the frequency distribution of recruitment has been found to be skewed with the log normal being a good empirical fit. An implication of this is that there are certain fish stocks in which the variability in recruitment is so high that they lead an ephemeral existence on the basis of occasional, very large year classes. Although pattern in recruitment variability is difficult to discern, within historic data sets there are examples of stocks that appear to depend (in the sense of supporting the continued existence of a spawning stock) on episodic recruitment. These, however, are from a wide range of unrelated areas, taxonomic groups, and ecological situations. The more striking examples are: the Atlanto-Scandian herring (15); the North Sea mackerel (4); perch in Swedish lakes (2); oysters in parts of Britain (29); bivalves in Normandy (11).

It is possible to combine the variation in recruitment with an estimate of the mortality suffered by the stock to arrive at some measure of the variation in stock size. Fish species higher in the trophic web and nearer the poles have in general a lower natural mortality, and hence for similar levels of recruitment variation will fluctuate less. However, where fishing intervenes and hence the total mortality increases, the stock will fluctuate more markedly. It should be emphasized that this effect is independent of any indication that exploitation leading to a reduction in stock size increases the variability of recruitment. The combination of a high natural mortality and medium to high recruitment variation exhibited in some small pelagic species, such as the Peruvian anchoveta, makes the management of such fish stocks on a sustainable basis extremely difficult and unwise; for example, it would be virtually impossible to stabilize the biomass, hence such a management objective should be avoided. This point is referred to later in this report.

Tropical pelagic species show some distinctive features: recruitment extends for several months and spawning grounds, feeding grounds, and nursery grounds all overlap (16). These features pose additional complicated management problems.

## EXPLOITED SYSTEMS

It is a plausible idea that the changes produced by exploitation are unlikely to have been mimicked in the evolutionary history of a species. Accordingly, it is reasonable to enquire whether such changes are likely to lead to a fundamental alteration in the state of the community. Viewed in the context of a single species, in the absence of exploitation, stock abundance will have some equilibrium probability distribution with characteristic mean and variance. This will be determined by the multitude of factors that affect its recruitment, growth, and mortality.

An interesting question is: following a reduction in the species produced by a fishery, do fundamental changes in that probability distribution occur? The evidence from the history of exploitation of a number of stocks is equivocal. The first problem is that, when there has been a major decline in the stock, the variability in the system makes the calculation of the expected timing and rate of recovery highly uncertain. However, there do appear to be examples where fish stocks have stayed at very low levels of abundance for long periods (the Californian sardine (24), the Japanese sardine (21), the Atlanto-Scandian herring (5)); while in other cases, following the closure of the fishery, recovery has apparently occurred towards the original levels of abundance, e.g., the North Sea (Downs stock) and Icelandic summer-spawning herring (5). Accordingly, the answer to the question is uncertain. Indeed, in the Northeast Pacific, different stocks of the same species of herring have showed totally different responses (20).

## SCIENTIFIC UNCERTAINTY

There are two levels of scientific uncertainty which determine the predictability of the response of fish stocks to exploitation. The first concerns the ability of scientists to monitor the abundance and trends in stock size. The second concerns their ability to discover and quantify the causal mechanisms that determine the productivity and abundance of the stock. Clearly, proper quantification of such causal mechanisms depends on the monitoring of the stock in the first place. Accordingly, it is reasonable to enquire what the limitations to such monitoring are.

It is obvious that greater scientific effort will tend to produce greater certainty. It is perhaps less obvious that the marginal return in knowledge per unit effort can be small and that there is a limit to the amount of information that can be gleaned from even the most competently designed program (Shepherd, this volume).

It should also be recognized that increased scientific work can lead to a recognition that uncertainty is greater than had previously been believed, for example, by the refutation of previously well established hypotheses. Hence, more science can lead to even more equivocal scientific advice. A further problem is that successful management aimed at stabilizing stock abundance can mean that uncertainty over the response of the system to stock changes increases with time.

One area where attempts have been made to increase the predictive power of scientific advice is by correlating environmental variables with the various components of fish dynamics, particularly recruitment (7, 8, 10, 26). For example, in the Moroccan pilchard fisheries a correlation seems to exist between the success of recruitment on the one hand, and the regime of trade winds and upwelling circulation, inasmuch as they affect the food supply and the rate of dispersion of food larvae. In addition, medium-term fluctuations have been observed in the atmospheric circulation and its effects on the upwelling regime. Inasmuch as the future evolution of these climatic manifestations can be forecast with some degree of probability, some prediction of strategic value could be made on the pilchard abundance and distribution fluctuation.

It is important to recognize the limitations of such approaches and the potential they have for misuse. For example, a functional relationship between temperature and recruitment cannot be used to assess potential variations in recruitment from the variations in temperature. To do so would be to ignore the variations in recruitment unexplained by the changes in temperature (Walters, this volume). Nevertheless, such analyses can be useful in two ways: they can be used to make short-term predictions and to gain time. For example, they can be used to predict the expected levels of recruitment from observations on temperature some months before recruitment to the fishery occurs. In addition, direct observations may be both difficult and expensive, and an environmental variable may be more readily or cheaply observed.

Typically, such analyses are generated from time series of observations on fish stocks. Many such observations appear to show auto-correlation, although formal statistical analysis of the data often fails to detect evidence for auto-correlation (Shepherd, personal communication). It is important to recognize that the assumption that auto-correlation is not present is highly optimistic, except in the very short term. Indeed, in general it may be said that to quantify the uncertainty of scientific predictions based on the assumption of white noise will underestimate that uncertainty. To that extent, it may be used as a "best case"; the result will usually be worse.

In the same way, scientific quantification of uncertainty places extra demands on data collection and analysis. The demand that such analysis should involve a proper investigation of the error structure of the models used is not mere statistical pedantry. The design of stochastic models, particularly those based on some underlying deterministic model, can be quite problematic. In particular, appraisal of the uncertainties associated with the predictions can depend markedly on the error structure (18, 25).

In addition to the problem of auto-correlation, a choice between multiplicative or additive error models can lead to very different results (3,

22). Many types of data in fisheries, for example, recruitment, catch per unit effort, or samples from acoustic surveys, present highly skewed frequency distributions. Accordingly, statistical analysis based on some standard technique which assumes normally distributed errors can be highly misleading (17, 23, 27).

The quantification of uncertainty also involves a consideration of the time scale involved. In the very short term, one year or less, uncertainty is largely produced by white noise and can be quantified. For example, Pope (28) has looked at the uncertainty associated with the calculations made of Total Allowable Catches (TAC's) typically performed by the working groups of ICES.

In the medium term, 2-10 years, predictions of average levels of catches, abundance, etc., can be made on the basis of standard models, but as indicated earlier, will be affected by any auto-correlations in the vital rates of the stock. However, of greatest importance in the medium and long term is the overall density-dependent response of the stock, particularly the stock and recruitment relationship and how well the models mimic this behavior.

In the long term, greater than 10 years, environmental changes can completely dominate the behavior of the system, and thus the uncertainty can depend on the ability to predict such environmental change.

The degree of predictability will depend on the natural time scale of the fish stock, set by the generation time, which will of course be affected by the level of exploitation. High levels of exploitation have a cost in lowered predictability.

There is a second important time scale which is set by the response time of the management system. How quickly can scientific advice be observed and utilized to alter the level of fishing? For example, where the reaction time is longer than, say, the time scale of natural fluctuations in the stock, major problems can occur. A basic way of approaching these problems is via decision theory, which permits an analysis to be made of the efficiency of the management system itself.

## THE FRAMING OF SCIENTIFIC ADVICE

Fishery science is by no means unique in the problems it faces in communicating scientific uncertainty to the various components of management. Essentially there are two extremes. One technique is where the uncertainty is laid out for all to see. Assumptions are explicitly stated, and the chances that particular events will occur are derived from the underlying equations which are specified. The other technique is to state the most likely event derived from the expectation of the model used. Although there are problems with both extremes, it seems clear that in almost all cases it is better for scientists to indicate some of the uncertainties involved in their calculations. In particular, the opportunity for interaction between scientists and different types of managers is enhanced if uncertainties are recognized. There has been a tendency for scientists to provide managers with quantitative information without sufficient reflection of uncertainty. This has probably arisen because of the need to have "firm" figures as a basis of negotiation. These figures also provide the manager with a convenient shield, for they are usually to a greater or lesser extent wrong. It is therefore most important that managers be provided with advice on such matters as: a) in what direction the stock should be moved, whether or not this can be supported with meaningful quantitative information, and how urgently action is required; b) what alternatives there are for achieving such a change; and c) what uncertainties are associated with the various pieces of information.

To do otherwise is to assume part of the managers' responsibility and also to fail to encourage the development of managers with an understanding of the complexities of fish stocks and the uncertain nature of the decisions required. It should be emphasized that uncertainty does not lead inevitably to a conservative management approach.

In other areas of activity there has emerged a group which solves the problem by interpreting the science for the managers and vice versa. To a certain extent, such groups may be observed within the community of fisheries scientists.

## IMPLICATIONS OF UNCERTAINTY AND MANAGEMENT METHODS FOR HANDLING IT

To a large extent the implications of scientific uncertainty are as variable as the components of the various institutions that depend on and interact with the fishery resource. Without seeking to extend unalterably the terms of reference of the group, it was felt important to emphasize that the resource and its harvesting is only one element of the fishery system which encompasses the processing, distribution, and marketing components as well as the institutions set up to regulate the system. Accordingly, while this report quite properly concentrates on biological aspects of fisheries, notice needs to be given to the technological, economic, social, and managerial dimensions, too. Typically, hundreds of individuals - separately and joined together in various institutional settings - affect the present and future state of a fishery resource. Indeed, the social ecology may be as challenging in its complexity as is the biological one (see (1)).

So then, what is fishery management? Most generally, it is the sum total of human acts affecting a given marine fishery. Some such will, of course, be quite evident and hence are well-known to most of us. But many others, not as proximate in space or time, may result in most consequential effects. And, because of interdependencies and synergies, apparently rational or well-meant decisions made in one sector or area may turn out to have unintended or even egregious consequences in others, such as a marine ecosystem. A major, general task confronting those with stewardship responsibilities over marine resources is to be aware of such eventualities so as to head off their most damaging effects, if possible, or to mitigate them if not. These ideas are discussed elsewhere in terms of preventive and reactive management, respectively (13).

A strong suspicion one gleans from even the most cursory observation of realistic managerial performance is that efforts to partition and reduce the "fishery problem" to its biological essentials are misguided. And, by extension, the "solutions" thereby determined and presented by biologists are too often suboptimal - if not irrelevant - to the actual needs of those trying to make the best use of the resource. It was therefore felt appropriate to consider the ways in which other institutions dealt with uncertainty.

## UNCERTAINTY IN OTHER SECTORS

Thinking about uncertainty as both unexpected and unwanted changes in an area of interest or responsibility and then asking how others cope with such opens up several intriguing prospects.

In any competitive endeavor certain, select few, individuals will through experience, intuition, and luck "rise to the top" - at least suggesting that uncertainty may be dealt with successfully, albeit by a limited number of decision-makers. Thus there exists an opportunity to investigate "success stories," individual instances where on average beneficial choices have been made. With specific reference to fishery management, who are the "good" or "successful" decision-makers? Why? What attributes and criteria define "good" and "success"? (It may also be well worth the time to repeat the exercise for successful scientists and analysts.)

As a general matter, success elsewhere stems from several sources. Two of these stand out: generating and synthesizing information about the decision setting and simultaneously working to control it.

Information comes from many sources and is created through many quite different means. Scientific analysis is but one of these, and in given circumstances it may not even be the most important or relevant. Formal and informal networks of communication exist and are heavily relied on to alert decision-makers to changes in the environment and to give them a sense of competitive intentions (professional associations, clubs, family connections and ties, "insider information," and industrial espionage are all illustrative). With respect to scientific information, in the energy field particularly, there has evolved advocacy and competitive modeling (the Energy Modeling Forum at Stanford University, and the Energy Models Assessment Group at the Massachusetts Institute of Technology), the purpose of which is to allow diverse and selected scientific perspectives on energy matters to be advanced, examined, and assessed as to relevance, accuracy, and worth in the decision setting. Besides helping to expose the inevitable biases and limitations that inhabit any analysis, this procedure also allows decision-makers an opportunity to "see" their problems from several different standpoints - an eventuality foregone in the one-model, take-it-or-leave-it case.

It is also worthwhile to point out that the multiple model approach to information generation and synthesis, hedges risk to the decision-maker by implicating scientists and analysts directly. If there is weak consensus (or dispute), the decision-maker has the opportunity to exploit this, doing as he wishes and/or placing blame for subsequent poor performance on the advisor groups employed.

Hedging bets and dispersing responsibility for poor performance are two constant strategies, neither of which necessarily invites sinister or negative connotations. For instance, reliance on futures markets (commodities), buying insurance, or seeking other forms of guarantees to indemnification, and working to share risk through reliance on public resources are all time-worn and time-honored means to these general ends. Protection for infant industries, tariffs and trade barriers, tax and loan guarantees, bail-outs, bankruptcy provisions, regulatory and adjudicatory processes, and many others are all exemplary.

Both in agriculture and forestry, uncertainties in supply are particularly important and hence are of particular interest. In both of these industries diversification of crops has been a standard method of limiting risks. Such activities potentially have parallels in multispecies fisheries. However, perhaps the closest analogy to fisheries management lies in the management of rangelands. Unfortunately, this is a cautionary tale rather than a prescriptive one.

## RANGELANDS ANALOGY

The problem for rangeland managers is to determine carrying capacity of the ranges so that an optimal number of livestock can graze without destroying the rangeland's resources. That carrying capacity will change from year to year, and strategies and tactics have to be devised so that livestock can be removed from the ranges if the rains prove to be poor, or added to the ranges in the case of rains that produce an abundance of vegetation beyond what the current herd can use.

The West African rangelands, for example, have unlimited access by herders whose goal is to keep as many livestock as possible for economic, political, and cultural reasons. Herds expand to large numbers during favorable times (that is, when good rains come, primary production increases sharply). When the rains are favorable for a few years, herds tend to expand in size, encouraging herders to increase the size of their family (there is an assumption that it takes at least five cattle to support each person on the rangelands).

Thus, in wet periods, when there is sharp increase in primary productivity, human perceptions about the carrying capacity of the vegetative cover change, and there is a rapid buildup of herds, which in turn often stimulates an irreversible increase in human population dependent on that increase in vegetation.

Rainfall in arid areas is highly variable in time and space and is skewed to dryness. When the seasonal rains drop below the long-term average (keeping in mind that the rainfall and primary productivity time series for West Africa is short), vegetation on the rangelands becomes overtaxed: too little vegetation and too many animals. As a result, the land surface becomes stripped of preferred vegetation and either unpreferred (less palatable) vegetation moves in to replace it or, because of the reduced density of vegetation, the land's productivity is reduced because of wind and water erosion.

Overgrazing in the West African Sahel for the past 20 years has been blamed in large measure for the conversion of productive rangelands into an area with desert-like conditions. The exacerbation of the problem by the introduction of artificial water supplies or other forms of support is familiar in the context of fisheries management (14).

## THE TOOLS AVAILABLE TO THE FISHERY MANAGER

All tools may, in the hands of the ingenious, be used in more than one way. Nevertheless, the methods available to fishery managers for dealing with uncertainty may be classified broadly into three main patterns of usage: for controlling the level or types of mortality on the fish resource and hence the risks of depletion or collapse, for controlling the income of the fishery, and for reducing uncertainty about the resource behavior or potential.

In the first case are included the traditional methods of quota control, effort control, and a variety of measures such as closed seasons, areas, or restrictions on mesh or gear type.

In the second, there are inter alia direct regulation of fishing capacity, e.g., buy-back schemes and allocation of property rights to the fishermen, and indirectly, insurance schemes and subsidies. In the final category there are a variety of methods for scientific monitoring, including the experimental manipulation of the resource (see Walters, this volume). Clearly the use of certain tools for one purpose can either enhance or interfere with the use of another tool for another purpose. Yet it is bordering on the trite to remark that the choice of tools depends on the situation and the management objectives.

Bearing that in mind, a short list of the important characteristics of different situations was prepared and discussed:

1 - The level of resource fluctuation was obviously felt to be important. This has two components, the fluctuation in space due to alterations in migration habits, or in time due to variations in recruitment, growth, or mortality, but primarily recruitment.

2 - The level of reliability of scientific information about the resource, which can also usefully be thought of as consisting primarily of information on the abundance and potential yield. A complementary characteristic to the scientific uncertainty about the stock is that concerning the interaction of the resource with the fishing gear. To take a well-known example, the abundance of pelagic stocks is not reflected in the catch rates of the fishery. This is a special case of a more

general situation. For example, rapidly increasing fishing efficiency due to either technological innovation or simply to an increase in efficiency for other reasons may leave catch rates independent of stock abundance.

3 - The degree of economic development of the fishery and its associated industrial infrastructure. Framed as a question: what is the degree of overcapacity, if any?

4 - Finally, the potential for flexibility of the fishing fleet, the processing industry, the market, and management. This involves several types of question - are there other suitable species available for capture? Can the fleet operate on such species in different areas, i.e., how specialized is the fishing gear or processing equipment? What is the market demand?

The group shrank from the obvious task of producing a catalogue of situations and the appropriate tools for them. Instead, it decided to consider selected tools and examples of their use from amongst the large number of possible situations.

Catch quotas can regulate the mortality imposed on a fish stock, and hence the risks of depletion or collapse, yet there are a number of associated problems. If the scientific information on the state of the resource is poor, the levels set must be low compared to the productivity of the resource. If the catch levels are to be altered annually to attempt to obtain a greater yield, demands on scientific information become much greater. Without some restricted access to the fishing, as in situations of overcapitalization, fishermen's incomes will be low, enforcement difficult, and accurate catch statistics more difficult to obtain. All these problems are exacerbated if the resource fluctuates significantly.

Effort control by licensing with the limitation of fishing capacity can control the mortality on the resource as long as the catch rate varies with the size of the resource. Where catch rates are independent of stock size, similar considerations to those for catch quotas apply. In other situations, licensing may produce fluctuations in fishermen's incomes in tune with the fluctuations in the size of the resource. An insurance system, in which proceeds from peak years may be utilized in the bad, affords an opportunity for mitigating such effects. Information requirements about the size of the stock become less important than information on changes in fleet efficiency.

Such methods of regulation are obviously best introduced in fisheries where there has been no overcapitalization. In these situations, an initial choice of the number of licenses depends on estimates of the potential of the resource. There are immediate trade-offs between the number of licenses that can be issued and the degree of uncertainty about the resource potential.

One method of fine tuning estimates of resource potential is the method of Dual Control (sensu Walters, this volume). This involves a deliberate manipulation of the stock size to assess the limits of density-dependent response and hence the stock's potential for sustaining a yield. It would appear to be most appropriate in situations where there are relatively minor variations around the stock and recruitment relationship and/or good chances of stock recovery following depletion.

To a large extent, the potential response of management to uncertainty considered in the above situations is essentially of a short- to medium-term nature only. Long time scales of change, even if predictable, tend to be beyond the interest of even the most socially committed manager or politician. If there is an appropriate managerial response to long-term uncertainty, it is likely to be found in the sort of analysis described by Brewer (this volume) as "worst case" analysis, although it is unlikely that any general presumption other than the need to preserve flexibility would emerge.

## REFERENCES

(1) ACMRR. 1983. Working Party on the Principles for Fisheries Management in the New Ocean Regime. FAO Fish. Report 299. Rome: FAO.

(2) Alm, G. 1952. Year class fluctuations and span of life of perch. Inst. Freshwat. Res. Drottningholm Report 33: 17-38.

(3) Anon. 1981. Report of the ad hoc working group on the use of effort data in assessments. ICES C.M. 1981/G: 5 (Mimeo).

(4) Anon. 1982. Report of the ICES Advisory Committee on Fishery Management. Copenhagen: ICES.

(5) Anon. 1983. Report of the ICES Advisory Committee on Fishery Management. Copenhagen: ICES.

(6) Arntz, W.E., and Rumohr, H. 1982. An experimental study of

macrobenthic colonization and succession, and the importance of seasonal variation in temperate latitudes. J. Exp. Mar. Biol. Ecol. 64: 17-45.

(7) Bakun, A., and Parrish, R.H. 1980. Environmental inputs to fishery population models for eastern boundary current regions. In Workshop on the Effects of Environmental Variation on the Survival of Larval Pelagic Fishes, ed. G.D. Sharp, pp. 67-104. IOC Workshop Report 28. Paris: UNESCO.

(8) Bakun, A., and Parrish, R.H. 1982. Turbulence, transport and pelagic fish in the California and Peru current systems. CA Coop. Ocean. Fish. Invest. Report 23: 99-112.

(9) Beddington, J.R., and Cooke, J.G. 1984. The Potential Yield of Fish Stocks. FAO Tech. Paper 242. Rome: FAO.

(10) Belvèze, H. 1984. Biologie et dynamique des populations de sardines (Sardina pilchardus) Walbaum peuplant les côtes Atlantiques marocaines et proposition pour un aménagement des pêcheries. Thèse de doctorat d'Etat, Université de Bretagne Occidentale (Mimeo).

(11) Berthou, P. 1981. Contribution à l'étude du stock de praires (Venus verrucosa, Linné) du Golfe Normano-breton. Thèse de Doc. spéc., Université de Bretagne Occidentale.

(12) Boesch, D.F. 1974. Diversity, stability and response to human disturbance in estuary marine ecosystems. In Proceedings of the First International Congress of Ecologists, Pudoc, pp. 109-114. Wageningen, The Netherlands.

(13) Brewer, G.D. 1983. The management challenges of world fisheries. In Global Fisheries: Perspectives for the 1980s, ed. B.J. Rothschild, pp. 195-210. New York: Springer-Verlag.

(14) Brochmann, B.S. 1984. Financial measures to regulate effort. FAO Fish. Report Suppl. 2: 167-172.

(15) Dragesund, O. 1980. Biology and population dynamics of the Norwegian spring spawning herring. Rapp. P.-v. Reun. CIEM 177: 43-71.

(16) Dutt, S. 1981. A comparison of some ecological features of the populations of clupeid fishes of cold and tropical waters. Proc. Symp. Ecol. Anim. Pop. Zool. Surv. India Part 1: 149-245.

(17) Elliot, J.M. 1971. Some methods for the statistical analysis of

samples of benthic invertebrates. Scient. Pub. Freshwat. Biol. Ass. 25: 144.

(18) Gudmunsson, G. 1982. Statistical analysis of catch observation. In Nordic Symposium in Applied Statistics and Data Processing, ed. A. Nöskuldsson, pp. 15-45. Copenhagen.

(19) Hassell, M.P. et al. 1982. Impact of infectious diseases on host populations. In Population Biology of Infectious Diseases, eds. R.M. Anderson and R.M. May, pp. 15-35. Dahlem Konferenzen. Berlin, Heidelberg, New York: Springer-Verlag.

(20) Kasahara, H. 1960. Pacific herring. In H.R. MacMillan Lectures in Fisheries. Part 1. Fisheries Resources of the North Pacific Ocean, pp. 49-63. Vancouver: Institute of Fisheries, University of British Columbia.

(21) Kondo, K. 1980. The recovery of the Japanese sardine - the biological basis of stock-size regulations. Rapp. P.-v. Reun. CIEM 177: 332-354.

(22) Laurec, A., and Le Gall, J.Y. 1977. The seasonalizing of the abundance index of a species. Application to the albacore (Thunnus alalunga) monthly catch per unit of effort (cpue) by the Atlantic Japanese longline fishery. Bull. Far Seas Res. Lab. 1977: 145-169.

(23) Lockwood, S.J.; Baster, I.G.; Gueguen, J.C.; Joakimsson, G.; Grainger, R.; Eltink, A.; and Coombs, S.H. 1981. The western mackerel spawning stock estimate for 1980. ICES C.M. 1981/H: 13.

(24) MacCall, A.D. 1980. Population models for the northern anchovy (Engraulis mordax). Rapp. P.-v. Reun. CIEM 177: 292-306.

(25) Nielsen, N.A. 1982. Estimation of the relation between nominal effort and fishing mortality in the fishery for sandeels. ICES C.M. 1982/G: 49-65.

(26) Parrish, R.H.; Nelson, C.S.; and Bakun, A. 1981. Transport mechanisms and reproductive success of fish in the California current. Biol. Oceanogr. 1(12): 175-203.

(27) Pennington, M. 1983. Efficient estimators of abundance for fish and plankton surveys. Biometrics 39: 281-286.

(28) Pope, J.G. 1982. Analogies to the status quo TACs: their nature and variance. In Fisheries and Oceans Canada Workshop on Sampling Assessment Data, Ottawa, Canada.

(29) Waugh, G.D. 1957. Oyster production in the Rivers Crouch and Roach, Essex, from 1950 to 1954. Fish. Invest. (London) Ser. II, vol. 21(1).

Exploitation of Marine Communities, ed. R.M. May, pp. 245-262. Dahlem Konferenzen 1984. Berlin, Heidelberg, New York, Tokyo: Springer-Verlag.

# Kinds of Variability and Uncertainty Affecting Fisheries

J.H. Steele
Woods Hole Oceanographic Institution
Woods Hole, MA 02543, USA

## INTRODUCTION

For the exploitation of all living natural resources, management involves a large element of uncertainty. As an initial simplification, the uncertainty can be attributed to a) variability in the external forces acting on the set of resources, and b) the internal complexity of the system in its response to external variations. Both aspects have one feature in common. The external variability and the internal interactions occur across a very wide range of space and time scales - from questions of initiation of larval feeding (hours/meters) to the appearance and disappearance of regional breeding stocks ($10^2$ years/$10^3$ kms).

There is an analogy, and probably a direct relation, with the purely physical theories for the atmosphere where there are defined limits to weather prediction (about ten days) because of the generation of uncertainty within the system. Thus, even where the basic physics are known and the fundamental equations can be written down, approximations must be made, essentially by defining a finite range of space and time scales and then parameterizing the other parts of the range of scales, or using observations to force the system.

The same process is necessary for fish population dynamics, and the success of the yield per recruit models arises both from their elimination of ecological interactions and their restriction to a narrow range of time scales, usually 1-5 years, with an emphasis on annual forecasts. Some of the failures with this approach have arisen from a correspondingly narrow range of spatial scales where each species is aggregated into a number of separable stocks. The neglect of smaller-scale behavioral responses in pelagic fish such as herring can have a severe effect on stock estimation and management criteria.

Improvements in management for single species or extensions to several species generally require not only some greater internal complexity in the model but are aimed at extending the range of time scales under consideration. Much of the science is concerned with improved understanding of the larval and juvenile phases before annual recruitment. There is also a need to provide longer-term forecasts of fish abundance to match the decadal time scales of boats and processing plants. Socially, there are the usual questions about the acceptability and utility of forecasts with a very high variance.

This review will consider the relevance of small-scale structure, the nature of the longer-term variability, and the consequences of combining these in a complex system.

**FINE STRUCTURE: TEMPORAL**

A major study over many years has been made of the anchovy and sardine populations off California. In particular, there have been detailed studies by Lasker (16, 17) of the feeding of anchovy larvae in relation to variations in the physical environment. At first feeding, anchovy larvae appear to require relatively high concentrations of phytoplankton in the size range 30-50 um. Appropriate concentrations can occur in mid-water chlorophyll maxima with species such as the naked dinoflagellate Gymnodiuium splendens. Lasker (16) reported that "a storm which caused extensive mixing of the top 20 m of water obliterated the chlorophyll maximum layer and effectively destroyed this feeding ground of the larval anchovy." In a later paper (17), upwelling in the feeding area was reported to have caused the dinoflagellates to be replaced by smaller diatoms which laboratory experiments have shown do not provide the caloric requirements of first feeding larvae. However, for adult fish, upwelling is usually considered a necessary condition to fuel the general food webs on which they depend.

There are complex but discernable relations between upwelling, longshore water movements, and zooplankton abundance (4), but no clear connection with fish stocks. Recent studies of El Niño and the southern oscillation (3) show the ocean-wide connections between fluctuations in the Pacific, atmospheric perturbations, and marked changes in other stocks such as Peruvian anchovy (2). These studies also show that we are still some way from an understanding of the coupled ocean-atmospheric system which would permit predictions, even six months ahead, of the oceanic fluctuations.

These studies illustrate the significance of temporal variability at scales beween "weather" ( < 10 days) and interannual periods. Such results suggest that an understanding of these types of physical/biological interactions would be a necessary but probably not a sufficient condition for a description of the processes underlying variable recruitment to the exploited phase.

If larval survival could be related to observable physical factors and to recruitment (a concept which is not generally acceptable), there would be two potential uses of such information. Determination of any recruitment level could be advanced by about one year and the data requirements could be obtained very much more easily and cheaply. These would be highly significant advantages for the applied scientist concerned with setting quotas on a year-by-year basis. The second potential use relates to longer-term forecasts. Could an estimation of the processes be parameterized to provide an input to useable longer-term assessments? This is the problem of deriving a stock/recruitment formulation, which can be combined with an age-structured yield per recruit formulation. There are several formulations, but, as Shepherd (26) has pointed out, the crucial term is the slope of the asymptote at the origin, i.e., $\underline{a}$ in

$$\text{recruitment rate} = a \times \text{stock size } S$$

as stock size $\rightarrow 0$.

In turn, the critical question is, what are the variations of $\underline{a}$ with environmental processes? Since there are very few values for $\underline{a}$ and these have wide limits, there is essentially no observation of environmental variation in $\underline{a}$. Furthermore, since the relation is with population density, laboratory experiments cannot provide direct evidence. If we accept the evidence of Lasker (16, 17) that physical variability at a range of temporal scales can affect larval survival, then one assumption would be that the statistical character of this variability could be regarded as a longer-term input. The relevance and usefulness of such an assumption then depends on the ability to describe the internal processes on which such variability will operate.

## FINE STRUCTURE: SPATIAL

The basic unit in fish population dynamics is the breeding stock. The usual and simplest assumption is that, for a given fishing effort b, the catch per unit effort (CPUE) Y is linearly related to stock size S; Y = bS (Fig. 1a). This implies, ecologically, that changes in total stock will result in corresponding changes in the density distribution. Technically and economically it is assumed that for heavily fished stocks, the catching rate is the dominant component, and processing of the catch on board or ashore is less significant. Thus, although a hyperbolic relation (12) may be appropriate as a general concept (Fig. 1b), the technical and economic factors imply that we operate sufficiently close to the origin that a linear approximation is acceptable.

The problems which have arisen concern the ecological assumption of a simple dependence of density distribution on stock size. It has been shown for various herring stocks (34) that catchability b increases as stocks decrease so that the catch per unit effort (CPUE) remains roughly constant over quite a wide range of stock sizes - rather than decreasing with stock. Although this phenomenon does depend on the newer technologies of sonar searching and purse seining, the basic reasons for this appear to lie in the small-scale behavior of the fish. With decreasing stock size, the individual shoal size remains nearly constant. Also, the area occupied by the stock decreases so that distance between shoals does not decrease in proportion to stock size (25). Obviously at some very low stock density this relation breaks down and the predator, man, disengages for economic or statutory reasons. The consequences in terms of the CPUE-stock relation is an almost flat response down to some low level of stock with a rapid descent to the origin. The overall shape

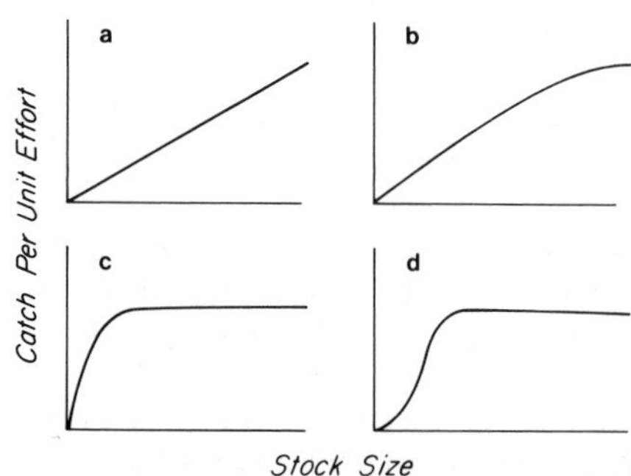


FIG. 1 - Four relations between stock and catch per unit effort (see text).

might be hyperbolic (Fig. 1c), or it may be "S-shaped," if there are alternative prey to which the predator can switch (Fig. 1d). The former relation (Fig. 1c) can cause severe managerial and ecological problems where there is only one available prey which can then be threatened with extinction. Both forms have the common feature that, without decreasing CPUE, there is not the expected evidence for management of decreasing stock size that can be used as a statutory or economic restraint on the predator.

On the other hand, the system derived from Fig. 1d can be regarded as ecologically efficient, giving relatively constant return over a wide range of stock size and so requiring little "management" by the predator so long as it can switch to an alternative prey or, in terms of the whole system, be replaced by another prey-predator pair. When man is the predator, the problems arise when such switching is technologically difficult or disadvantageous economically. With fishing fleets that are technically and geographically flexible, "pulse" fishing might be considered an appropriate response in a variable environment.

In a broader ecological context, there is considerable evidence (10) that the S-shaped curve of Fig. 1d is a general feature of predator-prey interactions. Furthermore, behavioral changes in stock distribution may be induced by environmental variations other than predation. During El Niño events off Peru, the anchovy populations are restricted to a narrower than normal coastal strip, thus permitting a higher predation rate at a time when recruitment is likely to be poor. It is probable that similar but less dramatic and so less easily observable changes can occur with other species subject to a wide range of environmental fluctuations. Thus, the spatial changes observed when man is the predator may not differ in kind from those occurring with "natural" fluctuations in the environment (including predators), particularly as these affect the larval and juvenile states which may be most subject to such fluctuations (including high natural mortalities).

There is a critical need to use new techniques such as multi-frequency acoustics to determine not only total population levels but the spatial patterns within the populations and changes in these patterns with population size.

## VARIABILITY IN ECOSYSTEM STRUCTURE

There is accumulating evidence of the variability in community structure at all time and space scales. Using methods such as in situ fluorometry and particle counting, the small-scale spatial patchiness of phytoplankton and zooplankton was studied extensively during the 70s (8, 24). More recently, Mackas (20) has pointed out that zooplankton community structure has much longer (circa 50 km) coherence scales than has total biomass (circa 5 km). At larger spatial scales, the Plankton Recorder Program (5) has documented trends over decades in community structure within the North Sea. The possible relations with corresponding changes in the physical environment have been discussed (e.g., (6)). What effects do these changes in ecosystem structure have on fish populations?

The most obvious changes are demographic - changes in weight at a given age and in age at sexual maturity. Cushing (7) lists the consequences of an increase in abundance of the large copepod Calanus upon North Sea herring as a) an individual weight increment of 10 - 20%, b) a reduction in the age of recruitment, and c) a concentration on the nursery ground, which may have stimulated the industrial fishery. Such demographic changes may also be induced by population decreases due to overfishing as with certain other herring stocks and with quite different species such as whales (18).

Yet, the surprising feature for fish is that the changes in growth rate are relatively small given the very large changes that can be induced experimentally with different feeding regimes, particularly in the juvenile stages (14). Thus, the assumption is that changes in the availability or suitability of food results in variations in mortality so that the end effect is fluctuation in numbers of recruits to the adult stock. The mediating factors may be changes in behavior such as the timing of descent to the bottom for feeding in juvenile demersal fish (1). Furthermore, during this juvenile phase, total food requirements of the cohort may be at a maximum (28) and so most critical in terms of interactions with prey density, with competitors, and with predators on these juveniles.

This summary discussion of ecosystem structure is intended to display the complexity of the paths from a range of physical factors through the induced spatial or temporal variability in community structure and abundance, to changes in abundance of several fish stocks within the system. The intent is to illustrate several features:

1. Community or size structure of lower trophic levels may be more important than simpler measures, such as total biomass, in determining fish species distributions.

2. Since the patchiness is relevant to intake rates and can affect behavior patterns, the smaller-scale interactions with food supplies may determine the mortality rates of juveniles in a manner analogous to that postulated for fishing predation on adults.

3. It seems inherently unlikely that, in the long term, there will be simple linear (or low-order) relations between physical factors and recruitment of particular species.

## LONG-TERM VARIABILITY: FISH STOCKS

The previous sections have examined some of the shorter-term and smaller-scale factors which play significant roles in causing these fluctuations, and whether we can elucidate patterns or processes which can be used to model, conceptually or numerically, multi-year fluctuations. But, the essential problem concerns long-range forecasting of fish stock changes. Can we discern any regularities in the long-term records which could be useful for direct empirical application or for indirect testing of such models?

The general subject of long-term variations and their possible relation to climate has been reviewed on several occasions. As might be expected, there is no single regular pattern nor any explicit relation with longer-term atmospheric changes, although there are suggestions that general warming or cooling trends in the North Atlantic may be associated with changes in distribution (11).

One frequently quoted example (Fig. 2) from fish scales in varved sediments off California shows a fairly regular 70-year appearance and disappearance of sardines, but it also shows much longer time scales for anchovy and hake. Another well-known example concerns the similar appearance and disappearance of herring stocks around Scandinavia (Fig. 3) with a comparable 50-year interval. A problem with such data in recent years is the obvious added effect of human predation on the stocks.

There are other cases involving possible interactions of species such as herring and mackerel off Nova Scotia (29) (Fig. 4). These data and Skud's interpretation (29) suggest alternating periods of dominance by one or the other species, possibly related to environmental changes and, again, occurring at about 50-year intervals. But the best documented case concerns the sequence of species dominance in the English Channel, where pilchards replaced herring in about 1920 and in turn were replaced by mackerel in about 1970. Evidence from a long series of plankton

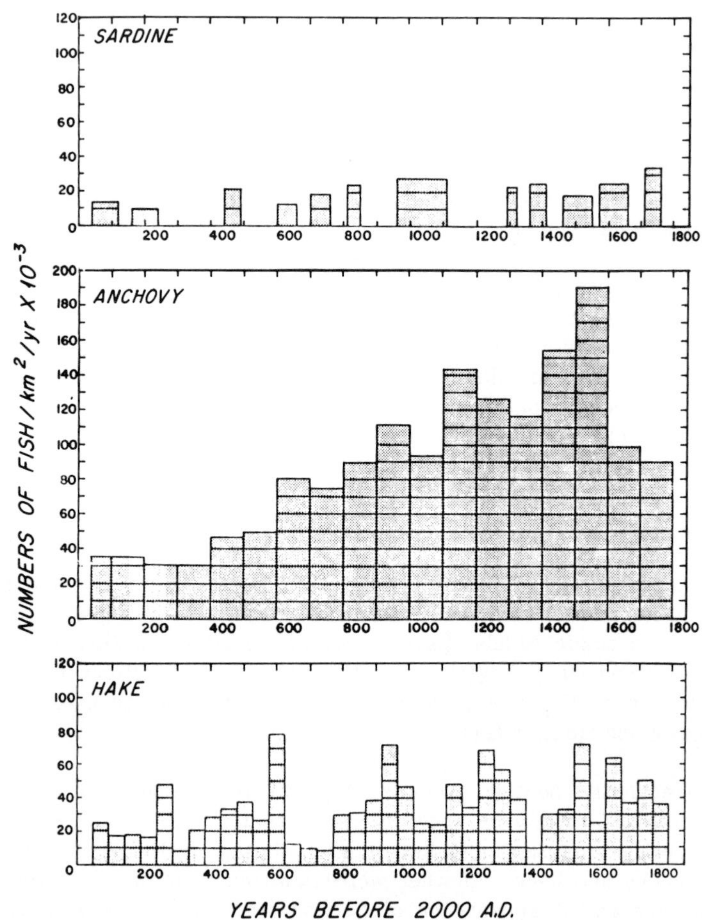


FIG. 2 - Relative population estimates based on fish scales from varved cores off southern California (30).

samples at one position (31) indicates that these changes were quite abrupt and involved a range of other species within the food web (Fig. 5).

The most recent and most controversial changes have been in the North Sea with the virtual elimination of the pelagic species, herring and mackerel, probably because of excessive fishing. This corresponded

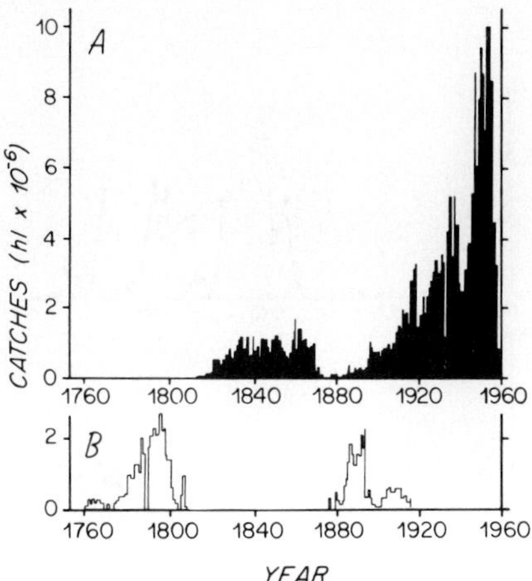


FIG. 3 - Catches of winter herring from the Swedish Bohuslan fishery and from the Norwegian fishery from 1760 to 1960 (9).

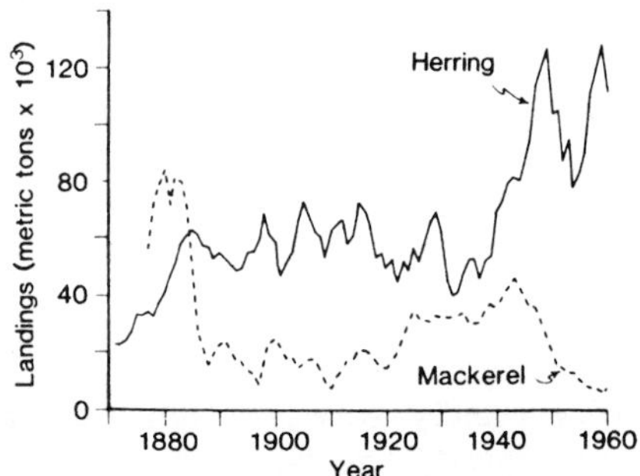


FIG. 4 - Fluctuations in herring and mackerel stocks off Nova Scotia (29).

in time with very large increases in several demersal gadoid species, particularly haddock. The argument (7, 15) is about whether there is a direct causal link involving transfer of energy within the food web, or whether the latter increase was due to environmental/climate factors.

This rapid survey indicates that although there is no simple predictable pattern, historical evidence suggests that regional fish stocks can change

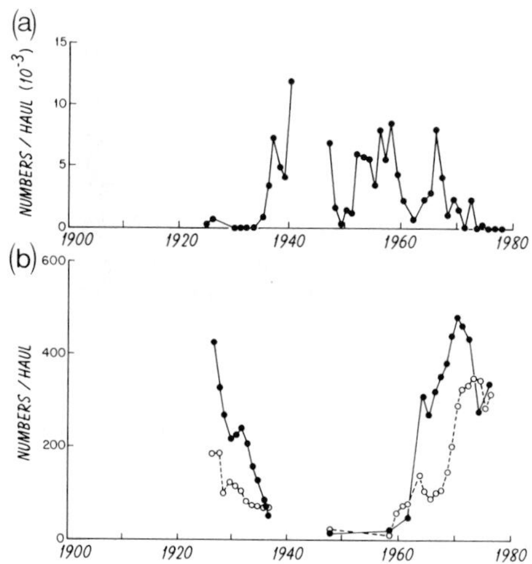


FIG. 5 - Examples of the changes in fish populations off Plymouth, England, based on weekly samples with a plankton net. (a) Numbers of eggs of pilchard (Sardina pilchardus (Walbaum)) expressed as the average per haul during the spring/summer spawning season from April to July inclusive. (b) Five-year running means of postlarval teleosts (excluding clupeids). The young fish are divided into spring-spawners (●), taken as the sum of the monthly means for March to June, and summer-spawners (o), taken as the sum of the monthly means for July to September (31).

very markedly and very rapidly between high and very low levels of abundance at intervals of around 50 years. Such abrupt natural fluctuations need to be considered in the context of longer-term management. We should especially consider the interactions of such "jumps" with the stresses imposed by heavy fishing mortality.

It may be that much of the principles of fishery management, which assume a single underlying long-term equilibrium, were developed during a period (1920-1970) when there was a relatively stable situation. The historical data suggest that for regional ecosystems there may be more than one equilibrium state. The questions arising concern the relation of such alternating states, particularly their periodicity, to some combination of environmental variability and fishing stress.

## LONG-TERM VARIABILITY: PHYSICAL

There are very significant differences in the temporal character of atmospheric and oceanic variability. If we remove the regular cycles (diurnal, lunar, and seasonal), then the variance of atmospheric temperature as a function of frequency is relatively constant from days to a few decades (23) and can be considered as "white" noise. We know that the oceans, because of their great heat capacity, can ameliorate the shorter-term variability in the atmosphere, but correspondingly, the longer-term interannual or decadal variations can be much larger. For these reasons, variance per unit frequency in the oceans increases rapidly with period (called "red" noise by analogy with visual spectra). This can be observed in deep-sea temperature (Fig. 6) (35) or in near-shore sea level (22) over time scales of available observations up to decades. As Chelton et al. (4) have shown, sea level can be a good indicator of changes in offshore biology. This difference in terrestrial and marine habitats can be expected to lead to significant differences in the system response. The dominance of poikilotherms in the ocean

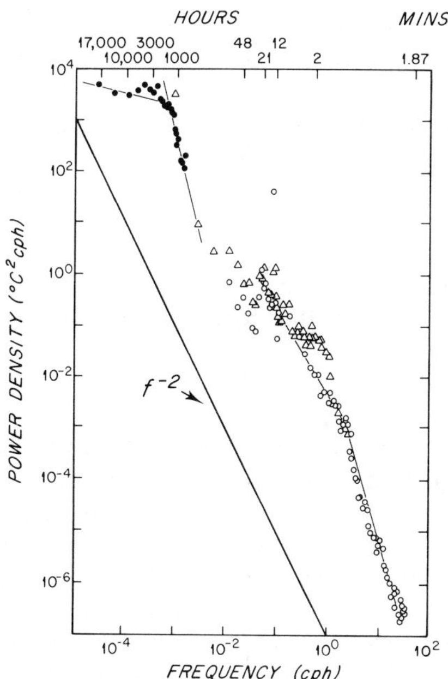


FIG. 6 - Spectrum of temperature at Bermuda in the main thermocline from record duration of 13 years.

may be an indication of the general ability to absorb the smaller short-term variability but in turn can make these marine systems more sensitive to the longer and larger trends.

The basic statistical character of the variability in the physical system in the oceans suggests that the increasing amplitude of change with time could trigger structural changes in particular regional ecosystems. For most species there are upper and lower tolerances to temperature or salinity, but these are usually much broader than the observed variability. Furthermore, the available evidence, although not voluminous, indicates that changes can occur simultaneously in several components of the food web. In some sense, the response to increasing variance occurs in the ecosystem as a whole, even if it is triggered by one or two keystone species near the top of their food chains. Thus the response in the ocean to the increasing variance with period may be to force regional systems to jump from one equilibrium state to another. Because of the essential unpredictability in these longer-term variations and because of the internal complexity of the system, there is unlikely to be any simple method of forecasting such events. However, the available historical observations indicate a possible occurrence for certain pelagic stocks at about 50-year intervals in such jumps (or a period of 100 years if there were cyclical responses). A very simple single-species model (33) combining a logistic growth curve and s-shaped predation (21) with stochastic forcing by "red" noise gives the right order of time scales for such jumps. Furthermore, this model suggests that with increasing fishing the frequency of the jumps might be increased. The much more rapid changes in relative species abundance observed in the yield from heavily fished stocks (Fig. 7) might be a "natural" consequence of increased stress on systems with multiple equilibrium states rather than being solely the result of inadequate control by management.

## DISCUSSION

As a means of summarizing the interactions which have been reviewed, I shall reuse a very simple diagram (32) which emphasizes the space and time scale relations at the expense of the food web complexities. Essentially, Fig. 8a portrays the spatial scales for the patchiness of phytoplankton, P, herbivorous zooplankton, Z, and pelagic fish stocks, F, such as herring in relation to their life span. It is intended to stress the large range of scales involved in their interrelation. Figure 8a implies that as one goes up the trophic ladder both scales increase providing, at each level, interactions with a wider set of food species over longer periods of time. Thus each predator has to absorb temporal variability

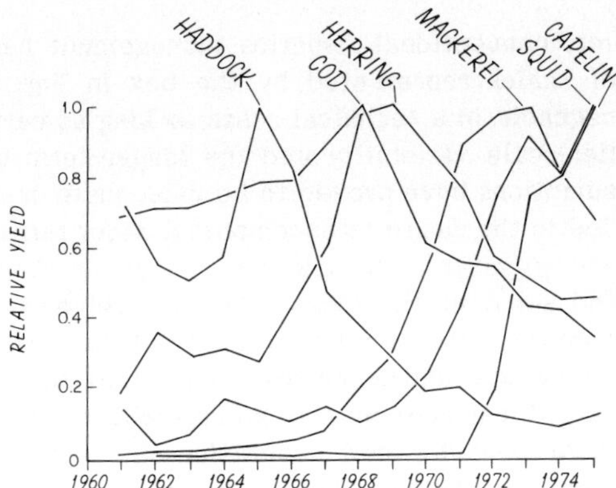


FIG. 7 - The relative yield of fish and squid from the area covered by the International Commission for Northwest Atlantic Fisheries (13).

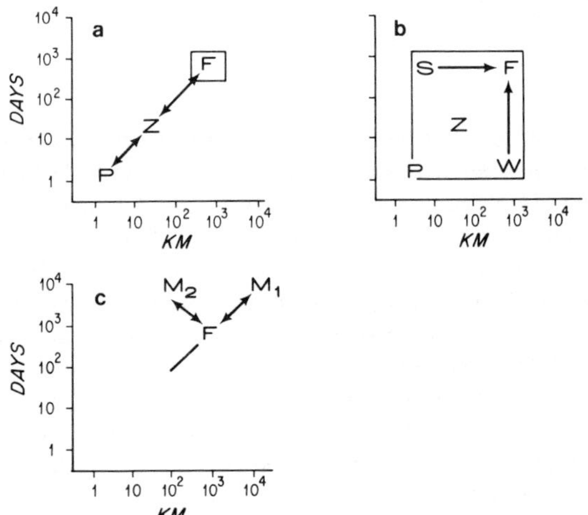


FIG. 8 - Relations between the planktonic food web, physical factors, and man as a predator (see text).

in its shorter-lived prey but has a larger ambit than its prey populations.

As stated earlier, conventional fisheries management has chosen a very narrow range of scales represented by the box in Fig. 8a. This makes the system manageable in a technical sense so long as certain assumptions about the smaller-scale variability and the longer-term equilibrium hold. Both sets of assumptions have proved to be inadequate or invalid in certain cases and have led to the desire to encompass a wider range of scales.

I have illustrated some of the ideas, and the problems, associated with attempts to incorporate the smaller scales (Fig. 8b). The smaller spatial scale behavior of pelagic and demersal stocks with changes in stock size can have a major influence on the stock/yield relation. An example is change in shoal behavior and separation (S). Possibly an increased understanding of these behavioral patterns could improve the parameterization of "catchability."

A second arena for much research and speculation is the year-to-year fluctuation in recruitment which may be significantly affected by short-term weather (W) events. The timing and intensity of these may, in turn, depend on interannual climatic variations whose oceanic expression may display certain statistical features, called "red" noise. Figure 8b illustrates the conceptual and technical problems in combining physical and biological complexities. It is not clear how increased knowledge of the system within this box could be extrapolated to scales outside it.

When we add man as a predator who now interacts significantly with the fish stocks, we can ask where and how he could be fitted into this simplified space-time framework. The time periods are known. In terms of fishing technology (the life span of a boat), of processing, and of social structures, the times are on the order of 30-50 years. On the ecological basis portrayed in these diagrams, one would expect this component of the system to have correspondingly large spatial scales $M_1$, on the order of $10^4$ km or greater (Fig. 8c). Such an approach, exemplified by the traditionally successful fishing nations such as Japan and Russia, depends on mobility, free access, and inherent economic restraints. In an ideal structure, such "pulse" fishing would correspond to the usual ecological practice of feeding on prey while they are at peak abundance or catchability and then disengaging. We know that nearly all prey-predator systems would be energetically impossible if they had to feed on the mean densities rather than on the variance. Man uses this factor in day-to-day fishing operations, but its real application, for the relevant time scales, should be decadal. Such an extrapolation would take account not only of the direct predator effects but also could absorb the natural flips in system structure which appear to occur on these time scales.

Social and political constraints have removed much of the freedom for such large-scale competitive interactions among human predators. It has proven unacceptable for a variety of reasons - the possible extinction of certain species such as whales - but mainly because of local or regional economic factors. Thus, while technical developments tend to increase the time scales of a particular fishing method, the ambit of operations are reduced to similar or smaller-scales $M_2$, than the fish populations (Fig. 8c). At the same time, the intensity of effort has increased. Under such conditions, particularly with economic subsidy, there is an enhanced likelihood of increasing frequency of such flips in particular stocks and a decreased probability of disengagement. Even under these conditions, the response of open sea systems and even large, enclosed bodies such as Lake Erie (19), appears ecologically adequate, with replacement species entering the system (27) or unexpected recoveries (herring in the North Sea). The problems are, immediately, economic, but in the long term are social and political.

The assumption of a single state provides forecasts which will degrade in reliability with increasing time but in a regular manner. Using short-term recruitment predictions, the forecasts are updated with the intent that the steady state will be maintained, providing a long-term basis for technical and economic development. This approach has not worked in practice, and its failure is usually blamed on inadequate fisheries data and the inability to apply suitable management techniques. The implication is that more data and stricter management control is required. If, in fact, the assumptions are inappropriate, as suggested by historical data and ecological speculation, then the inherent ecological uncertainties may require a different approach.

We should replace the single equilibrium assumption by the recognition of possible multiple states, each markedly different from the others, with the changes between them occurring rapidly, and with the frequency of change increasing with increased predatory fishing pressure. The expected variability is then very different from noise about some single steady state. The external or environmental fluctuations are responsible not only for variations about a particular equilibrium state but contribute significantly to the changes in state. These changes depend on the internal complexity of the ecological system, and it is this aspect which gives the inherent uncertainty to the resultant output - an uncertainty which should be incorporated into the management process.

**Acknowledgement.** I would like to thank the Coastal Research Center of the Woods Hole Oceanographic Institution for support.

## REFERENCES

(1) Bailey, R.S. 1975. Observations on diet behaviour patterns of North Sea gadoids in the pelagic phase. J. Mar. Biol. Ass. U.K. 55: 133-142.

(2) Barber, R.T., and Chavez, F.P. 1983. Biological consequences of El Nino. Science 222(4629): 1203-1210.

(3) Cane, M.A. 1983. Oceanographic events during El Nino. Science 222(4629): 1189-1194.

(4) Chelton, D.B.; Bernal, P.A.; and McGowan, J.A. 1982. Large-scale interannual physical and biological interaction in the California current. J. Mar. Res. 40: 1095-1125.

(5) Colebrook, J.M. 1978. Changes in the zooplankton of the North Sea, 1948-1973. Rapp. P.-v. Reun. Cons. Int. Explor. Mer 172: 390-396.

(6) Cushing, D.H. 1975. Marine Ecology and Fisheries. Cambridge University Press.

(7) Cushing, D.H. 1983. Climate and Fisheries. London: Academic Press.

(8) Denman, K.L., and Mackas, D.L. 1978. Collection and Analysis of Underway Data and Related Physical Measurements. Spatial Pattern in Plankton Communities. New York: Plenum Press.

(9) Devold, R. 1963. The life history of the Atlanto-Scandian herring. Rapp. P.-v. Reun. Cons. Int. Explor. Mer 154: 98-108.

(10) Hassell, M.P.; Lawton, J.H.; and Beddington, J.R. 1977. Sigmoid functional responses by invertebrate predators and parasitoids. J. Anim. Ecol. 46: 249-262.

(11) Hempel, G., ed. 1978. North Sea fish stocks - Recent changes and their causes. Rapp. P.-v. Reun. Cons. Int. Explor. Mer 172: 445-449.

(12) Holling, C.S. 1965. The functional response of predators to prey density and its role in mimicry and population regulation. Mem. Entom. Soc. Can. 45: 1-60.

(13) Horwood, J.W. 1981. Management and models of marine multi-species complexes. In Dynamics of Large Mammal Populations, eds. C.W. Fowler and T.D. Smith, pp. 339-360. New York: John Wiley & Sons, Inc.

(14) Jones, R. 1978. Estimates of food consumption of haddock (Melanogrammus aeglefinus) and cod (Gadusmorhua). J. Cons. Int. Explor. Mer 38(1): 18-27.

(15) Jones, R., and Richards, J. 1976. Some observations on the interrelationships between the major fish species in the North Sea. ICES C.M. 1976/F: 35.

(16) Lasker, R. 1975. Field criteria for survival of anchovy larvae: The relation between inshore chlorophyll maximum layers and successful first feeding. Fish. Bull. 73(3): 453-462.

(17) Lasker, R. 1978. The relation between oceanographic conditions and larval anchovy food in the California current: identification of factors contributing to recruitment failure. Rapp. P.-v. Reun. Cons. Int. Explor. Mer 173: 212-230.

(18) Laws, R.M. 1977. Seals and whales of Southern Ocean. Phil. Trans. Roy. Soc. Lond. B 279: 81-96.

(19) Leach, J.H., and Nepszy, S.J. 1976. The fish community in Lake Erie. J. Fish. Res. Board Can. 33(3): 622-638.

(20) Mackas, D. 1984. Spatial autocorrelation of plankton community composition in a continental shelf ecosystem. J. Mar. Res., in press.

(21) May, R.M. 1977. Thresholds and breakpoints in ecosystems with a multiplicity of stable states. Nature 269(5628): 471-477.

(22) Munk, W.H., and Cartwright, D.E. 1966. Tidal spectroscopy and prediction. Phil. Trans. Roy. Soc. Lond. A 259: 533-581.

(23) National Academy of Sciences. 1975. Understanding Climatic Change: A Program for Action. US Committee for the Global Atmospheric Research Program.

(24) Platt, T., and Denman, K.L. 1975. The physical environment and spatial structure of phytoplankton populations. Mem. Soc. Roy.

des Sci. de Liege, Serie 6, 7: 9-17.

(25) Saville, A., and Bailey, R.S. 1980. The assessment and management of the herring stocks in the North Sea and to the west of Scotland. Rapp. P.-v. Reun. Cons. Int. Explor. Mer 177: 112-142.

(26) Shepherd, J.G. 1982. A versatile new stock-recruitment relationship for fisheries and the construction of sustainable yield curves. J. Cons. Int. Explor. Mer 40(1): 67-75.

(27) Sherman, K.; Jones, C.; Sullivan, L.; Smith, W.; Berrien, P.; and Ejsymont, L. 1981. Congruent shifts in sand eel abundance in western and eastern North Atlantic ecosystems. Nature 291(5815): 486-489.

(28) Sissenwine, M.P.; Cohen, E.B.; and Grosslein, M.D. 1984. Structure of the Georges Bank ecosystem. Rapp. P.-v. Reun. Cons. Int. Explor. Mer, in press.

(29) Skud, B.E. 1982. Dominance in fishes: the relation between environment and abundance. Science 216: 144-149.

(30) Souter, A., and Isaacs, J.D. 1974. Abundance of pelagic fish during the 19th and 20th Centuries as recorded in anaerobic sediments off California. Fish. Bull. US Fish Wildlife Serv. 72: 257-275.

(31) Southward, A.J. 1980. The western English Channel - an inconstant ecosystem. Nature 285(5784): 361-366.

(32) Steele, J.H. 1978. Some comments on plankton patches. In Spatial Pattern in Plankton Communities, ed. J.H. Steele. New York: Plenum Press.

(33) Steele, J.H., and Henderson, E.W. 1984. Modeling long-term fluctuations in fish stock. Science, in press.

(34) Ulltang, O. 1980. Factors affecting the reaction of pelagic fish stocks to exploitation and requiring a new approach to assessment and management. Rapp. P.-v. Reun. Cons. Int. Explor. Mer 177: 489-504.

(35) Wunsch, C. 1981. Low-frequency variability of the sea. In Evolution of Physical Oceanography, eds. B.A. Warren and C. Wunsch, pp. 342-374. Cambridge, MA: MIT Press.

Exploitation of Marine Communities, ed. R.M. May, pp. 263-274. Dahlem Konferenzen 1984. Berlin, Heidelberg, New York, Tokyo: Springer-Verlag.

# Managing Fisheries under Biological Uncertainty

C.J. Walters
Institute of Animal Resource Ecology
University of British Columbia
Vancouver, BC V6T 1W5, Canada

**Abstract.** Fisheries management must face three types of uncertainties: environmental variations that will continue to be unpredictable, errors in production model parameter estimates and structures that can in principle be reduced over time, and persistent changes in parameters and structure due to a host of possible factors that are mostly unpredictable even in principle. Thus we must learn to live with continuing uncertainty as a basic feature of fisheries policy design. Uncertainties of the first type can be dealt with nicely without ever pretending to forecast responses accurately, by using robust feedback policies. Reduction in uncertainties of the second type depends fundamentally on what management policies are used, and here the idea of dual control or experimental policies appears very promising. Contrary to what some optimists have supposed, uncertainties of the third type cannot in general be sorted out through statistical procedures for adaptively reestimating parameters, since various effects are generally confounded with one another. Detailed process research and understanding may be of some help, but ultimately we may be forced to use the old scientific idea of replicated experimental units that are deliberately treated in different ways.

## INTRODUCTION

Management under uncertainty has become a favorite research topic for fisheries theoreticians, and there have been a number of surprising discoveries in the last decade. Even in the late 1940s, workers like W.E. Ricker were beginning to suggest that it is important to have feedback harvest policies that deal with unpredictable variation as it arises. A recent surprise (due to Reed (7, 8)) is that the best policy might be very simple, i.e., try to allow a fixed escapement (stock after harvest) every year. Discoveries about feedback policies have raised another question: really how important is it to have models that will predict dynamic changes over time? Provided it is possible to monitor changes and respond rapidly as they arise and to assign odds to various long-term outcomes, some simulation studies such as Buckingham and Walters' (3) suggest that forecasting is not important at all, except as a planning service for the fishing industry. But the ability to monitor and regulate rather tightly stock sizes creates a potentially more serious problem for the long term, namely, a reduction in informative variation in stock sizes. Without permitting or deliberately encouraging stock size variation through management experiments, we have no way to determine empirically what levels are best to try and maintain through feedback regulation. There have been a number of dramatic improvements since 1975 in statistical models for stock assessment and parameter estimation, and these methods indicate that mortality and recruitment parameters are very poorly determined from most fisheries data sets; the difficulty is sometimes due to large measurement errors, but more often there has simply not been enough contrast in stock sizes and policies over time to allow accurate estimation, even if abundance and production rates were measured with no error at all. Also, some long-term data sets lead to the nasty suspicion that a favorite theoretical assumption of a fixed "production function" corrupted by environmental "noise" may be fundamentally unsound; the quantities that we like to call parameters may change quite substantially over time, with effects that are confounded so as to make it impossible to track the changes by statistical methods.

So on the one hand there are some very encouraging results about the essential simplicity of good feedback policies, and on the other hand some serious concerns about the effect of such policies on long-term adaptive performance. Most fisheries modellers have shrugged off these concerns, arguing for a kind of passive adaptation in which management is advised to proceed cautiously or as though the "best" current model were correct and wait for nature to tell us about errors. The trouble with passive adaptation is that it allows the management system to become locked in on some sustainable and apparently optimum regime, when in fact there are large opportunities for increased harvest by managing the system differently. Such opportunities are likely to be detected only by deliberately "probing" system responses under more extreme conditions than would arise naturally or through conservative (cautious, risk averse) policies.

The following sections will cover three topics. First, I will review in a bit more detail the types of uncertainties that are faced in harvest policy design and the state of the art about how to deal with them. Second, I will look at the rather controversial question of how important it is to understand causes of natural "noise" around production functions and to use this understanding in the construction of more accurate temporal predictions. Finally, I will discuss what I consider to be the most fundamental issue in management under uncertainty, namely, the conflict between behaving cautiously to avoid unnecessary risks versus deliberate probing to uncover opportunities for higher long-term performance.

## TYPES OF UNCERTAINTIES

Fisheries managers must face four basic kinds of uncertainties in the interpretation of data from exploited systems. Taken together, these uncertainties provide a formidable barrier to the assessment of even what has happened historically, let alone what opportunities might exist for the future.

### Measurement of States and Rates

It has long been recognized that the analysis of stock dynamics using fishery data (catches, effort, etc.) can lead to a variety of biases in estimates of states and rates. Thus there has been a recent tendency to invest more in direct monitoring (survey fishing, etc.) by management agencies. Such monitoring programs may lead to unbiased estimates of variables (e.g., stock biomass), but it is seldom feasible to make the estimates very precise. Random sampling errors are thought to be less a problem than various biases.

However, stock assessment is concerned also with the measurement of functional relationships between production rate components (growth, mortality, recruitment) and the most basic management variable, stock size. It has recently been emphasized (13) that random errors in the measurement of stock size can lead to gross biases in the apparent form (and parameter estimates) of such functional relationships. So efforts to reduce bias in stock assessments may just be leading to an equally bad (or worse) result, when the survey data are used in functional modelling. Statisticians know this as the "errors in variables" problem: errors in independent variable measurement destroy the actual functional relationship and replace it with the deceptive appearance of independence between the variables. Probably this effect was responsible for the early conclusion in fisheries that recruitment rates are usually independent of stock size.

**Environmental Effects**

No production model can account for all factors that influence stock size and production over time, and we accept "process errors" that have no obvious functional pattern (relation to stock size) or temporal persistence as being an inevitable source of management problems. It is wise to continue searching for the causes of deviation, in hopes that they will be factors that are predictable or controllable.

The biggest "random effects" are in recruitment rates and often seem to follow a lognormal pattern (most points near recruitment curve, but a few very high outliers). Unfortunately, we expect precisely this pattern when there is <u>no</u> small set of major factors that cause variation, and the variation is instead due to many small random effects operating at various life stages prior to recruitment. If survival to recruitment is a product of many survival rates $s_1\ s_2\ s_3\ \dots\ s_n$, each varying a little independently of the others, then the logarithm of total survival is the sum of logarithms of the component rates; thus it will be approximately normally distributed (central limit theorem), with a variance equal to the sum of variances of the component logarithm rates. Thus it can have very high variance (and its antilog even higher) even when the component rates each vary only a little.

Therefore, I predict that most efforts to explain recruitment variation will end in frustrated failure at best, or a collection of spurious correlations at worst. The investments that management agencies now make into research on causes of recruitment variation might be better spent on the design and implementation of better control systems for more timely detection of and response to that variation.

**Uncertain Parameters**

The measurement and process errors just mentioned lead inevitably to errors in estimation of parameters for production functions, even if we can find the "right" functions to use. Much of the literature on fisheries stock assessment uses some very foolish assumptions (such as continuous equilibrium in surplus production analysis) in order to achieve results that "look reasonable" or fit nicely to the data. When these assumptions are discarded, as for example in Schnute's (9) method for surplus production assessment or Paloheimo's (6) method for estimating mortality and recruitment rates, we see that in fact most data sets contain very little information about dynamic model parameters.

Even using methods that assume no measurement errors at all, we usually find that several critical parameters are highly confounded. That is, the data can equally well be "explained" by a variety of parameter combinations. These discouraging results can mean either that process errors are very large, or that there has been insufficient contrast in states (stock size) and inputs (fishing effort) in the historical record. To make matters worse, data sets with informative contrasts are usually those from the early part of fishery development, when we expect some key "parameters" (such as the catchability coefficient) to change rapidly and before reliable monitoring systems are in place.

So we are much more uncertain about the production parameters for most stocks than the published literature would indicate, and this state of affairs will not be significantly improved just by investments in better monitoring systems. Nor is there much hope for better statistical recipes to coax more information out of existing data; modern recipes using likelihood and Bayesian formulations are already pressing information theoretic limits on parameter estimation performance.

**Parameters That Aren't**

Most exploited resources are not nice, homogeneous unit stocks that theoreticians have assumed for convenience. There are complex spatial structures, and variations among individual animals in heritable attributes ranging from fecundity to net avoidance behavior. Stocks are embedded in ecosystems, which are certainly not constant over time.

Thus when we examine how practically any aggregate parameter of stock response is determined by events or processes operating on individual animals that are not all alike, we find good cause to expect the parameter to vary over time. Harvesting progressively erodes stock structure, favoring the most productive and inaccessible substocks. There is strong selection for ability to avoid fishing gear. Reduction in densities due to harvesting may affect selection pressure for resistance to epidemic disease. The list goes on: any competent biologist or economist should be able to rattle off at least twenty factors that will result in slow changes in productivity and harvesting parameters.

There have recently been some Monte Carlo simulation experiments, where fake data without measurement errors are generated with known parameter changes over time, to see if changes can be detected and tracked by using filtering methods from the theory of system identification (12). These experiments have been quite discouraging; unless it is known in advance exactly which parameters are changing and roughly how fast, and unless the stock size is strongly perturbed over time by varying harvest rates, the estimation methods simply cannot tell one source of deviation (prediction error) from another and thus end up not tracking any changes correctly. In other words, the patterns we see in typical historical data sets can equally well be ascribed to a variety of different hypotheses about parameter changes, and we are not going to track the changes successfully into the future just by being careful with monitoring systems and statistical methods.

Parameter changes over time imply that there is some level of informative management disturbance below which we will actually become more uncertain over time about parameter values. That is, we cannot rely on gradual improvement in estimates as experience accumulates, because the older experience is not informative about the current state of affairs. Considering that there are also upper bounds on management disturbances, set by socioeconomic factors and by risks of irreversible damage, we should view good management as operating with a "disturbance doughnut." The hole in the doughnut represents levels of variation that are too small to be informative about changing parameters, and the outside of the doughnut represents "safe" limits of variation.

## PREDICTION, CAUSATION, AND FEEDBACK POLICIES

Recent surveys and workshops on the state of fisheries science have emphasized the need for increased investment in research on causes of variation, especially in recruitment rates. It has been taken for granted that a) understanding the causes will be helpful in making predictions about future stock changes, and b) prediction is important for good management. In the following paragraphs I will examine each of these assumptions, in light of recent discoveries about feedback policy design. My basic conclusion is that it is no longer wise to take either of them for granted.

It will quite likely be discovered that recruitment variation is driven largely by marine climate factors such as upwelling patterns. With luck it may even be found that there are a few dominant factors rather than a multitude of small effects with a lognormal result (see above). Now suppose we construct a population model that correctly represents the effects of at least the major factors, and we use this model to assess a probability distribution for future variation (it would obviously be foolish to construct a single prediction, since the climate factors are not in themselves predictable). This probability distribution might then be used in the design of robust management policies for meeting objectives such as minimizing the chance of a major collapse, or for examining the trade-off between mean and variance of catches.

Now, how would we establish the credibility of a probability distribution for future variations due to known environmental factors? The answer is, of course, to look at the historical pattern of variation at least to see whether the known factors will generate realistically high amounts of variation and realistic temporal patterns. Then the questions arise: why bother to generate a causal model for the variation in the first place, when the basis for policy design will be a future probability distribution that can be estimated directly from historical variations (and therefore be immediately credible) without making any special assumptions? Why risk using a causal model that may easily underestimate future variation (and not include some factors)?

So I conclude that in the very likely event that the causal factors will themselves be unpredictable, the justification for studying them must be sought elsewhere than in arguments about forecasting or placing odds on future outcomes. But there are at least two valid justifications: a) we may be lucky and find that the factors are predictable and/or controllable, and b) by sorting out their effects from the historical data, one may be able to see more clearly the effects of such variables as stock size that are controllable. However, there is a little snag in the second justification: factors that cause large variation in productivity will also indirectly cause variations in stock size; resulting correlations between stock size and the factors may make it impossible (statistically and logically) to decide what was actually causing the changes in productivity.

Let us turn now to the second assumption, that prediction is necessary for good management. A case can be made for this assumption in situations where decision-making involves large unit investments that take a long time to bear fruit, as in energy development or even large-scale aquaculture. But for most decision-making, it is perfectly sufficient to have a clear perception of possible outcomes and mechanisms for responding rapidly when the actual outcome makes itself evident (i.e., to have a feedback policy based on reasonably good monitoring). For example, in salmon management where the short-run objective each year is to regulate harvest so as to reach a preset escapement goal (estimated from analysis of long-term data), prior forecasts of stock size are of little value compared to information gathered from the fishery as harvesting proceeds. The salmon fishermen and the processing industry would certainly value good forecasts, but it is not at all clear that the public management agency should subsidize them by investing in a good forecasting system.

One argument for forecasting is that management agencies often lack the flexibility to respond rapidly, especially to declines in stock size. Again and again we see the old story that legal mandates to restrict fishing just do not carry much weight when the fishing industry is aroused through fear that immediate livelihoods are at stake. However, one must be a real optimist to think that we will ever have good enough long-term forecasts to avoid being trapped occasionally into the need for some hard regulation. A better approach is to seek increased flexibility in the economic system, for example, through taxation policies, insurance programs, and capital investment strategies that make the industry better able to absorb variation in catches.

The concept of managing for economic flexibility rather than stability of harvests is not a new one, and it has been practiced for decades with other resources. In agriculture, crop insurance programs help the industry during bad times; it is a small step from those programs to less costly schemes for the public, such as wealth buffering plans in which the fishermen pay money into a savings pool during good years and withdraw in bad ones. In many cases, we know enough about the likely variation in future catches to provide sound advice about how much to save and when to withdraw, provided the fishery is also managed with a limited entry scheme that prevents general overcapitalization and spreading of earnings across such a large industry that no one can afford to save even in the good years.

A very interesting policy for providing economic stability when there is the possibility of artificially enhancing production (hatcheries, etc.) has recently been discovered by Ray Hilborn at the University of British Columbia. He examined situations such as salmon management, where natural and "hatchery" stocks are harvested together in mixed fishing areas. His idea is simply to manage the mixed fishery for a fixed catch quota (which would ordinarily be disastrous in the face of variable natural stocks), while increasing hatchery production so as to increase the total stock available in the area. The exploitation rate (quota/stock) must then fall (shorter fishing seasons, etc.), and natural stock sizes should increase and contribute further to reducing the exploitation rate. So in the end, natural stock sizes recover and part of the hatchery production is still available, and the industry does not have to forego catches during the recovery period. I suspect that imaginative plans like this can be devised for a lot of fisheries, even using buffer systems other than hatchery production, and that investment in the search for such plans will be far more productive than research on better forecasting.

## ACTIVELY ADAPTIVE MANAGEMENT

When we examine long-term data sets using various procedures for estimating dynamic parameters, it always turns out that the estimates depend critically on only a few informative data points obtained during the fishery development or later "accidents" (like World War II) that greatly perturb fishing effort. Thus what initially seem like very large bodies of data turn out in fact to be very small indeed. We are led immediately to wonder whether it is worthwhile deliberately to introduce informative variation through experimental management policies.

In the general literature of stochastic control, active adaptation and experimentation are discussed under the heading of "dual control problems." When controlling actions can influence both immediate system performance and also learning rates about unknown parameters, these so-called dual effects are usually in conflict: the best action for learning may be a disruptive one for short-term performance. It has been found (1, 2) that long-term performance can generally be approximated as a sum of three value components: a) a "nominal" (deterministic) component based on the best single prediction forward in time, b) a "caution" component that accounts for uncontrollable and unpredictable process variations in the future, and c) a "probing" or information value component. All three value components depend on the immediate decision (control, action) and at any moment define which decision is best (imagine plotting long-term yield as a function of harvest rate, with the yield curve broken into three value components). The caution component varies so as to favor more conservative (rather than nominal) decisions, while the probing component usually favors decisions that are far from the nominal (or at least far from the historical average choice). In sum, the components usually indicate that it is best to behave either somewhat more cautiously than indicated by the nominal component or else make a substantial and informative probe. Small probing tests ("dithering") are never favored, since they deteriorate short-term performance without substantially helping in parameter estimation (for examples of fisheries calculations along these lines, see Walters (12) and Ludwig and Walters (5)).

Thus recent work in stochastic control theory does not support two common recommendations about how to deal with uncertainty in fisheries, namely, to be very conservative in setting harvest policies and to regulate development so it proceeds slowly (and stock sizes remain near equilibrium). I consider it downright irresponsible that these recommendations continue to be voiced loudly by scientists who will happily argue about how important it is to construct models carefully, yet are not willing to put that same care into thinking about the implications of uncertainty. At this point, the few calculations that have been made about fishing policies in the face of functional uncertainty (parameters, model form) indicate that it is best to use occasional probing experiments interspersed with longer periods of cautious management (4, 10, 11).

Conditions that favor a switch from cautious to probing actions include: a) high stock productivity and management flexibility, which both reduce the costs of "mistakes" (recovery from collapse, etc.); b) high uncertainty about parameters that determine equilibrium stock size in the absence of harvesting (i.e., uncertainty about whether the stock is currently far from an optimum level for long-term production); and c) an accurate monitoring system so that responses to the probing actions are clear (and therefore rapid learning is possible).

Probing decisions are generally not optimal when the management agency is required to act in a risk averse manner, i.e., as though it were trying to maximize a risk averse utility function for which maintaining a good average catch is not so important as avoiding low catches. Here we must be careful to distinguish between agency mandates and personal preferences. Most biologists are implicitly very risk averse and never stop to question whether others share their preferences. Some fishermen, for example, might well prefer to behave in a risk neutral way (maximize average catch without regard to variability) while counting on opportunities to fish several stocks (or even take other employment) as a means to even out their earnings. For governments that manage many stocks supposedly for general public benefit, it is difficult to show that the public will be best served by treating every stock in a risk averse manner, especially considering that a few informative disasters may be immensely valuable in demonstrating how the majority of stocks should be managed.

**CONCLUSION**

In this paper I have questioned several prescriptions that are commonly made by fisheries scientists, ranging from the value of having survey sampling independent of the fishery through to the notion that uncertainty always favors cautious policies. I have come to see these presumptions as being not the products of careful analysis, but rather the result of intuition in which issues of valid public concern have become muddled with personal feelings about risk taking and with some plain wishful thinking about how informative future data are likely to be. Thus I have one main conclusion: it is past time for fisheries scientists to start acting like scientists and to become much more imaginative and critical in examining how best to deal with uncertainty.

## REFERENCES

(1) Bar-Shalom, Y. 1976. Caution, probing, and the value of information in the control of uncertain systems. Ann. Econ. Soc. Meas. 5: 323-337.

(2) Bar-Shalom, Y. 1981. Stochastic dynamic programming: caution and probing. IEEE Trans. Autom. Contr. Ac-26: 1184-1195.

(3) Buckingham, S.L., and Walters, C.J. 1975. A control system for intraseason salmon management. Proceedings of the IIASA Workshop on Salmon Management, CP-75-2, pp. 105-137. Laxenburg, Austria.

(4) Ludwig, D., and Hilborn, R. 1983. Adaptive probing strategies for age-structured fish stocks. Can. J. Fish. Aquat. Sci. 40: 559-569.

(5) Ludwig, D., and Walters, C.J. 1982. Optimal harvesting with imprecise parameter estimates. Ecol. Mod. 14: 273-292.

(6) Paloheimo, J.E. 1980. Estimation of mortality rates in fish populations. Trans. Am. Fish. Soc. 109: 378-386.

(7) Reed, W.J. 1974. A stochastic model for the economic management of a renewable resource. Math. Biosci. 22: 313-337.

(8) Reed, W.J. 1979. Optimum escapement levels in stochastic and deterministic harvesting models. J. Env. Econ. Mgmt. 6: 350-363.

(9) Schnute, J. 1977. Improved estimates from the Schaefer production model: theoretical considerations. J. Fish. Res. Bd. Can. 34: 583-603.

(10) Smith, A.D.M., and Walters, C.J. 1981. Adaptive management of stock-recruitment systems. Can. J. Fish. Aquat. Sci. 38: 690-703.

(11) Walters, C.J. 1981. Optimum escapements in the face of alternative recruitment hypotheses. Can. J. Fish. Aquat. Sci. 38: 678-689.

(12) Walters, C.J. 1984. Adaptive Policy Design in Renewable Resource Management. Book Series. Laxenburg, Austria: International Institute for Applied Systems Analysis, in press.

(13) Walters, C.J., and Ludwig, D. 1981. Effects of measurement errors on the assessment of stock-recruitment relationships. Can. J. Fish. Aquat. Sci. 38: 704-710.

Exploitation of Marine Communities, ed. R.M. May, pp. 275-285. Dahlem Konferenzen 1984. Berlin, Heidelberg, New York, Tokyo: Springer-Verlag.

# The Wider Dimensions of Management Uncertainty in World Fisheries

G.D. Brewer
School of Organization and Management
Yale University, New Haven, CT 06520, USA

**Abstract.** Natural variability of fish stocks is only one of many possible sources of uncertainty in fishery management, and in given circumstances, it may not even be the most consequential. Failure to recognize and accommodate these other sources, including intellectual, political, legal, and administrative ones, poses grave threats to effective resource use.

## INTRODUCTION

Most of us are well acquainted with the planning and decision-making dilemmas resulting from the natural variability of fish stocks around the world. Less well-known are uncertainties that derive from many other sources, both apparent and subtle. Let us consider these broader contributors to uncertainty in fishery management.

A recent report of the Food and Agriculture Organization's Advisory Committee on Marine Resources Research (ACMRR) ((2), p. vii) pointed out that:

> Major changes in the conditions affecting fisheries development and in the perceptions of fisheries problems have occurred since the early 1970s.... These changes have led to wide-spread frustration and dissatisfaction with the performance of fisheries management and policy and to a sense of crisis.

Frustration and dissatisfaction, in turn, caused the ACMRR Working

Party to prescribe some remedies. Improving understanding and policy-making figured prominently ((2), pp. viii, ix).

> Policies should be expressed as precisely as possible. Policies should be sufficiently broad to take account of all factors relevant to fisheries management. Policies should be based on comprehensive analyses, particularly with regard to what is at stake, who gains and who loses. Policies should be developed in such a way that affected individuals understand how they are affected. Those who are affected should have opportunities to express their views and to participate in policy development. Models dealing with variability and uncertainty need to be improved.

The goals are clear enough and hard to fault. The problem is figuring out how to reach them.

One way is to start with familiar territory. Here we explore the basic relations of scientists, with their commitment to objective truth, and managers, whose decisions are often governed by the imperatives of politics. Uncertainties flow from that relationship. It is further complicated when analysts enter the picture - their profession bridging science and politics. Uncertainties also flow from the inordinate complexity of the setting of fishery management. We cite several examples while arguing that intellectual and practical means are inadequate for coping with the problems created. We conclude by pondering the implications for the practice of fishery management.

**SCIENTISTS AND MANAGERS**

As practitioners of science, many biologically trained fisheries specialists have had success in understanding and solving selected problems. But many other problems fall outside the narrow bounds of rational hypothesizing and experimental procedures supporting the scientific edifice.

Science seeks the discovery of theory by generating and testing hypotheses that confirm, refine, and enlarge the common understanding of events. By theorizing and testing, small infusions of evidence can be marshaled to yield predictions about the future. However, science's success owes partly to the care with which practitioners select their problems. It also results from the consistency of the "client" for the work: the disciplines, as represented by their adherents.

For practical problems that demand policy responses, such as the management of specific fisheries, scientific rules are not as applicable. Just for a start, consider the complexities inherent in "target switching," by-catches and discards, multispecies and multinational predation on different kinds of fish, implications of economic interdependencies on fishing (e.g., regional and worldwide fluctuations in prices of alternative forms of protein), and chaotic oscillations in population abundance. Science simply does not suggest means for coping with these problems (8).

For example, no theory is capable of predicting the social, economic, and political consequences of El Niño. That natural event, arising off the west coast of South America, resulted in the collapse of one of the largest fisheries in the world. Scientists even dispute the underlying mechanisms of El Niño, when and whether the anchoveta will ever return to abundance, and what the fishery's future may hold (3, 7, 13). However, these are precisely the kinds of questions that managers must deal with. Managers likewise gain little guidance from elegant economic theories that "solve" their problems by advocating efficiency through open markets with centralized ownership of the world's fish by a nonexistent international authority (9). Despite its scientific rigor, that solution is neither practical nor politically feasible. Any recommendation, scientific or analytic, that lacks political feasibility is about as good as no recommendation at all.

Thus, managerial uncertainty turns out to have a deep intellectual root. Science is unable to explain many of the most consequential events facing the manager, nor do science's disciplined and rational principles go far in surmounting the raucous and passionate play of politics over conflicts large and small, present and potential.

## ANALYSIS: THE TWO FACES OF JANUS

Analysis seeks to discover the contours and dimensions of a problem in systematic and logical fashion. A phenomenon is resolved into its constituent parts, and each is distinguished individually and in relation to the whole. Analysis is partly adduction and partly inference: facts, evidence, and arguments are offered in support of a position, and the facts are used in proceeding through a series of judgments to a conclusion about what might be done to solve a problem (5).

To those identified with science, analysis offers a road to rational management and policymaking. But to those identified with politics, analysis may seem only another ingredient in the black art of forging consensus and keeping society safe and intact. The two-faced character of analysis and its distinctiveness from science are poorly understood.

Analysis and science share a common approach, including certain tools, terminology, and logical means. But analysis has a political side that is foreign to science. It provides a language for politics and a paradigm for political discourse. It is not a sterile or encapsulated activity, as scientific investigation can be. It is a process that defines, propagates, interacts, and diffuses policymaking processes. It also causes tensions among the participants.

Policymaking requires consensus, but the very act of exploring options and trying to attain consensus exposes the many interests and values at stake in political decisions. As a result, it becomes nearly impossible to produce a logically consistent set of policies. Attention to the "big picture" gives way "to fragmented and personalized interests and values, difficult to reconcile with integrated and rational planning and foresight" (6). Different interest groups interpret and act on the same analytic results in different ways, according to their values, interests, and perceptions. And the recommendations an analysis contains may not be politically palatable.

In fact, even good analysis can impede political accommodation, something difficult for many analysts to accept. Rationally, more and better analysis should only be helpful, but it often is a hindrance. When an analysis (or the analyst) goes public - explicitly identifying alternatives, arguments, and trade-offs - problems can be created for the decision-maker. Displaying your hand before the bidding begins in delicate negotiations can undermine your bargaining position and limit your political maneuverability.

Politicians instinctively grasp the difference between rationality, as embodied in science, and analysis. Scientists often do not. And managers are caught somewhere in between. We must keep that difference in mind if we are to appreciate the intellectual uncertainties that complicate effective fishery management.

## POLITICAL UNCERTAINTIES

The "fishery problem" must be recognized as political and institutional as well as biological. Moreover, it is the complex interaction of these quite different entities that defines and determines real world outcomes. Such interaction is a root source of most stunning uncertainty. Politics thus has its own logic and factual basis, both of which are as challenging as any problem a scientist might confront.

One fact is that politics is only partly a rational process, at least by scientific standards. Decisions and the analyses done in their support are always imperfect because the "events" involved are literally unreal in scientific terms. Problems are intricate and changeable, and they often mean very different things to the different people they affect. Nor do problems and their possible solutions stand still - they evolve naturally and in response to efforts to understand and master them. Thus, a decision to regulate a certain stock according to a rational abstraction, such as maximum sustainable yield (MSY), turns out to be far from a final solution when the nonstop bargaining and compromising that always attend regulation begins. Regulation is politics in action, which annoys many of a rational cast of mind (10). Furthermore, though the disciplines seem to provide tempting "optimal" or "best" solutions, the information used to reach them is selective and addresses only what has been, not the what-will-be that most concerns policy and decision-makers. Descriptive knowledge cannot encompass the future, but policy and managerial actions must always look toward it.

Of course, uncertainty flows from decision-making in a policy setting. How could it be otherwise? Politics is the realm of clashing interests. Its interpersonal dynamic has the character of chemistry, not deductive logic. Politics deals in pressure, power, and negotiation, not optimization or rational deduction. Its instrumentalities are coalitions, compromise, and reciprocity, not decision rules, computer models, or payoff functions.

How do these insights affect managers? One important implication is that the "effective policy" prescribed above by the well-meaning ACMRR working party will probably never happen. The realistic setting of management - the "system" - simply does not work in ways to achieve it. But how close does the reality permit us to get?

A large part of the policy problem, including its definition, analysis, and resolution, lies in identifying specific interests and settings (1). The problems managers confront nearly always involve disparate interests; no two individuals or groups comprehend or value an event the same way. The practical problem is that all relevant perspectives matter, each illuminating the problem differently in angle and intensity. In composite, the various views offer a better sense of the complexity of the whole - social as well as biological - complexity (11, 14).

## COMPLEXITY AND UNCERTAINTY

Fundamental uncertainties arise out of the complex settings in which fishery management decisions are made. To cite just one aspect, consider the low priority that fishery issues usually receive - they tend to be driven by, rather than drivers of, other policies. Thus, decisions taken in other arenas often have very consequential impacts on fisheries that are not under the manager's control. Let us explore several instances.

As exclusive economic zones (EEZs) are turned more to national rather than foreign advantage, pay heed to the reactions of the nations denied fishing access. Will they retaliate and how? Might they decide to rigidly enforce selective pollution or navigation and overflight regulations in their own EEZs? What will they do with the newly unemployed human and capital investments? Can the fishermen be beached and the vessels decommissioned without a violent reaction? Redeploying the fleets to other fishing grounds seems more palatable until one realizes that the world may already be overcapitalized and overfished. One nation's "phase in" is truly another's "phase out," and in our tightly connected world, how are these phasing decisions reached and their repercussions compensated for?

Many believe world energy markets to be precariously balanced. In their view, the energy crisis persists just under the surface, silent and lulling, ready to erupt in fury with slight provocation. If so, how might desperate efforts to secure reliable petroleum supplies affect fisheries? Georges Bank is an obvious case, as are decisions about the use of the outer continental shelf. Fisheries have served as bargaining chips in political decisions before, e.g., to resolve border, trade, and national security disputes. No doubt they will be so used again, but with what detrimental consequences?

Higher priority issues need not have harmful consequences for fisheries. The United States and Spain are currently planning the development of the fishery contiguous to the Canary Islands. The American contribution is neither altruistic nor motivated simply by a desire to develop the marine resource. The United States wishes to maintain military base rights on Spanish territory: furthering Spanish interests in other ways is a perfectly above-board quid pro quo.

National security figures into the manager's scope in other ways beyond his control. For instance, enforcement of regulations largely falls to the coast guards and navies of the fishing nations of the world - not directly to fishery managers. Even in the wealthy nations such as the United States, security forces used for fishery purposes are poorly trained and easily diverted from this task (15). The administrative complexities, and hence uncertainties, of planning and carrying out the most ordinary managerial responsibilities sometimes challenge comprehension.

Defining and using information effectively present comparable difficulties, especially when management's needs for data are poorly understood by others who control access to data collection (12). The problem transcends bureaucratic rivalry. The world's marine scientists, a primary source of essential biological data, are concerned about their continuing ability to work in foreign EEZs. UNCLOS adds the administrative demand to secure consent before mounting a scientific expedition, which subjects scientists' plans to the whims of inefficient bureaucracies. The increased uncertainties thus introduced by UNCLOS could have a chilling, if not parochializing, effect on the collection of basic fisheries information.

Reliable basic data could also diminish within EEZs as a consequence of foreign fishing phase outs. In the United States zone, for example, foreign vessels are more carefully observed and tightly controlled. They often have observers stationed on board to supervise and count the catch. American vessels are not so closely observed nor are their own reports believed to be as accurate as they might be. The precision and reliability of data on catch and effort will decline as the foreign fleets are replaced by U.S. counterparts. This problem may be hard to correct, considering the number of U.S. fishermen and their ability to exert domestic political pressure.

Many of the uncertainties arising from complexity can be understood in terms of the different clocks that guide scientists and politicians. Politicians seldom plan ahead but rather react to the crisis of the moment. Scientists can afford to take a longer view. They warn, for example, of grave problems resulting from the accretion of carbon dioxide three or four decades from now, to which political authorities hardly respond; none of them will be standing for election when that doomsday is at hand.

Though less urgent, many such important problems come to mind: recurrence of energy shortages, determination of safe means and locations for the disposal of nuclear and conventional wastes, development of new institutions to mediate conflicts without resort to courts of law, and understanding ocean circulation and transport systems and effects.

When such problems are not considered deliberately in advance, one is left with crisis decision-making yielding outcomes that satisfy hardly anyone. But it is nearly impossible to obtain funding for long-term, comprehensive research and analysis. And the scientific research that is being done is poorly integrated with policy needs.

Consider, for example, what information fishery managers need worldwide concerning the complex economic factors driving markets. In deciding how to allocate a scarce and endangered resource such as striped bass in America's New England and mid-Atlantic regions, managers need much more than simple statistics on landings and market value of the catch. Intelligent choices about quotas, size, seasons, and limits also require a full socioeconomic accounting of net contributions to public welfare, including estimates of the value added to ancillary industries such as boat sales, real estate, and bait and tackle. Current decisions affecting the bass are a crazy quilt reflecting the relative political strength of contending groups and personal whim more than anything else. One looks in vain for empirically based studies of the interconnections between fish products and competitive commodities such as beef, soybeans, and fertilizer. Expressed in economics jargon: what are the cross elasticities of demand for various primary uses of fish and related products? Knowledge of demand elasticities could become important in a world where product substitutions are likely to occur as prices reflecting supply and demand move dramatically.

Perhaps the most troublesome class of managerial uncertainty stems from the inevitable conflicts that surround decisions about resource use. It is one thing to advocate involving and informing all affected parties, as does the ACMRR working party. It is quite another to contend with the fearsome energies thereby unleashed - as managers have to do. As a rough measure of the resulting contention and irresolution, witness the number of manager's decisions that end up in court. The courts are valuable for resolving otherwise unresolvable conflicts, but they are not an ideal forum for issues of increasing technical content.

In my view the most critical uncertainty of all is whether individuals and institutions will assume responsibility for the many actions urgently needed to improve and secure fishery management around the world. The sensible steps defined in meetings, conferences, working groups, and private forums are no longer enough. We need responsible authorities with the will and wit to take action.

## WAITING FOR GODOT

In the meantime, what courses are open to those in the trenches? Preparing for the worst in various ways is certainly compelling, especially if one believes that fishery issues will continue to be subordinated to others and that political authorities will only move on crises of the moment.

Elsewhere I have distinguished between three kinds of management: preventive, where problems are identified before they reach crisis proportion and while enough time remains to intervene effectively; active, the most common present-day managerial style; and reactive, where current decisions and practices are assessed for their consequences and necessary remediation is undertaken (4). The third has special appeal.

The poorly developed state of reactive management is a distinct shortcoming if the abuse of fish as a resource continues, as many fear. Contingency and fallback plans need to be prepared before disasters large and small strike. There simply will not be sufficient time otherwise. In the good years we need to accumulate resources to cushion the blow of bad years; they certainly will not be as available when stocks diminish or fail. Creative intelligence is needed to devise incentives limiting overcapitalization before it occurs, not when its pernicious effects are felt and the resource is ruined. An attractive suggestion in this vein is the levying of a "decommissioning fee" before a new vessel is constructed. The amount would be scaled according to best estimates of the abundance and condition of stocks at risk - the more threatened, the greater the fee. The monies could be allocated to unemployment, retraining, relocation, and similar social expenses engendered by a stock depletion or collapse. Suppose a fishery is in jeopardy. Is rehabilitation feasible, and if so, who pays for it? The question is pertinent when stocks are shared by nations, as many are. Whose job is it to facilitate the reallocation of human and capital resources when bad years continue or when the assertion of one nation's rights deprives another of access to habitual foreign fishing grounds?

Who is raising such questions and worrying about the demands encompassed in the answers? Where in the world are repositories of knowledge available to ease the reactive and damage-limiting tasks of fishery management?

Why are we waiting for Godot?

## RECAPITULATION

Biological uncertainty will probably always exist. We therefore need to be more defensive and reactive in the care and use of our fishery resource. We lack enough scientific knowledge about even the most basic factors affecting fisheries, particularly for those regions of the world where the resource is at risk from overexploitation and misuse.

A more comprehensive and longer-term view of fishery problems needs to be adopted to avoid the pitfall of crisis decision-making that in turn yields unsatisfactory results. Multidisciplinary, large-scale, and long-term policy studies are a partial solution, if obstacles to securing adequate financial support can be overcome.

We all need to accept and disseminate a more sophisticated appreciation of the interconnected and complex nature of fishery management. The harmful consequences of piecemeal consideration and treatment of the resource are all too evident. The difficulty in adopting the wider view is itself complex and deeply rooted in the intellectual, professional, and conceptual habits of scientists, analysts, and politicians.

## REFERENCES

(1) Advisory Committee on Marine Resources Research. 1980. The Scientific Basis for Determining Management Measures. Fisheries Report No. 236. Rome: FAO.

(2) Advisory Committee on Marine Resources Research. 1983. Working Party on the Principles for Fisheries Management in the New Ocean Regime. Fisheries Report No. 299. Rome: FAO.

(3) Barber, R.T., and Chavez, F.P. 1983. Biological consequences of El Niño. Science 222: 1203-1210.

(4) Brewer, G.D. 1983. The management challenges of world fisheries. In Global Fisheries: Perspectives for the 1980s, ed. B.J. Rothschild, pp. 195-210. New York: Springer-Verlag.

(5) Brewer, G.D., and deLeon, P. 1983. The Foundations of Policy Analysis, Ch. 1. Homewood, IL: Dorsey Press.

(6) Brooks, H. 1979. Technology: Hope or catastrophe? Technol. Soc. 1: 15.

(7) Cane, M.A. 1983. Oceanographic events during El Niño. Science 222: 1189-1195.

(8) Clark, C. 1981. Bioeconomics of the ocean. BioScience 31: 231-237.

(9) Cooper, R.N. 1974. An economist's view of the ocean. In Perspectives on Ocean Policy, ed. R.E. Osgood, pp. 145-165. Baltimore, MD: Johns Hopkins University Press.

(10) Freeman, A.M. III, and Haveman, R.H. 1972. Clean rhetoric and dirty water. Pub. Inter. 28: 57.

(11) LaPorte, T.R., ed. 1975. Organized Social Complexity: Challenge to Politics and Policy. Princeton, NJ: Princeton University Press.

(12) Lieberman, W. 1983. The Use and Abuse of Information in Fishery Management. New Haven: Yale University, School of Organization and Management.

(13) Rasmusson, E.M., and Wallace, J.M. 1983. Meteorological aspects of the El Niño/southern oscillation. Science 222: 1195-1202.

(14) Simon, H.A. 1969. The Sciences of the Artificial. Cambridge, MA: MIT Press.

(15) Young, O.R. 1981. Natural Resources and the State, Ch. 4. Berkeley: University of California Press.

Standing, left to right:
Simon Levin, Daniel Pauly, Veravat Hongskul, Peter Larkin, Niels Daan

Seated, left to right:
Harald Rosenthal, Someschwar Dutt, Günter Radach, Colin Clark, Garth Newman

Exploitation of Marine Communities, ed. R.M. May, pp. 287-301. Dahlem Konferenzen 1984. Berlin, Heidelberg, New York, Tokyo: Springer-Verlag.

# Strategies for Multispecies Management
## Group Report

P.A. Larkin, Rapporteur
C.W. Clark
N. Daan
S. Dutt
V. Hongskul
S.A. Levin
G.G. Newman
D.M. Pauly
G. Radach
H.K. Rosenthal

### INTRODUCTION

Much of the literature of fisheries science is concerned with the management of single-species populations of fish. But most of the world's fish populations do not live in isolation from other fish populations. On the contrary, most live in association with many other species, and many are caught in association with others in various kinds of fishing gear.

It is against this background that many different kinds of fisheries are grouped under the heading of the term "multispecies." The papers of Clark, Newman, and Gulland and Garcia (all this volume) each allude to some of the considerations involved. It seemed logical for our group first to describe a typical situation, then to suggest how the various kinds of multispecies fisheries might be classified and what management measures might be taken to achieve certain objectives, and finally to consider the implications of the foregoing for research.

#### What Is a Typical "Multispecies Fishery" Situation?

A typical and somewhat stylized example of a multispecies fishery as described by Pauly (personal communication) involves artisanal fishermen who take many of the same species plus some other species with which they may interact. Characteristically, the fishery is prosecuted initially

(Phase A, Fig. 1) by artisanal fishermen, the total catch is not large, and the abundances of only some of the species in the catch are substantially influenced by fishing (2). With the introduction of large-scale trawling (Phase B), resulting in competition with artisanal fishing, total fishing effort steadily increases and total catch quickly increases until it more or less levels off (2-4, 6). As total fishing effort further increases, total catch may be maintained but there are changes in the species composition of the catch, the higher valued species often being the first to disappear, and the artisanal fishermen take a smaller proportion of the total catch (1, 5). With still further increases in effort, total catch fluctuates, and with each decline in catch there are strong pressures from the artisanal fishermen for intervention on their behalf (Phase C).

This pattern has been followed in fisheries in many parts of the world

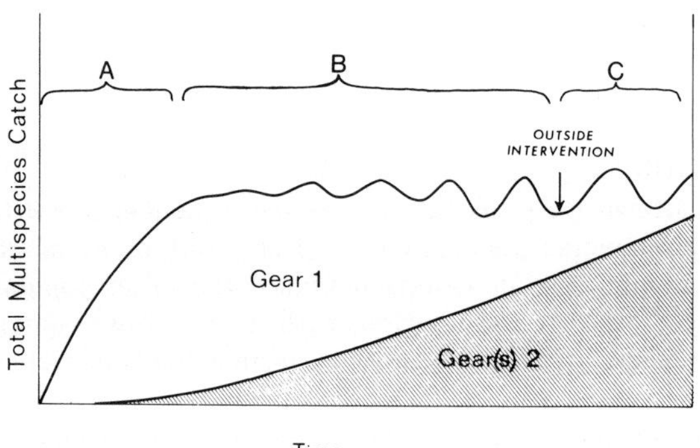


FIG. 1 - Three phases in the development of a multispecies fishery. A: "Development" of fishery, based on a given type of gear (1); B: transfer of an increasing proportion of the total multispecies catch from gear 1 to gear(s) 2, which may be more efficient, more capital intensive, subsidized, etc. The model assumes more or less constant total multispecies catch (massive changes in species and size composition are implicit) and indicates the reason for outside intervention at times when one sector of the fishery (using gear 1) turns to politicians for support. This model applies to competition between artisanal vs. modern, small vs. large, sports vs. commercial, local vs. foreign, subsidized vs. non-subsidized (etc.) fisheries. More than two gears may be incorporated either in temporal succession, or simultaneously competing with each other. (Derived from material in (1-5), and from (6), especially Figs. 16 and 17.)

and is especially common throughout the tropical seas. With only slight modification, the same pattern may be applied to a very wide variety of fisheries for which there would be the expectation that with an increased rate of exploitation there would be a change in the species composition of the catch, and in the absence of intervention, a change in the "species composition" of the "catchers." It is not essential that the catch of each kind of gear should comprise all of the available species, for it is the aggregate effect of all of the vessels on all of the species that decides the overall character of the fishery.

The essential features of this scheme are also readily extended to several types of gear, which in rapid transitions are successively squeezed as they become relatively ineffectual, and as well to the situation in which recreational and commercial fishermen compete for the same resource. It will not, of course, apply to those rare situations in which several species continue to be harvested with only a single type of gear, but even here it may be observed that intensive fishing induces changes in species composition.

In responding to the complaints of the operators of gear types that are squeezed, the manager's key problem is, having assumed a level of total catch and species mix, how to allocate the resource amongst the variety of users. Assuming that "the manager" is abstract and omnipotent, rather than a bureaucrat, the obvious choices are a) do nothing, b) set a biological objective such as maintaining all of the species or maintaining the total weight of catch, c) set a social objective such as the maintenance of the various groups of fishermen, d) set an economic objective such as improvement of the benefit/cost ratio where costs include the expenditures on research and management, or e) some combination of alternatives b, c, and d.

With this brief introduction we may turn to considering the classification of "multispecies" fisheries.

**THE CLASSIFICATION OF MULTISPECIES FISHERIES**

Classification of multispecies fisheries has little intrinsic importance but can serve to identify the nature of likely management problems. In the simplest cases that might come to mind, there would be no interactions or, at most, weak (or unknown?) interactions amongst the species, and single-purpose vessels would fish without targeting with particular effectiveness on any particular species ($W_1$ of Table 1). Alternatively (and more likely), various kinds of vessels would be targeting on a

TABLE 1 - Considerations in a scheme for classifying multispecies fisheries.

| Species interaction \ Fishing operation | Single-purpose vessels nontargeted fishing | Single-purpose vessels targeted fishing | Multipurpose vessels targeted fishing |
|---|---|---|---|
| weak interaction | $W_1$ | $W_2$ | $W_3$ |
| strong interaction | $S_1$ | $S_2$ | $S_3$ |

Elements of the matrix could be further divided by numbers of species, onshore/offshore stocks, habitats, etc.

particular species or a group of species ($W_2$). Finally, multipurpose vessels may be able to target on several species over the course of a fishing trip or over a fishing season ($W_3$).

A similar group of three kinds of fishing operation ($S_{1,2,3}$) can be distinguished for situations in which there are strong interactions amongst species. Thus, although one might imagine that some vessels catch only cod and others only herring, together they would constitute a "multispecies fishery," because cod eat herring (and vice versa).

Different types of multispecies fisheries can be readily accommodated in the classification of Table 1. For instance, the simple predator-prey system of Gulland and Garcia (this volume) would fall into category $S_2$, their multi-single-target fishery would fit within $W_3$, and their single-multi-target fishery could be classified as $S_3$. Clark (this volume) reorganizes fisheries within each of the columns of the table, and Newman (this volume) discusses types of fisheries which could be allocated to categories $W_1$, $W_3$, and $S_2$, respectively.

It should also be noted that mixes of the elements defined by the fishing operation are also possible within one fishery. From this kind of reasoning it becomes apparent that virtually all fisheries may be classed as multispecies fisheries either because they take several species in one kind of gear or because in taking one species they influence the abundance of species taken in other fisheries. Multispecies fisheries management could even be made virtually synonymous with regional fisheries management, with all that is implied in the definition of boundaries of regions.

Particularly for those who theorize, it may be useful to distinguish fisheries on the basis of the number of species, with categories such as one species, two species (predator-prey, competitors, symbionts, etc.), three species (one predator, two prey) and many species. From an empirical point of view, it may also be useful to distinguish situations in which there are from four to ten species from those with several hundred species.

In particular situations it may also be desirable to think in terms of stocks rather than species as the units ("multistock" fisheries), and while anadromous salmonids may come first to mind, there are many other species for which stocks are probably the most appropriate units to consider. There may also be value in classifying on the basis of habitat (pelagic, demersal, coral reef, and so on) or on the basis of which and how many trophic levels are harvested.

In any event, the classification should be based on the catches rather than on the landings, because the latter, residue after discards, do not reflect the true nature of the fishery. In a similar vein, it should be remembered that the use of some kinds of gear may have effects on the environment that are harmful to species other than those being caught, an "indirect technological effect" that may be the equivalent of being caught and discarded.

An implication of any scheme of classification is that the different classes might require different kinds of management. For example, a fishery which harvests a group of competing species at one trophic level may pose quite different questions for the manager than a fishery which harvests a hierarchy of superpredator, predator, and prey. If there is a variety of single-purpose vessels, each type targeting on a particular species, the management options may be more clear-cut than if several types of vessel are fishing the same mix of species. There are many obvious variations that would illustrate that management practices would not be the same for the various types of multispecies fisheries.

## OBJECTIVES OF MANAGEMENT

While it is difficult to establish a single objective for multispecies fishery management in general, it is likely that the objective for a particular fishery will comprise a mix of biological, economic, social, and political elements. On the biological side, by contrast with single-species fisheries, there may be less concern for short-term prediction of abundance and the setting of annual regulations. Because the total catch in a multispecies fishery may be more or less constant over the short term, the emphasis may be shifted to guarding against sudden and unforeseen changes in species composition that reflect a profound structural change in the fish community. It may be only rarely that fisheries threaten species with extinction, but the possibility should not be shrugged off. Multispecies fisheries management should aim for preservation of the full ensemble of species that occur naturally, as well as for optimal utilization of the complex of species of greatest commercial interest, although it is recognized that in some cases these goals may be incompatible.

The tendency of most contemporary managers, for reasons that stem from political sources, is to preserve the status quo, at least in the short term. In the longer term it might be desirable to diversify the fishery over as full a range of species as possible, both to avoid the waste of discarding and to foster maintenance of a more stable species composition. It may also be desirable to maintain or encourage diversification of a fishery to ensure that it can respond quickly, selectively, and opportunistically to unpredictable local increases in the abundance of particular species. This, incidentally, is an excellent rationale for taking measures to ensure that a substantial proportion of the total fishing effort resides with artisanal fishermen. Thus, although it will almost certainly be desirable in many instances to take a large portion of the total catch in a more cost-efficient manner, the ostensibly inefficient (but often actually very efficient) smaller fishing units should also be seen as a useful part of the total fishery mix.

In considering the economics of multispecies fisheries, it is important to include the full spectrum of social benefits and social costs, including the costs of research and management that are often conveniently forgotten. For some fisheries it is even doubtful that the value of the catch is equal to the cost of research and management. This should not be cause for concern if the expenditures on research are seen as an investment which will reap rich rewards in the future. At the same time, it must be acknowledged that for some fisheries (Canada's, for example) the costs of research have been high for many years but the fisheries are currently no more rewarding financially than they were twenty years ago. To the extent that this kind of situation might arise from the failure to apply the findings of research to management, it is like running hospitals for diagnosis but not treatment of patients.

The social and political objectives of multispecies fishery management are as varied as the social philosophies of the various countries that have fisheries, but they have this in common: they must all recognize that many fishermen may not easily find other employment. This is a central fact of all contemporary fisheries management and is particularly to the fore in multispecies fisheries in which there is such intense competition amongst fishermen using the different kinds of gear.

**MEASURES OF MANAGEMENT**

There is a wide range of measures that may be taken by a fisheries manager, and these measures can be conveniently considered as manipulations of the resource, or of the fishery, or of the infrastructure.

The resource manipulations involve the maintenance and creation of habitat as well as direct measures to propagate desired species. One of the most obvious kinds of habitat manipulation is the construction of artificial reefs, but there may be more than meets the eye in their deployment. Some claim that such constructions provide a good way of creating habitat for reef fishes, even though initially the effect may be more to attract fish than to increase production. The ulterior motive in building reefs may be to make trawling too risky, thus establishing a refuge for other kinds of fishermen.

Propagation of young fish for release in the wild may in some circumstances be highly effective (Pacific salmon, for example). In the circumstances of most multispecies fisheries, however, there is often considerable skepticism about such measures. Young released improperly may be nothing but a tasty snack for predators, but recently developed stocking strategies may hold promise for successful mass propagation. Young may be released initially into semi-wild conditions, thereby reducing initial mortality. As shown by the Japanese experience with Paneaus japonicus, such activities can be successful if combined with the proper choice of location for release, construction of artificial coastal nursery grounds, partial regulation of tidal water level, and control of predator populations on releasing grounds prior to stocking (mainly juvenile fish). Japanese stocking is carried out with the aim of supporting small-scale, local fishermen. Since prawns tend to migrate offshore as they grow larger, the mesh size for inshore fishermen was reduced (Rosenthal, personal communication).

The control of predators to enhance the survival of their prey species is frequently proposed as a device of resource manipulation. In situations in which the prey may be temporarily highly concentrated and easily captured, local control of predators may be effective, but as a more general practice, predator control is far more expensive than is usually justifiable. A complicating factor is that many of the top predators evoke strong and emotional support from conservation groups. It may take a strong-minded manager to carry on with a control program on marine birds such as gulls or marine mammals such as seals.

Many other examples could be cited of resource development in general and propagation techniques in particular. Without exception, they should be viewed as experimental until there has been an adequate scientific evaluation. Performance frequently falls far short of promise, especially in the context of multispecies fisheries. The need for research and evaluation at each step of the way will be especially important in the forthcoming age of biotechnology.

The introduction of exotic species is another technique of resource manipulation which is commonly mooted if not attempted. The introductions of Pacific salmon to the Atlantic and the Antarctic are the best-known examples, and it can be expected that many more such ventures involving salmon as well as other species will be launched in the future. The effects of such introductions cannot be predicted. It can only be said that most attempted introductions are failures, and the successes have often proved to have surprising consequences for indigenous species.

One hundred years hence, it is conceivable that much of the world's food production from the sea will come from sea ranching and aquaculture rather than from "hunting expeditions" for fish. The world has already witnessed some steps in that direction, but enthusiasm for further development should not override precautions for adequate protection of the integrity of natural ecosystems. It will probably never be possible to bring the sea under the same degree of control as the land, and natural production should always be fostered. The ICES code of practice on transfer and introductions of a non-indigenous species should be strictly followed.

The manipulation of the fishery sector is a far more complex and comprehensive way of influencing multispecies fisheries. Of all of the possible measures of regulation, the control of effort by simultaneously limiting entry and the fishing power of vessels is seen as being potentially the most effective. Unfortunately, this kind of measure is not usually implemented in the development stages of a fishery, and in consequence, there are too many overcapitalized units in the fishery and the downward adjustment of effort must depend in large part on the attrition of the fleet as it ages. It should be noted in passing that in limited-entry fisheries it is difficult to hold to a non-replacement policy.

In general, the practice of establishing a total allowable catch (TAC) for each individual species is neither practical nor sensible with or without an overriding total allowable catch for all species that is less than the sum of the individual TAC's. To implement individual TAC's it is necessary not only to have real time catch data, but also to assume a level of inspection of catches (or reporting of catches by fishermen and processors) that is unrealistic. These problems multiply with the number of species and in tropical seas are virtually insurmountable. An additional problem arises when the harvest of some species is predominantly fish of less than one year of age, and TAC's are perforce on a very short time basis. In the words of one of the group, at best TAC's control landings, not catches; at worst they are meaningless. Moreover, TAC's must be negotiated every year, while a reduction in effort has a lasting impact.

If TAC systems are not generally practical, it seems reasonable to ask why they have been so widely used in attempts to regulate fisheries. The answer seems to be that they have been used as second-best solutions to the problem of limiting effort. Thus, in ICNAF, from which much of the present practice of setting TAC's was derived (spawned?), quotas were far more readily negotiated than effort. Though TAC's have served a useful conservation purpose, they have not solved the economic problem of too much fishing effort chasing too few fish.

The Atlantic fisheries of Canada may be developing as an exception to the general rule. TAC's for all major commercial species have been in effect for several years. In the offshore groundfish fisheries of Newfoundland, TAC's for nine species are now allocated, on a proportional basis, to individual fishing companies under a system of "enterprise allocation" (Canada Dept. of Fisheries and Oceans, 1983). Early experience indicates that this system could effect significant savings in fuel and other costs because the companies would no longer be forced into a competitive scramble for the limited total catches.

It should be kept in mind in attempting to control fishing effort that there is an inevitable trend to technological improvement and accumulation of experience. Thus it always pays to limit entry to less than the number that one would speculate as appropriate. The restriction of effort is somewhat more difficult when there is more than one kind of gear, for the fishermen using each gear type do not wish to be differentially restricted. The safest approach is to reduce each by the same proportion, rather than on the basis of a calculation of relative fishing power.

One of the major shortcomings of controlling effort rather than catch is the difficulty of preventing the harvest of depleted species. It may be possible by arranging a judicious mixture of gear types and area and season closures to provide some measure of protection, but some species will probably have to be virtually written off in the interests of reaching an optimum compromise of gear types and species caught.

The regulation of mesh size has the merit of simplicity and can compensate for excessive effort. The disadvantages are the common failure to consider recruitment overfishing and the difficulty of enforcement. Mesh size regulation is much easier to enforce if it is coupled with measures to limit entry. In some instances, seasonal and area closures may be as effective as mesh size regulation in preventing recruitment overfishing. The use of variable effort coupled with variable mesh size has not been adequately explored as a management technique for multispecies fisheries.

The manipulation of infrastructures is a powerful tool in the management of multispecies fisheries. The development of new markets can encourage the use of by-catches that might otherwise be discarded or used for fish meal production. The development of processing facilities can give added value to catches and expand markets for fish products. The species composition of the catch can be influenced by taxing landings of some species and subsidizing landings of others. In these and other ways the infrastructure can be used to influence the fishery. The converse is also true. When a fishery is in trouble it can create political pressure for subsidies of various kinds, and indeed, it has often been remarked that in some situations the fishermen fish for subsidies rather than for fish.

It is relevant to observe that fishermen may not be socially mobile, perhaps for reasons of temperament as well as for lack of alternative opportunities. Whatever the reasons, it is often not easy to reduce the number of participants in a fishery, and the creation of new alternatives by infrastructure manipulation can be a constructive adjunct or alternative to other management measures.

## REQUIREMENTS FOR INFORMATION

The various measures that may be taken by management imply the need for different kinds of information. In general, the amounts of information that seem to be required are functions of the number of fisheries and, to a lesser degree, the number of species. It should be noted that if these are exponential functions, or even linear functions, the requirements become so enormous as to suggest the need for looking at the whole complex at a more encompassing level of aggregation than species. It is widely agreed that one of the prime requirements for multispecies modelling is measurement of the strong interaction coefficients between species. Because the potential interactions increase as the square of the number of species, it is essential either to identify a few strong dominating species interactions, if indeed there are only a few, or to find some other means of depicting the dynamics of the association.

There is a wide range of views on the best course for management and with each view there is an associated need for information. The most pragmatic emphasize that when a fishery is evidently in a sad state, the amount of information that is needed may be very small. Where there is a will for a change, the prophet who simply says that fewer boats will mean more fish per boat may well be heard. More commonly there is a desire to preserve the status quo, and even very large amounts of information fitted to highly sophisticated models may fail to convince fishermen and their managers that changes should be made. If a fishery is not in desperate straits, it may well be that the small changes that are politically acceptable will eventually allow the fishery to creep to some optimum. But the optimum may be local, and to discover if there are other and better optima may require bold experimentation that takes risks to gain knowledge. Even if it is known that the fishery can be moved quickly to a desired optimum state (bang-bang), the disruption to the fishery, however short-lived, may be seen as intolerable.

Such considerations lead directly to the question "what kind of model of the fishery should the manager use to guide his lily-livered or bold decisions, and how much and what kinds of information are needed for what kinds of model?" A widely accepted rule of thumb has been that the more parameters a model contains, the more precise it should be and the more information is needed. But there are many exceptions to this general rule, the simpler model often describing the relevant processes better than the more detailed "correct model." Leaving those considerations to his expert staff, the manager must nevertheless first decide how much precision he can afford.

A second consideration is what the model is designed to predict. The interests of management are likely to focus on such matters as total catch, catch per unit effort, and biomass, and since some will be easier (and cheaper) to predict than others, it pays to be precise about what and how precisely you wish to predict. A particular need of multispecies fisheries managers is for models that will predict the likelihood of changes in species composition because the changes have large consequences.

It should be emphasized that a large number of models is available and that there is a wide variety of alternatives in modeling. The "thoroughly modern manager" should almost certainly a) consider several models for addressing each of the aspects of his multispecies fishery, b) continually evaluate their cost effectiveness, c) continually develop new models to replace the old, and d) both create and exploit opportunities to critically evaluate whatever models he is currently using. At the present time there is a very wide spectrum of opinion about the merits of various kinds of models for describing multispecies fisheries dynamics. This the group found encouraging but they could agree on little more, except perhaps that between the extremes of overly simple and horrendously complex there was a vague but promising middle range of models characterized by "appropriate" aggregation procedures.

## IMPLICATIONS FOR RESEARCH

In the course of our discussions we were frequently reminded that much of the conventional wisdom concerning multispecies fisheries is based on anecdotal evidence. There is a dearth of thorough monographic studies on multispecies fisheries that could serve as the foundation for developing generalizations. It may perhaps be too much to expect that each monograph would include a review of the biology, the management decisions, the industry, and the environment, but it is this kind of comprehensiveness that is sorely needed.

Research should be concentrated on the problem of predicting major changes in species composition, particularly when there are large changes in fishing mortality. It may be unrealistic to make predictions at the species level, except for communities with few dominating species. There is therefore a related research need on the criteria for aggregating the elements in a fish community. Is it best to group by size, by food habits, by taxonomic similarity? Do different groupings reveal different things about the species assemblages?

Regardless of the grouping criteria, the essential requirement is for understanding of the mechanism of interaction among groups. It may well be that the key interactions take place amongst the larval and early juvenile stages, and further research into the dynamics of these associations is needed.

To expose fully the dimensions of multispecies fisheries management, if possible it would be useful to couple models of the biological systems with models of the economic systems. Economic models of multispecies fisheries are particularly needed. Many of the existing models are based on equilibrium assumptions and have the potential for being very misleading when applied to nonequilibrium situations. Research on models that describe the behavior of fishermen would also be valuable as they illustrate the microeconomics that underlie the macroeconomics of the fisheries.

The selection of the "best" population dynamics model for management purposes may be influenced by statistical considerations. For example, Ludwig (unpublished) has used a cohort-structured model in a computer simulation to generate data which is then fed into parameter estimation algorithms for the original model as well as for a simple stock-recruitment model. Often, better estimates of optimal fishing mortality are obtained from the simple model than from the "correct" cohort model.

The group was not enthusiastic about "shotgun" oceanography which too often, at very great expense, accumulates large amounts of information of little immediate or short-term relevance to management. Such oceanographic studies as are undertaken for fisheries management purposes should be targeted to answer specific questions, such as the current movements from a spawning area to purported nursery areas.

The group agreed with the other groups in strongly recommending experimental procedures designed to give empirical information about the consequences of various management regimes and the seizing of opportunities that may be provided by "natural experiments."

Finally, the group stressed the importance of maintaining a modicum of basic research in the total program of research activities of management agencies. As one of the workshop moderators so aptly pointed out, there is a difference between data that are gathered to gain understanding and data that are gathered to enable prediction. It is to be hoped that the expenditures on basic research will be sufficient to ensure that the data necessary for gaining understanding will continue to accumulate for the mutual benefit of fisheries science everywhere in the world.

## SOME IMPLICATIONS FOR POLICYMAKERS

As a postscript to our deliberations we spent a few minutes reflecting on "what's gone wrong." There is enough knowledge at hand to provide much better management, yet multispecies fisheries are in difficulty in all parts of the world. Why? The answer lies largely in the "tragedy of the commons" and leads to the suggestion that where it is possible to give fishermen ownership rights, there is reason to hope for better management.

It is also apparent that overcapitalization gives the tragedy of the commons a special twist in fisheries, for the effects can persist for a long time. Any developing country that is ambitious to develop its fisheries should explore the option of renting vessels to harvest surpluses when they are available. Along the same line of thinking, it may be cheaper to contract for some of the chores of research and monitoring than to undertake doing it all with public servants. It should, however, be acknowledged that many programs of assistance to less developed countries are available to provide such services and that these programs have stressed the importance of building at least a core of national expertise as the best way of ensuring that external advisory services are used to optimal advantage.

Many multispecies fisheries are also multinational fisheries and therefrom stem many of the problems of management. International agreements are particularly valuable for exchange of scientific information and research findings. Wherever they might serve a useful purpose they should be developed, and indeed, many have. The decisions of management, however, are best not made by the body that is concerned with the exchange of information. The exchanges should not be jeopardized because the parties are unable to come to agreements on management. The disasters to fisheries that arise from failures of countries to agree on management must be documented, if only to serve as reminders of what to avoid in the future.

## REFERENCES

(1) Alagaraja, K.; Kurup, K.N.; Srinath, M.; and Balakrishnan, G. 1982. Analysis of marine fish landings in India: a new approach. Central Marine Fisheries Institute Special Publication No. 10, Cochin, India.

(2) FAO. 1978. Some scientific problems of multispecies fisheries. Report of the Expert Consultation on Management of Multispecies Fisheries. FAO Fish. Tech. Paper 181. Rome: FAO.

(3) Larkin, P.A. 1982. Direction of future research in tropical multispecies fisheries. In Theory and Management of Tropical Fisheries, eds. D. Pauly and G.I. Murphy, pp. 309-328. Manila: ICLARM.

(4) Marten, G.G., and Polovina, J.J. 1982. A comparative study of fish yields from various tropical ecosystems. In Theory and Management of Tropical Fisheries, eds. D. Pauly and G.I. Murphy, pp. 255-285. Manila: ICLARM.

(5) Pauly, D. 1982. History and status of the fisheries. In Small-scale Fisheries of San Miguel Bay, Philippines: Biology and Stock Assessment, eds. D. Pauly and A.N. Mines, pp. 95-124. Manila: ICLARM.

(6) Sissenwine, M.P. 1984. The uncertain environment of fishery scientists and managers. Mar. Res. Econ. 1(1): 1-30.

Exploitation of Marine Communities, ed. R.M. May, pp. 303-312. Dahlem Konferenzen 1984. Berlin, Heidelberg, New York, Tokyo: Springer-Verlag.

# Strategies for Multispecies Management: Objectives and Constraints

C.W. Clark
Dept. of Mathematics, University of British Columbia
Vancouver, BC V6T 1W5, Canada

**Abstract.** While current fishery management practice remains largely based on the single-species paradigm, most actual fisheries are multispecies fisheries in some sense. The objectives appropriate for multispecies management remain obscure. Inherent complexity and irreducible uncertainty of marine ecosystems have hindered progress towards a "scientific" basis for multispecies fisheries management. The use of a decision-theoretic apparatus, currently under consideration for the single-species case, has not yet been extended to the multispecies situation.

## TERMINOLOGY

In its usual meaning the phrase multispecies fishery denotes a marine ecosystem in which two or more interrelated species are subject to exploitation. There are two cases: a) Mixed catches. Several species are caught by the same gear, as in trawl, seine, or long-line fishing. A particular species may be the target (yellowfin tuna), while other species are by-catches (porpoises), or several species may be desired equally. Less desirable species (or sizes) may be discarded. b) Separated catches. The different species are captured in fisheries separated by time, space, or vessel/gear combination (whales and krill, cod and capelin).

A second meaning of multispecies fishery applies to vessels that switch species according to opportunities. In general, such flexibility is economically desirable, but the tendency is often towards overgeneralization (10).

In a multistock fishery, a single, exploited species consists of several genetically distinct stocks, which should ideally be managed separately, although this may not be feasible. For example, the British Columbia salmon fishery exploits five species consisting of more than 3000 genetically separate stocks, many of which may be intermingled in any catch (2). This fishery is thus both a multispecies and a multistock fishery.

## OBJECTIVES

The traditional fishery management objective, Maximum Sustainable Yield (MSY), in spite of recent disparagements (6), at least had the advantage of separating "clean" biology from "dirty" economics and politics. The biologist was not required to make value judgements. MSY is still the basic management objective for most single-species fisheries. It is usually acceptable on economic grounds, but when treated as the sole or primary objective it invariably leads to overcapacity of the fishing fleets.

In the multispecies context MSY may be ambiguous, self-contradictory, or economically absurd. Consider two ecologically interdependent species, $X_1$, $X_2$. The objective of MSY for each of $X_1$, $X_2$ separately is obviously logically inconsistent. If one defines yield as a weighted sum

$$Y = \Sigma p_i H_i, \tag{1}$$

where $H_i$ = yield of $X_i$ and $p_i$ is the "weight," then a value judgement pertaining to the weights (prices?) $p_i$ is unavoidable. In exceptional cases it may be reasonable to consider yield Y in terms of total biomass, but usually this will be economically unacceptable. For example, if $X_1$ is a predator on $X_2$, and $X_1$ is marketed for direct human consumption while $X_2$ is reduced to fish meal for animal feed, a direct biomass yield objective should at least take account of the conversion efficiency for $X_2$.

Many prey species (anchovies, herrings, krill) supply two very disparate markets, one for high-priced delicacies, the other for fish-meal reduction. Prices for raw fish may differ by two orders of magnitude, but usually the high-priced market is highly inelastic. If Eq. 1 is taken as the definition of economic yield, the price $p_i$ will then be a function of $H_i$ and its allocation to the different uses. In economic terms, supplying the high-priced market would obviously receive higher priority than supplying the (larger) low-priced market.

Another complication: Eq. 1 does not consider the distribution of catches $H_i$ over time; the discounting effect applied to future harvests can have serious implications for multispecies fisheries (1). Furthermore, in the case of mixed catches, there may be technological constraints relating the $H_i$, such as, for example,

$$H_i = q_i E X_i, \quad (i = 1,\dots,n), \tag{2}$$

where the effort E is the same for all species. All this is just to indicate the various complexities that may be involved in merely quantifying "yield" in the multispecies case.

Many objectives other than simply maximizing yield (or revenue, or utility, ...) can be conceived. These may be based more on a <u>protection</u> motif than on an optimal-use motif. Some species (marine mammals) may be considered sacrosanct by certain interest groups. The idea of maintaining the pristine ecosystem structure may appeal to others.

The objective of maintaining genetic diversity, and in particular of preventing extinction of any species, is widely accepted. Such objectives also have economic significance, although these are not automatically included in the simple yield formulation of Eq. 1. Again, the discounting effect may greatly reduce the incentive for genetic conservation, since this is predominantly a long-term objective.

Various operational difficulties arise in applying the MSY concept to actual fish stocks which normally experience significant random fluctuations. In the case of salmon stocks, the objective of management is typically the maintenance of a fixed "optimal" spawning escapement S, determined so as to maximize <u>average</u>, or expected, yield. Thus the annual catch H is given by

$$H = \begin{cases} X - S & \text{if } X > S \\ 0 & \text{otherwise} \end{cases}, \tag{3}$$

where X denotes recruitment. Given that X fluctuates, the yield H in Eq. 3 is far from sustained! (For some salmon stocks X may be negligible in off years.)

The use of a constant-escapement policy requires accurate monitoring of the fish stock, which is seldom feasible (Clark and Kirkwood, in preparation). Also, if successful, the constant-escapement policy serves to minimize the amount of information contained in the fishery data, and this may cause the fishery to become "trapped" at a suboptimal escapement level (7).

For groundfish stocks, the currently popular management objective is to maintain a constant level of fishing mortality F (e.g., "$F_{0.1}$"). Under such a policy, annual catches are proportional to the current stock level, so that catches track the fluctuation in the stock level. The resulting catch fluctuations are less pronounced than for the case of constant escapement (9). As before, constant-F policies depend upon fairly accurate and timely stock assessments, a difficult and expensive task.

Constant-effort policies may be considered as proxies for constant-F policies. Ideally, fishing effort E and fishing mortality F would be in direct proportion at all times, but in practice this simple relationship fails to hold, as a result of various spatial and temporal nonhomogeneities of fish stocks and fishing fleets. But since they do not require expensive on-line stock assessments, constant-effort policies can be much more cost-efficient than constant-F policies.

As far as I know, no one has made any attempt to analyze alternative harvest policies for stochastic multispecies fishery models, and I state this as an unsolved problem. It seems possible that a multispecies system could be managed so as to reduce the variance in total yield (1), relative to the variance of individual yields $H_i$.

Another problem that should be considered pertains to spatial distribution and heterogeneity of marine ecosystems. Could this be taken advantage of in order to maintain resilience, in face of the vast uncertainties involved? Such considerations would imply a departure from the "more is better" philosophy that really underlies MSY and the majority of more sophisticated economic optimization objectives. Emphasis would instead focus on heterogeneity, complexity, uncertainty, flexibility, adaptability, and so on.

Perhaps the central unsolved problem for multispecies fishery management is the problem of finding a saleable, operational catch-phrase to replace MSY, or if that proves impossible, an understandable series of basic management principles. For the record, May et al. (8) listed the following:

1 - For populations at the top of the trophic ladder, stocks should be kept at or above the level at which they provide the greatest net

annual increment.

2 - For other populations, ...stocks should not be depleted to a level such that the population's productivity, or that of other populations dependent on it, be significantly reduced.

3 - ...The slowest time scale should be used in monitoring a harvesting regime.

4 - ...Harvesting levels should be set conservatively, allowing for safety factors to guard against accidental overexploitation.

Applying these vague precepts to specific multispecies management regimes is a challenging task for the future.

## CONSTRAINTS

The preceding discussion indicates that, even if we had a complete understanding of the pertinent ecosystem, we might still be hard pressed to formulate appropriate management objectives. While this is itself a severe constraint to management strategy, an even more serious constraint lies in the plain fact that our knowledge and understanding of marine ecosystems is extremely fragmentary. This is a situation that is not going to change rapidly. Indeed, there are good scientific reasons to suppose that quantitatively predictive models of nonlinear systems of the complexity of marine ecosystems may be impossible in principle. If so, we are faced with a very basic unsolved philosophical problem - if long-term predictions are impossible in principle, what management strategies are implied by that fact? (The whole subject of the practical implications of gross uncertainty, irreducible or otherwise, to resource management is one that has hardly been seriously addressed as yet. At present, uncertainty usually results simply in an impasse - think of the acid rain controversy. The scientific community uniformly recommends caution, while the business world prefers to ignore unpleasant future possibilities until they have been "scientifically demonstrated." In the same way, the fishing industry demands "proof" that stocks are overfished before any reduction in catch quotas can be considered. Such proof is, of course, seldom possible in any scientific sense. What is possible, namely, a decision analysis of the basic uncertainties involved, is seldom performed, probably for reasons tied to the scientific education of most fisheries scientists.)

For multispecies fisheries, the scientific dilemma is nicely surveyed by Stainsbury (12) who first asserts that

> Confident management of a fishery ... requires that reliable predictions

> can be made of the consequences of alternative exploitation strategies, from which the strategy deemed 'best' can be selected. Amongst other things, this progression implies a sound knowledge of (1) how the community in question operates, and (2) how this operation is influenced by exploitation.

Stainsbury reviews the two main approaches to ecosystem modeling, the "cybernetic" and the "ecological niche" approaches. He is led to the conclusion that

> Examination of ecological and fishery theory indicates that at present there is no ecologically adequate model of community dynamics and no adequate method of estimating the parameters of the models that are available.

Stainsbury does not despair of this situation; appropriate study and experimentation will eventually lead to the necessary models and parameter values, so that "confident management" may proceed. This is doubtlessly the received wisdom in ecology and fishery science, but it is probably insupportable, both on fundamental scientific principles, and also on simple cost-benefit grounds. The basic unsolved problem is: are the paradigms of traditional science valid (or appropriate) for exploited marine ecosystems, and if not, what alternative paradigms are required? The possibility exists that models which are vastly simpler than the systems they are supposed to represent may be the most useful models (4). The current single-species models constitute the ultimate simplication in this sense, of course.

The difficulties resulting from the complexity and limited observability of marine ecosystems are greatly increased because of the powerful influence that the ocean environment exerts on population dynamics. Whereas detailed monitoring of oceanic variables (temperature, wave motion, currents, etc.) might ultimately lead to improved short-term forecasts of fish abundance, it is clear that long-term predictions will depend upon the accuracy of long-term predictions of environmental conditions. What if such predictions are themselves impossible in principle - or even only in practice? If these problems prove to be insurmountable, opportunistic management strategies may prove to be superior to any rigorous set of management principles.

Besides these fundamental scientific constraints, many severe <u>institutional</u> constraints exist. While the latter are by no means peculiar to the multispecies setting, it is clear that the lack of an accepted theoretical basis, and of any accepted management objective, can only exacerbate these institutional difficulties.

In extreme cases, no effective control of the fishery whatever may be possible. Piracy, for example, is still rampant in certain areas (5). The chances of enforcing fishery regulations under such circumstances are obviously minimal.

Although it is often conveniently overlooked, no fishery management program can hope to succeed without rigorous monitoring and enforcement of the regulations. It is virtually always to the benefit of the individual fisherman to thwart any imposed regulations - this applies equally to catch quotas, closed seasons, fishing boundaries, gear restrictions, and so on. It is relatively easy to think up management measures that would help conserve marine resources and improve long-run yields, but such mental exercises are of little use unless the proposed measures can be successfully implemented and enforced. In particular, highly complex systems of regulations (which may seem necessary for controlling a complex ecosystem) are unlikely to succeed.

In the USA, the Northeast Fisheries Council had adopted by 1981 a set of management regulations for the multispecies ground fishery off New England, consisting of over 600 separate regulations. When this system proved unworkable, the Council reverted to a single mesh size regulation. This may be a bit of an overreaction, but the message is clear.

The extent to which enforcement is possible depends on jurisdictional limits, and also on the geographical extent of the fishery, port facilities, etc. For international fisheries, regulations must be unanimously accepted, and then enforcement will rely upon the goodwill of the participating states. National fisheries within Exclusive Economic Zones are in principle much more subject to control and regulation, but there are still strict limits to what can be achieved in practice - and at a practical cost.

Controllability of catch quotas may be particularly weak in trawl and similar fisheries, where the fisherman has little control over the mix of species taken. It can be predicted on common-sense grounds (the prediction is supported by more rigorous analysis in (3)) that species with relatively low reproductive rates will progressively be eliminated in such mixed-catch fisheries. In certain situations (e.g., tropical fisheries), these species may be replaced with smaller fishes, usually of lower commercial value. In other cases (e.g., Pacific salmon), the less productive stocks may simply be lost, and overall production thereby reduced.

Ideally these losses could be reduced, if not eliminated, by adopting more selective harvesting methods. But this may not always be economically feasible. In the case of Pacific salmon, stocks could be separated if harvesting took place closer to the entrance to spawning streams, but the quality of fish may deteriorate as they approach spawning areas.

In some fisheries a complete change in fishing methods or gear may be required in order to control the catches of individual species. For example, commercial trawl fishermen in Lake Michigan were deemed responsible for the decline of lake trout, which were popular with sports fishermen. The problem was solved by requiring that trawls be replaced with fish traps, from which fishermen were required to release lake trout unharmed (13).

The example of porpoises killed incidentally in the Eastern Pacific yellow-fin tuna fishery is probably well enough known not to require elaboration here. U.S. fishermen were faced with numerous regulations designed to reduce the kill of porpoises - but hardly any fishing occurs within U.S. waters.

It is clear from these examples that technological constraints on controllability may be important, but that a variety of methods may be available to improve the selectivity of fishing. The feasibility and enforceability of such methods may be less obvious.

## RESEARCH PROBLEMS

Some general statement of management objectives seems necessary in fisheries, both as a guideline to actual policy, and also for the purposes of legislation and international treaty formulation. Currently formulated objectives (including "optimal yield") are vague and lack clear operational implications. Given the inherent complexity, variability, and uncertainty of multispecies fisheries, no single objective is likely to be appropriate in general. Research is needed to elucidate the implications of possible management objectives for various types of multispecies fisheries. If possible, a decision-theoretic approach should be developed for dealing with the gross and irreducible uncertainties characteristic of marine ecosystems.

Alternative harvest strategies for stochastic multispecies systems should be investigated by modellers. The possible advantages of opportunistic strategies, or strategies that vary over time and space, may be worth investigating. How important are reserves, or protected areas, for preserving vulnerable species and preventing overfishing by miscalculation?

Finally, the costs, benefits, and institutional constraints associated with alternative management strategies need to be considered.

## REFERENCES

(1) Clark, C.W. 1976. Mathematical Bioeconomics: The Optimal Management of Renewable Resources. New York: Wiley-Interscience.

(2) Healey, M.C. 1972. Multispecies, multistock aspects of Pacific salmon management. In Multispecies Approaches to Fisheries Management Advice, ed. M.C. Mercer. Can. Spec. Publ. Fish. Aquat. Sci. 59: 119-126.

(3) Jones, R. 1982. Ecosystems, food chains, and fish yields. In Theory and Management of Tropical Fisheries, eds. D. Pauly and G.I. Murphy, pp. 195-240. Manila: International Center for Living Aquatic Research Management (ICLARM).

(4) Kirkwood, G.P. 1982. Simple models of multispecies fisheries. In Theory and Management of Tropical Fisheries, eds. D. Pauly and G.I. Murphy, pp. 83-98. Manila: International Center for Living Aquatic Research Management (ICLARM).

(5) Khoo Khay Huat. 1980. Implementation of regulations for domestic fishermen. In Law of the Sea: Problems of Conflict and Management of Fisheries in Southeast Asia, ed. F.T. Christy, Jr., vol. 2, pp. 49-59. Manila: ICLARM.

(6) Larkin, P.A. 1977. An epitaph for the concept of maximum sustained yield. Trans. Am. Fish. Soc. 106: 1-11.

(7) Ludwig, D.A., and Walters, C.J. 1982. Optimal harvesting with imprecise parameter estimates. Ecol. Mod. 14: 273-292.

(8) May, R.M.; Beddington, J.R.; Clark, C.W.; Holt, S.J.; and Laws, R.M. 1979. Management of multispecies fisheries. Science 205: 267-277.

(9) May, R.M.; Beddington, J.R.; Horwood, J.W.; and Shepherd, J.A. 1978. Exploiting natural populations in an uncertain world. Math. Biosci. 42: 219-252.

(10) McKelvey, R.M. 1983. The fishery in a fluctuating environment; coexistence of specialist and generalized vessels. J. Env. Econ. Manag., in press.

(11) Reed, W.J. 1979. Optimal escapement levels in stochastic and deterministic harvesting models. J. Env. Econ. Manag. 6: 350-363.

(12) Stainsbury, K.J. 1982. The ecological basis of tropical fisheries management. In Theory and Management of Tropical Fisheries, eds. D. Pauly and G.I. Murphy, pp. 167-194. Manila: ICLARM.

(13) Talhelm, D.R. 1978. Limited entry in Michigan fisheries. In Limited Entry As a Fishery Management Tool, eds. R.B. Rettig and J.J.C. Ginter, pp. 300-316. Seattle: University of Washington Press.

Exploitation of Marine Communities, ed. R.M. May, pp. 313-333. Dahlem Konferenzen 1984. Berlin, Heidelberg, New York, Tokyo: Springer-Verlag.

# Management Techniques for Multispecies Fisheries

G.G. Newman
Fisheries and Wildlife Service
Dept. of Conservation, Forests and Lands
Melbourne, Victoria 3002, Australia

**Abstract.** Multispecies fisheries management decisions require resource information which is complex. In addition, the wider spectrum of users of such resources can further complicate the task of managers who have to take into account biological, economic, social, and political factors when selecting objectives for management. However, the management techniques applied to multispecies fisheries are an extension of those used in single-species fisheries. In this paper, three fisheries are examined to illustrate the range of issues associated with the management of multispecies fisheries. The provision of a full suite of information required for management is likely to be beyond the scope of the organizations with responsibility for the fisheries concerned. Selection of key information is therefore critical, and a format which could aid managers achieve this selection is discussed.

## INTRODUCTION

The techniques available for managing multispecies fisheries are essentially similar to those which have been applied to single-species fisheries over a number of years. These techniques aim basically at manipulating fishing mortality to meet biological, economic, or social objectives. An extension of the techniques to multispecies resources is complex and makes demands and places constraints on both resource managers and researchers. The objective of this paper is to examine the special problems associated with management of multispecies fisheries.

It would not seem appropriate in a review of this kind simply to list the techniques available to resource managers and discuss the advantages and disadvantages of each as they apply to multispecies resources. An attempt is therefore made to establish a format within which selected examples of management of multispecies resources can be examined so that factors which influence the selection of management techniques can be highlighted and an indication given of the scientific requirements for management decisions. The examples have also been selected to illustrate the limitations of the currently available techniques.

The first element within the format relates to the nature or category of the multispecies resource, and in this paper the term multispecies is interpreted in its broadest sense. Multispecies resources are considered to be those in which a group of species either are caught by a specific fleet which may or may not be diversified or are harvested by different fleets but interact biologically and therefore have to be managed as an entity. The multispecies nature of the fishery may thus be determined by either the nature of the resource, the nature of the fishing fleet, or both of these factors.

The second element in the format comprises the range of techniques available to managers. The application of specific techniques or mixes of techniques must be appropriate to management objectives which are in turn influenced by biological, economic, and social factors. Thus management objectives may be regarded as a third element. In the case of multispecies fisheries, agreement on objectives can be complex.

## CATEGORIES OF MULTISPECIES RESOURCES

The term "multispecies fishery resource" is most usually applied to an association of species which is harvested by a uniform fleet of fishing vessels. In such fisheries it is usually difficult to determine target species, and the composition of the catch reflects the abundance or availability of the component species at the particular time, rather than preferences on the part of the fishermen or processors. Examples of this type of fishery are to be found in some upwelling regions which support pelagic fisheries based on a number of species. However, from the resource manager's viewpoint, this type of fishery reflects one extreme of a spectrum of fisheries which depend to a greater or lesser extent on more than one species.

There are fisheries in which diversified fleets of vessels are supported by groups of species which are not necessarily closely associated in terms of their biology or habitat. In this instance, the management problem relates more to the nature of the vessel and its ability to deploy a range of gears than to associations between the components of the total resource exploited by the vessel. Nevertheless, the fishery has to be managed in an integrated fashion to ensure the welfare of the resources as well as of the industry. Diversification allows fishermen to take advantage of some resources which may fluctuate widely, while maintaining alternatives during periods in which those components of the fishery are less abundant. An example of this type of fishery occurs in southeast Australia where multipurpose vessels harvest scallop, rock lobster, and shark resources.

Finally, there are fisheries in which the component species may interact strongly but are nevertheless harvested by separate vessels or fleets which capture target species. It is important that the management of these fisheries be integrated in view of the interdependence of the resources. A topical example of this type of fishery would be the interaction of harvesting of krill and its impact on the recovery of whale resources.

## MANAGEMENT TECHNIQUES

The two major techniques available for managing fisheries resources are catch quotas, which may be global or allocated to individual operators, or limits placed upon the total fishing capacity of a fleet utilizing a particular resource. The latter is achieved by limiting entry of vessels to the fishery and controlling vessel replacement. Either one or both of these techniques can be applied to a particular fishery, and its appropriateness would depend on the nature of the resource, the level of understanding of its dynamics, and the scope for effectively negotiating the introduction of the measure with participants in the fishery.

For instance, it is more feasible to negotiate annually on quota levels if fishing operations are centrally planned and catches can be monitored. Changing quota levels to reflect the state of the stocks is facilitated if vessels have some mobility and can be deployed elsewhere to maintain an economic operation. Negotiations about quota levels require good information on the status of the stock, and the models used by scientists must be comprehensible to both managers and fishermen so that the need for any sacrifices can be clearly understood. Acceptance of catch quotas is also more easily achieved in homogeneous fisheries where the participants take a proprietary interest in the resource and are in a position to weigh the longer-term gains against the immediate sacrifices which good management may require.

Effort restrictions, in the form of limited entry and constraints on vessel upgrading and deployment, are not as demanding as quotas in terms of the continuous information requirements and the need for ongoing and possibly sophisticated negotiation. These restrictions are therefore often applied to smaller fisheries in which the fleets do not operate in a corporate fashion. Such measures are also more appropriate when scientific information on the state of the resources may be sufficient to indicate the order of magnitude of permissible fishing effort, but the acquisition of information required to refine such decisions is either too expensive or not technically feasible. In such fisheries there may be little opportunity to redeploy vessels or labor, and therefore fishing effort is approximately fixed and fluctuations in stock size are reflected by differences in catch rate and the returns to the fishermen during the particular season.

The application of both of these measures is therefore subject to a number of constraints which relate to the level of understanding of the stocks, the shorter- and longer-term biological and economic goals, as well as a number of social factors such as alternative employment, social values, and traditions, all of which also have political ramifications.

Ownership of the resource, and therefore responsibility for its management, can have a bearing on the complexity of negotiations and the degree to which managers can take decisive and effective action. The recent extension of the fishing zones of many coastal countries to encompass some important resources has effected a change in the manner in which these resources are managed. By assuming sole ownership, a number of governments have been able to break deadlocks and improve compromise solutions which were the best which could be achieved within international fisheries commissions comprised of countries with differing stages of fisheries development and social objectives for their fleets. Difficulties in achieving consensus in these forums have hampered managers' attempts to improve resources and in some cases have seriously eroded safeguards which would otherwise have been applied.

An example of this problem is afforded by the attempts which have been made by the International Commission for South East Atlantic Fisheries (ICSEAF) to manage the important hake resource in Divisions 1.3, 1.4, and 1.5 of the Convention area. The Commission has adopted total allowable catches since 1977 (2), but until recently these have far exceeded the recommendations of the Commission's Scientific Advisory Council (Table 1). As a result there has been no significant recovery of the resource in spite of what appears to be a concerted effort to study and

TABLE 1 - Recommended and adopted Total Allowable Catches (TAC's) for ICSEAF Division 1.3, 1.4, and 1.5 in tons (see (2)).

| YEAR | RECOMMENDED TAC | ADOPTED TAC |
|---|---|---|
| 1977 | 480,000 | 536,000 |
| 1978 | 405,000 | 480,300 |
| 1979 | 360,000 | 415,900 |
| 1980 | 300,000 | 320,000 |
| 1981 | | 215,641 |
| 1982 | 352,000 | 352,000 |

manage the stock by countries actively supporting the Commission. The lag time involved in implementing the recommendations of the Commission's scientists was due to difficulties experienced in persuading a large number of participating countries, with different scales of operation and catch requirements, to agree to a reduction in catch as well as to how the sacrifices should be distributed between nations.

Attempts to limit entry to fisheries are usually made at an advanced stage in development of the fishery when effort is already excessive. The problem of constraining effort is further exacerbated by the fact that, in the limited entry situation, transferability of licences must be accommodated, and this is most easily administered in national fisheries by allowing the sale of licences. Under these circumstances, the introduction of limited entry places a monetary value on the entitlement to fish, and numbers of fishing units which in the open entry system may have operated at a low level of effort now have the same potential as more active units, and they are likely to realize this potential because of the opportunity cost of the entitlement. In this manner, the introduction of limited entry can be followed by a subsequent increase in the actual effort deployed.

A number of other measures are available to managers to achieve limits on catch and/or effort, but in a less direct fashion. Closed seasons and areas provide managers with the facility to limit fishing mortality to some extent if the more direct controls on catches or the total number of fishing units are not feasible. Closing seasons or areas are thus another means of achieving limits on catch or effort and can steer fishing pressure away from resources at times or in areas in which exploitation is not advantageous.

Limitations on the age at first capture can be achieved by gear selection or minimum size limits applied to individual fish which are landed. These measures have been widely used to ensure that yields per recruit from particular fisheries are maintained at an adequate level. Other controls can take the form of restraints placed on gear efficiency and limits on the daily landings of particular vessels. These constraints usually lead to inefficiencies in the operation of the fishery, and they often reflect an inability on the part of managers to resolve the more fundamental problems of resource management.

An understanding of resource dynamics as well as the economic and social consequences of management plays an important role in the selection of an appropriate management technique for a particular resource. However, such measures must also take into account the ability of the management agencies to implement and enforce any measures which may be selected. It is not necessary to discuss in detail the range of enforcement techniques available to managers but simply to record that the cost, feasibility, and effectiveness of enforcement often constrain the range of management packages which can be considered by the resource manager.

## MANAGEMENT OBJECTIVES

The objectives adopted by resource managers are influenced by biological, economic, social, and political issues. Clearly, the more diverse the participants in a fishery, the more complex will be the resolution of conflicting management objectives.

In terms of the biology of the resource, alternative objectives could be simply to maintain or improve yields or catch rates or to manage resources to accommodate recreational or aesthetic needs, in which case the yield could be substantially less than the maximum or economic optimum. Examples of the latter two requirements would be the clash of interests between recreational and commercial fishermen leading to attempts to restrain Japanese tuna long-line fishermen operating off the east coast of Australia from taking live black marlin. Decisions by managers not to harvest certain marine mammals may not only be based on the state of the resource, but may have to take into account community values which do not accept such harvests.

Conflicting economic and social requirements also confront resource managers. Economic efficiencies sometimes have to be sacrificed to maintain more traditional life-styles in communities which may have little mobility or alternative forms of employment. On the broader, international scale, different levels of development with regard to fishing technology, market demands, affluence, and administrative infrastructure can complicate attempts to manage resources which are exploited by a wide range of participants. Issues can be further confounded if managers and participants within a fishery adopt different views on the time horizons for resource management programs. Operators within a fishery may only wish to maintain high catches over a short period to take advantage of the life span of their capital equipment, the replacement of which may not be justified. In other instances, operators within fisheries may wish to generate high cash flows over a limited period to finance diversification, possibly to the extent of allowing them to leave the fishery. These short-term objectives may be incompatible with those of management agencies whose responsibility it is to manage common property resources for the benefit of the community over the longer term.

Political factors have to be taken into account in the design of resource management packages. The manner in which benefits from fishery resources flow to the community can be perceived in different ways by agencies with different political persuasions. For instance, in centrally planned and operated systems, these benefits can accrue to the community by means of controlled pricing structures. Alternatively, managing agencies may extract benefits for the community by means of direct resource taxes, in the form of high licence fees or levies. Part of this revenue can be utilized to improve or maintain the fishery by means of research, management, and the provision of infrastructure, with surplus monies being placed in general revenue. In more free enterprise systems, agencies can elect to charge modest licence fees and levies simply as a fee for service, while allowing the benefits to the community to accrue via normal taxation procedures.

Finally, resource managers have to deal with the realities of political influence brought to bear by sectors in the community which include commercial fishermen, recreational fishermen, as well as conservation groups. The objectives of the latter two groups can differ significantly from the hitherto straightforward issues which have confronted fisheries resource managers in the past. The progressively more vocal demands of recreational fishermen and conservationists have added a complicating dimension to a number of management situations.

## SELECTED EXAMPLES OF MULTISPECIES MANAGEMENT PROBLEMS

In this section three examples are presented and discussed within the framework established earlier in the paper. The examples have been selected to illustrate the three categories of multispecies resources which have been identified. They do not purport to illustrate neat solutions, but rather they reflect the range of problems associated with each category.

### South African Pelagic Fishery

The South African pelagic fishery conducted in ICSEAF Divisions 1.6 and 2.2 has been subject to management constraints since the early 1950s, which have included total catch quotas, limitations on fleet hold capacity, and the establishment of fishing seasons. The fishery provides an example of the first multispecies resource category identified earlier in this paper in that a homogeneous fleet exploits a multispecies resource with little ability to select target species. Although not explicitly stated in published documents, the management objective of the fishery is biased towards maintaining biomass rather than accommodating complex economic or social goals.

The management strategy essentially treats the multispecies stock as one resource, by application of total catch quotas and certain vessel hold limits on a long-term basis. The application of these techniques has resulted in fairly stable total catches, more especially during the past decade, but with considerable fluctuations in species composition (Table 2). It should be noted that the catch has been especially stable over the past five years, during which time the quota has remained unchanged.

The essential problem facing the managers of this fishery relates to uncertainties with regard to the interdependence of component species. Even after considerable study there is doubt as to whether the changing sequence of dominant species has represented replacement of heavily fished species or, alternatively, whether better management could increase the total landings by improving the productivity of each of the components of the multispecies resource (4).

The quota was originally based on historical catches with some adjustment to take into account the fluctuations in catch of certain species in the fishery. In addition, the results of some modeling exercises have been taken into account, but in spite of considerable scientific effort the usefulness of the models is questionable.

TABLE 2 - Species composition of landings ('000 t) in the purse seine fishery in ICSEAF Divisions 1.6 and 2.2, 1960 - 1982.

| YEAR | PILCHARD | ANCHOVY | HORSE-MACKEREL | CHUB MACKEREL | ROUND HERRING | LANTERN FISH | TOTAL |
|---|---|---|---|---|---|---|---|
| 1960 | 318,0 | | 62,9 | 31,0 | 0,1 | | 412,0 |
| 1961 | 402,0 | | 38,9 | 49,7 | 0,1 | | 490,9 |
| 1962 | 410,2 | | 66,7 | 20,4 | 0,1 | | 497,3 |
| 1963 | 390,1 | 0,3 | 23,2 | 13,2 | 0,2 | | 427,0 |
| 1964 | 256,1 | 92,4 | 24,4 | 50,0 | 2,7 | | 425,6 |
| 1965 | 204,5 | 171,0 | 55,0 | 41,4 | 8,2 | | 480,1 |
| 1966 | 118,0 | 143,9 | 26,3 | 53,4 | 15,4 | | 357,1 |
| 1967 | 67,7 | 270,6 | 8,8 | 128,2 | 32,0 | | 509,3 |
| 1968 | 107,8 | 138,1 | 1,4 | 91,0 | 30,3 | 0,1 | 368,6 |
| 1969 | 56,1 | 149,2 | 26,8 | 91,7 | 23,3 | 4,9 | 352,0 |
| 1970 | 61,8 | 169,3 | 7,9 | 77,9 | 23,7 | 18,2 | 358,9 |
| 1971 | 87,6 | 157,3 | 2,2 | 54,2 | 21,6 | 2,0 | 324,9 |
| 1972 | 104,2 | 235,6 | 1,3 | 56,7 | 20,6 | 15,2 | 433,6 |
| 1973 | 69,0 | 250,9 | 1,6 | 58,8 | 28,7 | 42,4 | 451,4 |
| 1974 | 16,0 | 349,8 | 2,5 | 30,7 | 1,3 | 0,3 | 400,5 |
| 1975 | 89,2 | 223,6 | 1,6 | 69,3 | 23,6 | 0,1 | 407,4 |
| 1976 | 176,4 | 218,3 | 0,4 | 0,5 | 11,7 | 0,1 | 407,5 |
| 1977 | 57,8 | 235,5 | 1,9 | 21,3 | 35,0 | 5,6 | 357,1 |
| 1978 | 97,0 | 209,5 | 3,6 | 2,4 | 67,0 | 1,0 | 380,5 |
| 1979 | 52,9 | 291,4 | 4,3 | 2,7 | 21,0 | 8,7 | 381,0 |
| 1980 | 50,4 | 315,5 | 0,4 | 0,2 | 14,1 | 0,1 | 380,5 |
| 1981 | 46,2 | 292,0 | 6,1 | 0,3 | 24,3 | 10,3 | 379,2 |
| 1982 | 33,5 | 306,9 | 1,1 | 2,7 | 31,0 | 0,7 | 375,9 |

Production curves relating total multispecies landings to fishing effort have suggested that a yield of about 360,000 tonnes of all species could be sustained (5). This estimate assumes that production is independent of species composition, an assumption which is barely tenable. A more explicit model of the fishery was developed by Crawford (1). Crawford's model was deterministic and simulated the multispecies fishery by taking into account the dynamics of component species and the characteristics of the fishery operation. However, the model did not take into account parent stock-recruitment relationships, nor was there any facility to account for interaction between species. These two factors may not be significant, but uncertainty about these issues limits the model's usefulness to managers.

Under the circumstances managers have sought, until recently, to maintain the basic quota level which has resulted in a catch of 380,000 tonnes for the past five years, in spite of fairly dramatic changes in species composition. The total catch has, in fact, with one exception, fluctuated between 410,000 and 380,000 tonnes over the past decade. Fleet hold capacity limitations have been applied continuously and have probably prevented severe escalation in the fleet size, but innovations affecting fishing power have resulted in substantial increases in potential effort. For instance, over the period 1966-1972 the fleet hold capacity increased by 15%, but the commensurate increase in fishing power due to innovations was on the order of 80% (6).

There has also been some adjustment of fishing seasons and areas to avoid the capture of small fish, but fundamentally the quota remains the major management measure available to those responsible for administering this fishery.

To improve Crawford's model of the resource would require considerable additional finance and scientific effort. It is also to be expected that the uncertainty associated with models of this fishery would be considerable. An alternative option to further modeling would therefore be for managers to deploy funds and facilities to make more direct estimates of biomass, by means of acoustic surveys as well as egg and larval programs. Although the variance of these estimates could be extremely wide and the data only useful for short-term management decisions, such information is easily comprehended by both managers and the fishing industry, and it can be used persuasively during the course of negotiations.

Alternatively, in the face of the difficulties and expense entailed in collecting better information, managers could simply continue to implement a fixed quota, reducing it to provide a safety factor for the stock, and phase down further attempts to improve the information base upon which to manage the fishery. Essentially this would amount to paying a premium, in the form of potential catches foregone, to decrease the risk to the stock while effecting cost savings associated with expensive research which is perceived to have a low probability of success and an outcome associated with high variance. Such an approach would have severe shortcomings in relation to large fluctuations in environment which could influence the resource but not be taken into account by this static form of management.

The South African pelagic fishery therefore provides an example of a fishery which is comprised of species with complex population dynamics and interactions between species which are poorly understood. These characteristics present major problems to both scientists and managers. In other respects, the fishery is homogeneous and national, so there are no serious conflicts about management goals. The economic and social aspects of the fishery are fairly directly related to the biological condition of the stock, which determines yields, so the task of management is not complicated by these factors.

**Southeast Australian Scallop Fishery**

Recent developments in scallop fishing in southeast Australia provide an illustration of the problems which are associated with the management of a fishery belonging to the second category identified earlier in the paper, that is, a fishery which is at least in part conducted by a fleet of multipurpose vessels harvesting a range of species which have little biological interaction.

The fishery concerned is located in the Bass Strait and is exploited by vessels which are based in two states adjacent to the Strait, that is, Tasmania and Victoria. Until recently, the Victorian fleet was responsible for a major proportion of the total landings (Table 3), and at least this segment of the fishery could be regarded as a single-species fishery in that the majority of the Victorian fleet comprises specialist scallop vessels. Although fishing in Bass Strait takes place predominantly in waters controlled by the Commonwealth government, a policy of limited entry has been maintained by the Victorian government by means of a state-based licensing system. State licences are freely transferable at a market value which is largely dependent on the state of the resource at the time of transfer. The Victorian government's policy of limited entry has ensured

TABLE 3 - Bass Strait scallop flesh weights in tons 1970 - 1981.

| YEAR | FLESH WEIGHTS VICTORIA | TASMANIA | TOTAL |
|---|---|---|---|
| 1970/71 | 641 | - | 641 |
| 1971/72 | 1,016 | 7 | 1,023 |
| 1972/73 | 906 | 74 | 980 |
| 1973/74 | 294 | 165 | 459 |
| 1974/75 | 342 | 180 | 522 |
| 1975/76 | 664 | 99 | 763 |
| 1976/77 | 231 | 71 | 302 |
| 1977/78 | 774 | 57 | 831 |
| 1978/79 | 885 | 154 | 1,039 |
| 1979/80 | 965 | 547 | 1,512 |
| 1980/81 | 885 | 480 | 1,365 |

that the number of vessels within the fishery has remained at about 100 over the past decade, but there has been no constraint on the upgrading of vessels, and the fishing power of the fleet has increased over the period.

Scallop resources are known to fluctuate widely and the Bass Strait resource is no exception (Table 3). Victorian fisheries managers have therefore attempted to stabilize the industry by limiting entry, thereby hoping to avoid "boom or bust" situations which would be accentuated if there were unrestricted entry into the fishery.

Although in the past the Victorian fleet has fished mainly in the northern part of Bass Strait, close to Victoria, its operations have recently extended southwards to waters adjacent to Tasmania. The vessels are thus operating in areas which have hitherto been mainly exploited by Tasmanian boats which have caught sharks and rock lobsters in addition to scallops. The more diversified operation of the Tasmanian vessels has been due to their design and their easy access to resources other than scallops. Opportunities for access to these resources by Victorian vessels have been more limited, hence their specialization as scallop boats.

The attitude of Tasmanian fisheries managers until recently has been to allow unrestricted entry for Tasmanian residents to the shark and scallop fishery, with many of the vessels concerned also having access to the rock lobster resource.

The advent of Victorian vessels operating over a wider area, on grounds exploited by Tasmanian boats, as well as a marked increase in the number of Tasmanian vessels entering the fishery, have created concern about the total number of participants in the fishery and the further threat posed by upgrading of a number of smaller vessels which have hitherto had free access to the fishery. As a consequence, fisheries managers from both the Commonwealth government and the two states concerned have been examining how entry to the fishery can be limited and escalation in fishing effort contained.

This particular example highlights a number of management problems associated with a fleet of vessels dependent on a range of species which do not interact biologically. At the present time the population dynamics of the rock lobster and shark fisheries is not adequately understood, and all that is known about the Bass Strait scallop resource is that it fluctuates widely. The technical problems associated with improving the information base available to managers on shark and rock lobster dynamics would appear more tractable than is the case with a more complex multispecies pelagic resource discussed earlier. Thus, although the prediction of scallop abundance remains problematical, the likelihood of obtaining scientific advice for at least some of the component species would appear to be relatively good.

The major problem facing the managers of this fishery relates to the provision of a package which will reconcile the objectives of the two types of fleet engaged in the fishery and ensure the correct deployment of effort, more especially on the scallop resource.

Limitation of participants is the first and seemingly obvious step for managers to take, and this requires fair allocation of an entitlement based on past involvement in the fishery. The limited number of licences comprising the Victorian fleet, which has been involved in the scallop fishery for a number of years, would qualify for an entitlement, as would those Tasmanian vessels which have participated in scallop fishing over an agreed period of time. Vessels which have been built or specially acquired for the fishery prior to its closure but which have not commenced fishing would also have to be accommodated. The total number of participants which the fishery can sustain is not known, and although it is not possible to deny access to the fishery of individuals who have participated, measures need to be introduced to ensure that within the constraints of the number of vessels which are granted an entitlement, upgrading and specialization do not result in the focus of an excessive amount of effort on a particular stock. Such focus on the scallop fishery, by upgraded Tasmanian vessels becoming specialized scallop operators, would result in a considerable increase in effort and impact on the operations of the Victorian fleet which has for some time been subjected to restraint.

Maintenance of fishing capacity at present levels can, to some extent, be achieved by vessel replacement policies which allow vessels to be replaced in such a way as to contain the total fishing power of the fleet. It is equally important that replacement policies be such as not to hamper unduly improvements in efficiency. A second requirement relevant to multipurpose vessels is to prevent the separation of entitlements applicable to specific boats, that is, to obviate the replacement of one multipurpose vessel with a number of specialized boats.

More explicit control over the manner in which the total fleet effort is deployed in this fishery would be difficult to achieve. Negotiating a formula for deployment of effort with the fishermen would be difficult, as would subsequent monitoring of their activities. The assignment of annual quotas for different species is a further option, but it would require a level of stock assessment beyond the present capability of the management agencies. The monitoring and enforcement of such a quota, whether it be global or allocated to individual fishermen, would be difficult and expensive and is probably not feasible.

Under the circumstances, the managers of this resource are concentrating their efforts on restricting, as far as possible, the number of entrants to the fishery and the subsequent upgrading of vessels to constrain effort escalation. Within these effort constraints it would appear that nothing more than laissez-faire management is feasible. It might be possible to augment this by manipulation of fishing areas and minimum sizes to protect those species which would benefit.

To summarize, this fishery illustrates the problems which managers have to overcome in reconciling differing management philosophies held by two sectors of the fishing fleet and the agencies which have been responsible for administering them.

The major constraints on managers are thus more economic than biological, at least with regard to the immediate problems. The management package has to accommodate both specialized scallop vessels from the developed

Victorian fishery as well as the multipurpose vessels engaged in a more developing Tasmanian fishery.

A large number of participants in the fishery are fishermen who own their vessels and are not associated with larger companies. This can make negotiations with fishermen difficult. In addition, owner-operator fishermen are less mobile in terms of employment alternatives and the personal stake which they have in the fishery. These factors further constrain the manner in which managers can change fishing effort. Under the circumstances, initial controls on effort through entry need to be rigorously applied, for the only possibility of rectifying errors at a later stage would seem to be by means of government-arranged buy-back schemes which can prove expensive and unwieldy.

The biological constraints facing managers could be regarded as being more tractable, in that at least two of the three species concerned could be studied using single-species models which are not subject to the variance associated with multispecies models.

**Antarctic Resources**

The management of Antarctic marine living resources, and in particular krill and whale stocks, provides a topical example of the third category of multispecies resource, that is, one in which specialized fleets direct their efforts to an identifiable target species, the catches of which can be controlled. However, the strong interaction between component species requires that the fisheries be managed jointly because of the interdependence of the resources.

The possibility that Antarctic whaling will be phased out and the doubtful economics of krill harvesting at the present time are such that the need to manage these resources may not materialize in the near future. Nevertheless, the magnitude of the management problem, as well as of the resources, is such that a number of international organizations have considered the management of Antarctic resources. Under the auspices of SCAR, SCOR, IABO, and ACMRR, a group of specialists on Southern Ocean Ecosystems and Their Living Resources have been addressing problems associated with Antarctic stocks for a number of years and have initiated the BIOMASS program to provide better information on these resources (6). More recently, the Commission for the Conservation of Antarctic Marine Living Resources (CCAMLR) was established for the purpose of managing the living resources of the Antarctic, under a convention which is unique in that it requires that these resources be managed with regard to ecosystem principles.

In particular, Article 2 of the Convention states that "the principles of conservation to be followed when managing resources include:

> Prevention of decrease in the size of any harvested population to levels below those which ensure its stable recruitment. For this purpose its size should not be allowed to fall below a level close to that which ensures the greatest net annual recruitment.
>
> Maintenance of the ecological relationships between harvested, dependent and related populations of Antarctic marine living resources and the restoration of depleted populations to the levels defined in the above paragraph.
>
> Prevention of changes or minimisation of the risk of changes in the marine ecosystem which are not potentially reversible over two or three decades, taking into account the state of available knowledge of the direct and indirect impact of harvesting, the effect of the introduction of alien species, the effects of associated activities on the marine ecosystems and of the effect of environmental changes with the aim of making possible the sustained conservation of Antarctic marine living resources.

This Article has two serious implications for resource managers.

First, it implies that there is sufficient understanding of individual resources and of their interaction with other exploited and non-exploited species to assess the impact of fishing on recruitment and net annual increment, and to move depleted resources to agreed levels of recruitment. This is a formidable scientific task.

The International Whaling Commission has documented whale stock assessments over a number of years, and there have been some attempts recently to model conceptionally the resources of the Antarctic to identify the order of krill biomass which could be considered to be surplus in view of the reduction in consumption by whales. In addition, there have been some acoustic surveys, executed during the course of the BIOMASS program, to study the abundance and distribution of krill in limited areas (6). Although both IWC and the BIOMASS program are persisting with efforts to improve scientific information on resources, it is likely that in the foreseeable future management decisions concerned with Antarctic resources and their impact on the ecosystem will have to be made under conditions of high uncertainty. Scientists participating in the International Whaling Commission hold differing views on the catch of whales which can be sustained, and these differences have resulted in very rigorous examination of the assumptions which have to be made when modeling these resources. In the scientific committee of CCAMLR it is likely that an equal, if not wider, range of views on the impact of exploitation will be held, and scientific models and conclusions are likely to be subjected to very rigorous scrutiny. This should ensure good science, but it is likely that managers will be confronted with complex models with large numbers of associated caveats.

The second serious issue is that of reaching agreement with regard to management goals. In this respect the participating nations have a wide range of economic objectives and social values. Whereas countries with fishing fleets which are capable of economically harvesting Antarctic resources will be seeking to ensure quotas which allow for viable operations, some other nations with less direct or perhaps longer-term interest in the utilization of the resources will certainly be opting for management objectives which are more geared towards protecting the ecosystem.

At its second meeting the Technical Group on Data Statistics and Resource Evaluation of the SCAR, SCOR, IARBO, and the ACCMLR group of specialists on Southern Oceans Ecosystems and Their Living Resources noted that the interpretation of Article 2 3 of the Convention presents some difficulties for resource managers (3). For instance, the greatest net annual increment of prey species, which is suggested as a management objective in Article 2 3 a, might only be achieved if exploitation reduces predators to a low level. Likewise, it may not be possible to maximize the net annual increment of predators while significantly harvesting their prey. Furthermore, the interaction of resources at the same trophic level could make it difficult to achieve the goals which are suggested by strict interpretation of the Convention. For instance, the maximum net annual increment of a number of baleen whale species which have been heavily exploited in the past may not be achieved if species which are competitors have significantly increased. The problem of determining management objectives in relation to Article 2 of the Convention is further compounded because the Antarctic ecosystem is probably not in equilibrium. This factor makes it difficult to establish realistic and defensible objectives towards which fisheries can be directed by management.

Under these confused circumstances, the attitudes taken by various

CCAMLR countries on resource management will probably be influenced by the values of their society and the pressures that these bring to bear on managers.

A question which should concern managers of the Antarctic multispecies resource is the feasibility of collecting sufficient information to improve materially the current understanding of ecosystem dynamics. Scientific work in the Antarctic is expensive, and the possible cessation in whaling will effectively reduce opportunities for collecting further population data on this important component of the system. The loss of this opportunity relates not only to estimates of numbers of whales but also to such biological parameters which can reflect population status.

The BIOMASS program provides a good illustration of the logistic problems associated with collecting data on krill and the expense of such operations. At the present time, opportunities for collecting population data on other important components of the system appear to be remote. Under the circumstances, managers may have little alternative other than to base their decisions on models using existing data. Bearing in mind the high degrees of uncertainty attached to this information and to the models which have hitherto been available, agreement by participating nations with a wide range of attitudes to exploiting the Antarctic is likely to be extremely difficult to achieve.

In an attempt to address this problem at its first meeting in 1979, the BIOMASS Technical Group on Data and Resource Evaluation noted that there were two possible approaches to managing future fisheries for krill, that is, the predictive or the reactive approach (8). The former attempts to provide a comprehensive prediction of the effects of future management actions on both target and nontarget species. The group recognized not only that the theoretical basis for such predictions was insufficient, but also that important parameters required for such a predictive modeling exercise were lacking. They therefore concluded that at the present time it is only possible to make predictions of a very general, qualitative kind, which would relate to the effect of heavy exploitation of a key resource such as the krill.

The reactive approach, on the other hand, entails initiation and expansion of exploitation, with some monitoring of predetermined components, and the application of management measures if and when they are seen to be appropriate.

The group felt that harvesting of krill could be managed using an intermediate approach which would allow carefully controlled expansion of harvesting, while information was collected on its impact on the krill as well as on the reactions of other species in the system which may be associated with krill. In this manner development can take place under conditions of high uncertainty while some effort is made to contain risks to the ecosystem. Managers will thus be balancing the risk to and the return from the resource. This in itself will be difficult to reconcile amongst CCAMLR nations, few of whom would contemplate exploitation, with the majority having an interest in the welfare of the ecosystem.

**CONCLUSION**

The policy instruments available to managers of multispecies fisheries are essentially similar to those applied to single-species fisheries but with a number of additional, important dimensions added to the decision-making process. These are concerned with interactions between either the resources, the fishing units exploiting them, or a combination of both.

All exploited resources interact with other components of their ecosystems, but in adopting a single-species approach, fisheries managers and scientists have been able to conveniently express a number of interactions in a few, relatively simple parameters. When, however, the fishery impacts on more than one species, interactions cannot be dealt with simply, and the modeling process becomes complex as additional interventions have to be accommodated. The variance around such models is likely to be large, although it could be argued that the simpler models applied to single-species problems are less realistic and therefore provide a sense of precision which cannot be justified.

In multispecies fisheries the spectrum of exploiters is often wide, and this, together with the complexity of biologically modeling the system, suggests that a broad approach to the problem of management is required, with emphasis on management objectives which take account of biological, economic, social, and political factors. This contrasts with the more conventional single-species approach to fisheries management which has tended to rely heavily on quantitative information on resource dynamics, which in turn is used to achieve the objective of safeguarding yields from the resource. The multispecies problem has, by virtue of its scope, highlighted the real range of the problems which fisheries managers should be addressing.

The cases selected for analyses in this paper clearly illustrate the wide range of information required for management and the difficulty and expense entailed in its collection. In some instances, it is unlikely that the agencies responsible for management will have the capacity to obtain the information required or that they could justify its collection in relation to the monetary value of the decision concerned. Under the circumstances, a more holistic approach may be appropriate. More emphasis should be given to identifying and deciding upon management objectives and the key information requirements for management decisions, whether these be in biological, economic, social, or political spheres. This would better allow agencies to decide what information, within the full suite of desirable data, they must acquire, taking into account whether or not such acquisition is feasible.

A systematic approach is required, and the simple three-dimensional format discussed in this paper provides a possible example of such an approach. It identifies categories of resources, management techniques, and management objectives. The application of the format to three case studies attempts to illustrate the approach. It has, in addition, suggested that the selection of appropriate policy instruments, the principle subject of this paper, is one of the simpler tasks entailed in managing multispecies fisheries.

## REFERENCES

(1) Crawford, R.J.M. 1979. Implications of recruitment, distribution and availability of stocks for management of South Africa's purse-seine fishery. Ph.D. Thesis, University of Cape Town.

(2) ICSEAF. 1977. Proceedings Report of the Meeting of the International Commission on SE Atlantic Fish.

(3) Meeting of the Group of Specialists on Southern Ocean Ecosystems and Their Living Resources. 1982. Biomass Report Series 24.

(4) Newman, G.G., and Crawford, R.J.M. 1980. Population biology and management of mixed species pelagic stocks off South Africa. Rapp. P.-v. Reun. Cons. Int. Explor. Mer 177: 279-291.

(5) Newman, G.G.; Crawford, R.J.M.; and Centurier-Harris, O.M. 1974. Assessment of fishing effort, stock abundance and yield for the South African multispecies pelagic fishery. Coll. Scient. Pap. Intl. Comm. SE Atl. Fish 1: 163-170.

(6) Newman, G.G.; Crawford, R.J.M.; and Centurier-Harris, O.M. 1979. Fishing effort and factors affecting vessel performance in the South African purse-seine fishery, 1964-1972. Investl. Rep. Sea Fish. Brch. S. Afr. 120.

(7) Report of the Technical Group on Data, Statistics and Resource Evaluation. 1979. Biomass Report Series 4.

(8) Report of the Technical Group on Data, Statistics and Resource Evaluation. 1980. Biomass Report Series 16.

# Epilogue

J.A. Gulland
Dept. of Fisheries, FAO
Rome, Italy

An examination of today's fisheries throughout the world suggests that management lacks the necessary ability, or is not using it properly. Fisheries are not being managed well, though the degree of failure is not the same for all possible objectives. Despite some spectacular failures (e.g., Peruvian anchoveta), the total landings of marine fisheries continue to increase slowly, but the economic losses through mismanagement - overcapacity with accompanying excessive fuel and other operating costs, as well as very often far too high administrative and research costs associated with supporting and implementing inappropriate management policies - are widespread and very large. Probably most disturbing to the fishery administrator is the extent of conflict. This conflict is mostly political, but there is increasing physical violence, to the point approaching loss of government control. Conflict has been most visible between countries interested in the same stock, but despite their prominence in the world's news media, these types of conflict are of declining importance. The basic conflict has been resolved by the new Law of the Sea Convention in favor of the coastal state. The most serious conflicts are now within countries, particularly between growing industrial fisheries with relatively large vessels and traditional fishermen with small, often unpowered, boats.

What has been said and learnt at this Dahlem Workshop that could help resolve these problems? On the purely biological side (see Beverton et al. and Sugihara et al., both this volume), the main message is one of relative comfort. Those fishery scientists (as much in developed countries as in developing countries) who are struggling with obviously complex situations with limited data and with models which clearly greatly simplify the true situation have been assured that what they are doing is scientifically sensible and respectable. The simple single-species models that have been the basic tools of the fishery scientist for the past twenty years can be used to give advice with only a small number of unpleasant surprises (e.g., the collapse of the Peruvian anchovy), and advice given now, based on these single-species models, is much more useful to the manager than a situation in which he gets no advice until more sophisticated models are developed. In new fisheries, examination of the general characteristics of the fish and the fishery, and comparison with existing fisheries, may enable reasonable advice to be given to the managers in the advance of formal assessments. Where there have been surprises it seems that the current state of the art in more complex and more realistic modeling is not yet such that events such as the anchovy collapse can be predicted, even with the benefit of hindsight. What these models can tell us now is that collapses and other signs of instability (e.g., abrupt changes in species composition) can take place, not that they will take place, or when they will take place, if they do occur. In the long term more attention should be paid to the questions of stability, as well as to the possible genetic effects of usually selective fisheries (e.g., towards catching the larger and faster growing individuals).

On the management side (see Beddington et al. and Larkin et al., both this volume), the situation is less satisfactory and more could have been learnt from the workshop, regarding both the tools of management and the willingness of managers to use them. The latter two groups noted the disadvantages of catch quotas (need for annual reassessments, difficulties of enforcement, etc.). While quotas were probably the most appropriate tool in the 1960s and 1970s, in the age before the 200-mile limit when most management was done by international commissions, there seems little justification for the degree to which catch quotas dominate present-day discussions and practice of fishery management.

More attention should be paid to other management tools, both those that directly control the fishing effort (or fishing mortality), such as various forms of licensing and limited entry, and the less direct approaches such as various fiscal measures. At present many of the latter (e.g., subsidies), while designed to reduce short-term problems, only add to long-term difficulties. Although some of these techniques have their direct impact only on the shoreside problems, indirectly they can greatly improve the way the resource is managed (e.g., by reducing the pressure from industry for too high catch quotas, etc.).

To what extent will these conclusions lead to better management? Probably more than the somewhat bland summary above might suggest. Despite some weaknesses in the tools used (e.g., over-attention to catch quotas as the primary method), the main cause of poor management has been a failure to use the advice and tools that were available, rather than an absence of advice or a lack of suitable tools. Too often nothing is done until overfishing, and particularly overdevelopment in boats, gear, etc., has gone so far that any remedial action must, in the short run, be very painful. Lying behind this has been a general lack of appreciation among those with some of the greatest influence on the pattern of development of fisheries (bankers, senior planning officers, etc.), of the nature of fishery resources, and poor communication between these people and the biologists and others (e.g., economists) who should advise them. It has seemed that for each new fishery the message has had to be learnt afresh from painful experience that fish stocks are not unlimited and may not be stable, and that left to themselves in the usual open access conditions, fishermen and the fishing industry will tend to expand fishing capacity beyond the limit that the resources can withstand, with consequent biological, economic, and social problems.

The situation is now changing. There is a much wider appreciation among policymakers that fishery resources may not be unlimited - and there may not be enough fish to support development plans, however desirable they may be for economic, nutritional, or social purposes. There is also some appreciation - still small, but growing - that effective fishery management involves several disciplines and is only possible if there is good communication between biologists, economists, administrators, and others. If these trends can be continued and strengthened, they could lead to a much brighter future for fishery management - even more than purely technical progress in, say, stock assessment. This Dahlem Workshop has helped; it has confirmed that while much more needs to be learnt about the dynamics of fish stocks, particularly concerning their possible instability and the interactions between species, the current methods of biologist assessment do provide results that can be used by policymakers. It has also shown that there can be effective dialogues between biologists and the other disciplines concerned in fishery management.

# Geographical Glossary

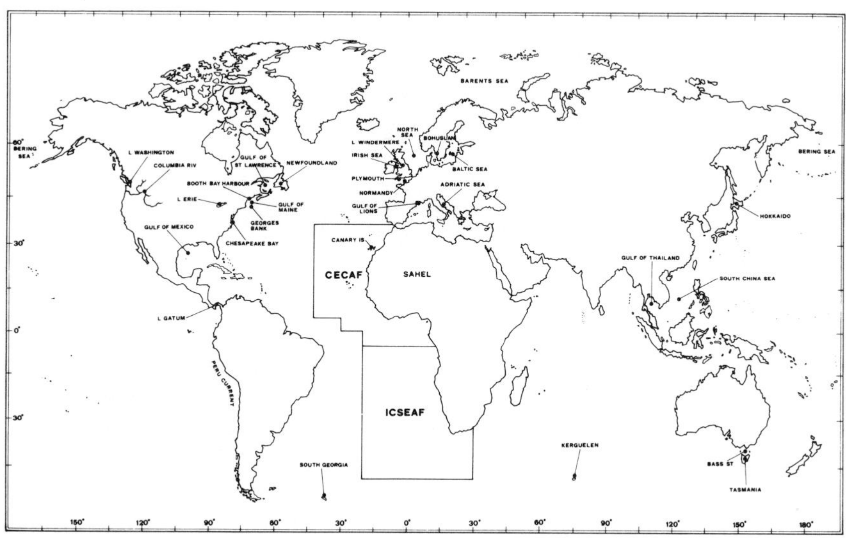


Map showing location of places mentioned in the text. Also shown are the areas of responsibility of the FAO's Fishery Committee for the Eastern Central Atlantic (CECAF) and the International Commission for the Southeast Atlantic Fisheries (ICSEAF).

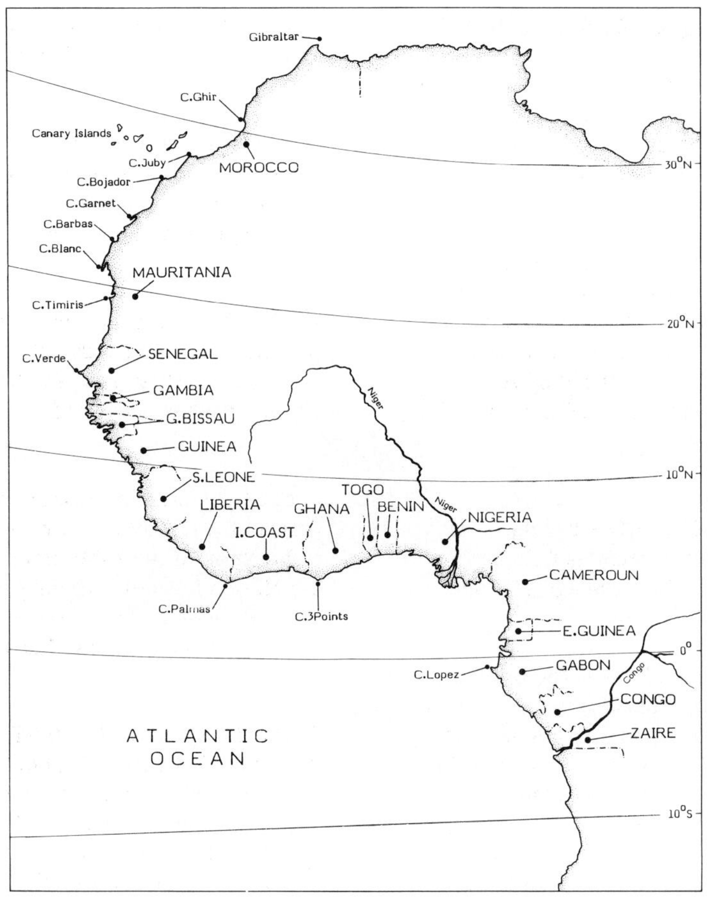


Map of West African Coast.

# Glossary of Technical Terms

**ADAPTIVE MANAGEMENT** - A form of management (qv) which involves continuous active response to new information about the behavior or current state of the resource.

**AGE CLASS** - A group of individuals of the same age range in a population. This term usually refers to a year class in long-lived annually breeding species, but shorter units of time are also used.

**AGE-SPECIFIC** - The dependence of a factor, such as fishing mortality (qv), on the age of fish.

**AGGREGATION** - The practice of incorporating groups of individuals or species into a single, indivisible component when building models of complex systems.

**ASSEMBLAGE** - A collection of species inhabiting a given area, the interactions between the species, if any, being unspecified.

**ASSESSMENT** - A judgement made by a scientist or scientific body on the state of a resource, such as a fish stock (e.g., size of the stock, potential yield, whether it is over- or underexploited), usually for the purpose of passing advice to a management (qv) authority.

**BY-CATCH** - Catches of a species in a fishery which is directed primarily at another species. These are sometimes discarded.

**CATCH** - Is usually expressed in terms of wet weight. It refers sometimes to the total amount caught, and sometimes only to the amount landed. The catches which are not landed are called discards.

**CATCH PER UNIT EFFORT (CPUE)** - The catch obtained by a vessel or fishery per unit of fishing effort (qv) expended. This term is often used as a measure of abundance of the target stock(s), with greater or lesser degrees of justification.

**CATCH RATE** - Means sometimes the amount of catch per unit time,

**CATCH RATE** (continued) - and sometimes the catch per unit effort.

**CATCHABILITY** - The fishing mortality on a stock generated by a unit of fishing effort (qv). It is usually denoted as q in the equation:

$F = qf,$

where F and f represent fishing mortality and fishing effort, respectively. q will depend on the habits of the fish as well as on the type and deployment of fishing gear; it may also depend on the abundance of the fish (less abundant fish may be more catchable due to less saturation of gear).

**COLLAPSE** - Reduction of a fish stock by fishing or other causes to levels at which the production is only a negligible proportion of its former levels. The word is normally used when the process is sudden compared with the likely time scale of recovery, if any, but is sometimes used melodramatically for any case of overfishing (qv).

**COMMUNITY** - The collection of organisms or species inhabiting an area. This term usually implies particular known or hypothesized interactions between the organisms (as opposed to ASSEMBLAGE).

**COMPENSATION** - A compensatory mechanism is a process by which the effect of one factor on a population tends to be counteracted or compensated for by a consequential change in another factor. For example, a reduction in the egg production of a stock may be compensated for by an increase in the survival rate of the eggs. Compensation can be partial or complete. In some cases, OVER-COMPENSATION can occur: a reduction in egg production could, for example, lead to an increase in the resultant production of larvae. Many density-dependent (qv) mechanisms are compensatory.

**COMPETITION** - The detrimental interaction between two or more organisms of the same or different species which utilize a common resource.

**CONCENTRATION PHASE** - A stage in the life cycle of a fish at which the individuals are particularly concentrated. For example, the adults of a species may inhabit an entire sea for most of the year, but the larvae may be confined to the beaches.

**DEMERSAL** - Inhabiting the sea

**DEMERSAL** (continued) - bottom. The application of the term to a species usually refers to the adult stage of the species.

**DENSITY-DEPENDENCE** - The dependence of a factor influencing population dynamics (such as survival rate or reproductive success) on population density. The effect is usually in the direction that contributes to the regulative capacity (qv) of a stock.

**DEPENSATION** - The opposite of COMPENSATION (qv).

**DETERMINISTIC MODEL** - A model whose behavior is fully specified by its form and parameters, unlike a stochastic model (qv).

**DISCOUNT RATES** - The rate at the relative weight attached to benefits or losses is reduced in proportion to their distance into the future. Discounting is often used when making investment or policy decisions and can have serious consequences for future generations. For example, a discount rate of 10% means that gains or losses occurring n years into the future are ascribed a weight equal to their nominal value multiplied by $0.9^n$.

**DIVERSITY** - The variety of species in a community, sometimes expressed by various quantitative measures which reflect not only the total number of species present but also the degree of domination of the system by a small number of species.

**ECOSYSTEM** - The sum total of biological populations and abiotic factors present in a region, and their relationships to each other. No ecosystem is a closed system, hence the precise meaning of the term varies according to the scale of the region to which it is applied.

**EFFORT** or **FISHING EFFORT** - This term is defined to varying levels of precision. It can be simply the total number of boats operating in a season, or the actual total number of net hauls per unit time. When different types of fishing gear are deployed, the amounts of effort expended by each are usually standardized according to their relative fishing power (qv) before being summed as an index of total effort.

**ESCAPEMENT** - That part of the stock which survives at the end of the fishing season.

**EXIT FISHERY** - A fishery directed at a transient phase in the life

**EXIT FISHERY** (continued) - cycle of the prey, for example, a fishery catching only the juveniles of a species.

**F** - See MORTALITY.

**FECUNDITY** - In fish this term usually refers to the production of eggs per individual.

**FISHING POWER** - The relative fishing power of two vessels or gear types is the ratio between the catches they would obtain per unit time for a given level of abundance of the resource.

**GROWTH RATE** - In fish this is often measured in terms of the parameter K of the von Bertalanffy curve for the mean weight as a function of age;
W = Wmax (1 - exp (-K.age)).

**HABITAT REFUGES** - A part of the range of a stock which is not accessible to fishing. If the species in question is not too diffusive, the existence of such refuges may enable it to persist under a higher level of fishing mortality than it otherwise would.

**ICES** - International Council for the Exploration of the Sea.

**INTERACTIONS** - Between species these are the processes by which the fate of one species in an ecosystem (qv) can influence another. Interactions can include predation, competition, etc.

**LEVEL OF EXPLOITATION** - This can mean the amount of catch or the level of fishing mortality (qv), or is sometimes used without any precise quantity in mind.

**LICENSING** - Restriction of the right to fish to those persons or vessels issued with licenses for the purpose.

**LIMITED ENTRY** - Restriction of the right to join a fishery, by the use of licenses or other means.

**LINKAGES** - See INTERACTIONS.

**LONG-LINING** - Fishing using baited hooks on lines.

**MANAGEMENT** - The art of taking measures affecting a resource and its exploitation with a view to achieving certain objectives, such as the maximization of the production of that resource. Management includes, for example, fishery regulations such as catch quotas or closed seasons. MANAGERS are

**MANAGEMENT** (continued) - those who practice management.

**MESH SIZE** - The size of holes in fishing net. Minimum mesh sizes are often prescribed by regulations in order to avoid the capture of the young of valuable species before they have reached their optimal size for capture.

**MORTALITY** - Is usually defined as an instantaneous death rate of fish, usually expressed in units of $(\text{years})^{-1}$. Thus a proportion exp(-Z) of a population would survive a constant mortality rate, Z, operating for one year. The mortality rate is divided into FISHING MORTALITY, usually denoted by the symbol F, and NATURAL MORTALITY, usually denoted by the symbol M. When both are expressed as instantaneous rates, the total mortality is simply the sum of these two. Natural mortality is usually taken to include not only mortality due to natural causes (predation, disease, etc.) but also mortality due to non-fishing artificial causes such as nuclear weapons testing or chemical waste dumping.

**OVERCAPITALIZATION** - The situation in which an excessive investment has been made in fishing or processing capacity in relation to the yield which the resource can sustain (also referred to as OVERCAPACITY).

**OVERFISHING** - In the wider sense, any level of fishing greater than some defined, optimal level. In the classical sense, a level of fishing effort or fishing mortality such that a reduction of this level would, in the medium term, lead to an increase in the total catch. Two distinct types of classical overfishing are recognized: GROWTH OVERFISHING is the situation where a reduction in the proportion of fish caught would be more than compensated for by an increase in their average size; RECRUITMENT OVERFISHING is the situation where a reduction in the proportion of fish caught would be more than compensated for by the increased number of recruits to the fishery that would accompany the increased escapement (qv) of mature fish.

**PELAGIC** - Inhabiting the open sea, not bottom-dwelling. Usually refers to the adult stage of a species when not otherwise stated.

**PERSISTENCE** - The tendency of a population (qv) to continue to exist in the long term, despite short-term fluctuations.

**POPULATION** - Strictly, a distinct

**POPULATION** (continued) - group of members of a species inhabiting a certain region, which are reproductively isolated from other populations. In loose usage, the term is often synonymous with STOCK (qv).

**PRODUCTIVITY** - Generally used loosely to refer to the capacity of a stock or ecosystem to provide a yield of some useful kind.

**PROPERTY RIGHTS** - (Varying degrees of) ownership of a resource by particular individuals or associations.

**QUOTA** - A limit placed by an authority on the amount of fish which may be caught, often applying to a particular species and/or area and/or fishing gear type.

**RECRUIT** - A young fish entering the exploitable stage of its life cycle. RECRUITMENT can mean either the rate of entry of recruits into the fishery or the process by which such recruits are generated. Recruitment can be defined as the attainment of a particular AGE AT RECRUITMENT, being the youngest age group which is considered to belong to the exploitable stock. The age at recruitment depends both on the biological characteristics of the fish itself and on the nature of the fishery (location, mesh size (qv), etc.). Alternatively, recruitment can be defined as the attainment of a certain size, or as the appearance on a particular fishing ground, or as the attainment of a particular level of catchability (qv) relative to that of older fish. PRERECRUITS are fish which have not yet reached the recruitment stage.

**RECRUITMENT OVERFISHING** - See OVERFISHING.

**REGULATIVE CAPACITY** - Is a population's tendency to revert towards some typical average level of abundance rather than to increase or decline indefinitely or to drift aimlessly. The REGULATIVE MECHANISMS by which this can be achieved include, for example, inverse dependence of survival rate and/or reproductive success on population density (often used synonymously with DENSITY-DEPENDENCE (qv) and sometimes called homeostasis).

**RELIABILITY** - The extent to which a resource, if managed properly, can be depended on to provide a reasonably constant yield (qv). An unreliable resource is one which may fail suddenly due to causes not obviously related to fishing. As used in Beverton et al. (this volume), this term is distinct from ROBUSTNESS (qv).

**RESILIENCE** - The tendency of a community or ecosystem not to change greatly despite heavy disturbances such as intensive fishing.

**REVERSIBILITY** - The extent to which a change in a stock or ecosystem induced by exploitation will reverse itself when the causative factor is removed. Extinction of a species is an example of an irreversible change.

**ROBUSTNESS** - The capacity of a population to persist in the presence of fishing. This depends on the existence of compensatory (qv) mechanisms.

**SINGLE-SPECIES MODEL** - A model describing the dynamics of a species which does not explicitly incorporate the effects of interactions with other species.

**SPATIAL HETEROGENEITY** - The nonhomogeneous nature of habitats or spatial distributions of organisms, often ignored in simple models.

**SPAWNING BIOMASS** - The total biomass of fish of reproductive age during the breeding season of a stock.

**SPAWNING SUBSTRATE** - The type of habitat required by a fish species for spawning.

**STABILITY** - This term is applied very loosely to ecosystems or communities, but it usually means their tendency to retain their essential characteristics in the shorter or longer term. It sometimes refers to the more particular capacity of systems to return to their original state following a disturbance.

**STATUS QUO** - Can mean the general current state of affairs in a fishery, but in certain fora, such as ICES (qv), it refers specifically to the current level of fishing mortality (qv).

**STOCHASTIC MODEL** - A model whose behavior is not fully specified by its form and parameters, but which contains an allowance for unexplained effects which are represented by random variables.

**STOCK** - In its strict sense, a distinct, reproductively isolated population (qv). In practice, the term is applied to the members of a species or group of species inhabiting any conveniently defined area, which is regarded as a discrete population for management purposes.

**STOCK-RECRUITMENT RELATIONSHIP** - The dependence of recruitment (qv) on the size of the parent breeding stock, the latter usually being measured in units of spawning biomass (qv). Such a relationship always exists in principle, in that the existence of a parent stock is a prerequisite for the generation of recruitment. However, in many cases there exist regulatory mechanisms (qv) such that the number of recruits is not strongly related to the parent stock size over the range of stock sizes observed: this situation is sometimes described as the absence of a stock-recruitment relationship, but is more logically described as a special case of a stock-recruitment relationship.

**STOCKING** - The practice of putting artificially reared young fish into a sea, lake, or river. These are subsequently caught, preferably at a larger size.

**UPWELLING** - An oceanographic phenomenon whereby deep, nutrient-rich water is forced up to the surface, often leading to an exceptionally productive area.

**VITAL RATES** - Rates (such as natural mortality, fecundity, and growth rates) affecting the

**VITAL RATES** (continued) - dynamics of a stock.

**WHITE NOISE** - A random effect in a stochastic model (qv), which is taken to be independent, identically distributed over the time intervals in which the model is expressed. Other types of noise such as red and blue noise are also recognized, but these terms are less common and should not be used without explanation.

**YIELD** - Sometimes synonymous with CATCH, but usually with the implication of a degree of sustainability, especially when potential yields are under discussion. The YIELD CURVE is the relationship between the expected yield and the level of fishing mortality or (sometimes) fishing effort.

- Compiled by J.G. Cooke

# List of Participants with Fields of Research

**ARNTZ, W.E.**
Institut für Meeresforschung
Am Handelshafen 12
2850 Bremerhaven
Federal Republic of Germany

*Structure, dynamics, and stability of marine ecosystems (special reference to "El Niño")*

**BAILEY, R.S.**
Marine Laboratory
P.O. Box 101
Victoria Road
Aberdeen AB9 8DB
Scotland

*Population dynamics and stock assessment of pelagic fish*

**BEDDINGTON, J.R.**
Marine Resources
Assessment Group
Center for Environmental
Technology
Imperial College
48, Prince's Gardens
London SW7 1LU
England

*Applied ecology*

**BEVERTON, R.J.H.**
Dept. of Applied Biology
UWIST
P.O. Box 13
Cardiff CF1 3XF
Wales

*Fisheries dynamics*

**BREWER, G.D.**
School of Organization and
Management
Yale University
Box 1A, Yale Station
New Haven, CT 06520
USA

*Applied social science, policy sciences (research and applications); natural resource policy and management*

**CLARK, C.W.**
Dept. of Mathematics
University of British Columbia
Vancouver, BC V6T 1W5
Canada

*Renewable resource economics*

**COOKE, J.G.**
Institute of Animal Resource Ecology
2204, Main Hall
University of British Columbia
Vancouver, BC V6T 1W5
Canada

*Estimation of abundance and yield of fish and marine mammal stocks*

**CSIRKE, J.B.**
FAO
Dept. of Fisheries
Via delle Terme di Caracalla
00100 Rome
Italy

*Fish population dynamics, stock assessment, fisheries management*

**DAAN, N.**
Netherlands Institute for
Fisheries Investigations
Postbox 68
1970 AB-IJmuiden
Netherlands

*Fisheries biology*

**DOYLE, R.W.**
Dept. of Biology
Dalhousie University
Halifax, Nova Scotia B3H 4JI
Canada

*Ecological genetics of marine organisms; aquaculture genetics*

**DUTT, S.**
Dept. of Marine Living Resources
Andhra University
Visakhapatnam 530 003
India

*Biology and ecology of tropical marine fish stocks*

**GARCIA, S.**
FAO
Dept. of Fisheries
Marine Resources Service
Via delle Terme di Caracalla
00100 Rome
Italy

*Stock assessment; advice on management; tropical stocks, Penaeid shrimps*

**GLANTZ, M.H.**
Environmental and Societal Impacts
Group
National Center for Atmospheric
Research
P.O. Box 3000
Boulder, CO 80307
USA

*Climate impacts and fisheries management*

**GULLAND, J.A.**
41, Eden Street
Cambridge CB1 1EL
England

*Fishery management*

**HEMPEL, G.**
Alfred-Wegener-Institut
für Polarforschung
Columbus Center
2850 Bremerhaven
Federal Republic of Germany

*Ecology of Antarctic krill*

**HOLT, S.J.**
2, Meryon Court
Rye, East Sussex TN31 7LY
England

*Management of whaling and sealing; role of marine resources in human nutrition*

**HONGSKUL, V.**
SEAFDEC Liaison Office
956 Olympia Building, 4th Floor
Rama IV Road
Bangkok 10500
Thailand

*Fishery dynamics: multispecies, multifleet exploitation in tropical areas; fishery management (marine)*

**LARKIN, P.A.**
Institute of Animal Resource Ecology
University of British Columbia
Vancouver, BC V6T 1W5
Canada

*Population biology of fishes*

**LAUREC, A.J.Y.**
Institut Scientifique et Technique des Pêches Maritimes
POB 1049
Rue de l'Ile d'Yeu
44037 Nantes Cedex
France

*Parameter estimation, management under uncertainty*

**LAWTON, J.H.**
Dept. of Biology
University of York
York YO1 5DD
England

*Population dynamics and community ecology*

**LEVIN, S.A.**
Ecology and Systematics
Corson Hall
Cornell University
Ithaca, NY 14853
USA

*Mathematical ecology*

**MacCALL, A.D.**
National Marine Fishery Service
Southwest Fisheries Center
P.O. Box 271
La Jolla, CA 92038
USA

*Population modeling for purposes of fishery management; development of harvesting strategies; population dynamics of density-dependent habitat selection*

**MASKE, H.**
Abt. Planktologie
Institut für Meereskunde an der Universität Kiel
Düsternbrooker Weg 20
2300 Kiel 1
Federal Republic of Germany

*In vivo fluorescence as a tool in remote sensing, photoinhibition - wavelength dependency, ecological significance*

**MAY, R.M.**
Dept. of Biology
Princeton University
Princeton, NJ 08544
USA

*Population biology, with particular emphasis on host-parasite systems*

**NELLEN, W.P.**
Abt. für Fischereibiologie
Institut für Meereskunde an der
Universität Kiel
Düsternbrooker Weg 20
2300 Kiel 1
Federal Republic of Germany

*Fish eggs and larvae distribution and ecology in the sea; experimental work on the physiological and ecological capacity of fish at different life stages with emphasis on changing salinities*

**NEWMAN, G.G.**
Fisheries and Wildlife Service
Dept. of Conservation, Forests
and Lands
250 Victoria Parade
P.O. Box 41
East Melbourne, Victoria 3002
Australia

*Resource management*

**PAINE, R.T.**
Dept. of Zoology
University of Washington
Seattle, WA 98195
USA

*Marine ecology, especially dynamics and organization of rocky intertidal communities; major focus is on interspecific interactions as studied by experimental manipulations*

**PAULY, D.M.**
ICLARM
MCC. P.O. Box 1501
Makati, Metro Manila
Philippines

*Population dynamics of tropical fish*

**PLATT, T.**
Marine Ecology Laboratory
Bedford Institute of Oceanography
P.O. Box 1006
Dartmouth, Nova Scotia B2Y 4A2
Canada

*Structure and function of marine ecosystems; physiological ecology of microbial plankton*

**POLICANSKY, D.J.**
National Research Council
National Academy of Sciences 347
2101 Constitution Avenue NW
Washington, DC 20418
USA

*Life-history patterns, especially transitions; ecological and physiological correlates of sexual maturation, metamorphosis in fishes, sex change in plants and animals*

**RACHOR, E.**
Institut für Meeresforschung
Am Handelshafen 12
2850 Bremerhaven
Federal Republic of Germany

*Marine ecology; dynamics and productivity of bottom communities (invertebrates)*

**RADACH, G.**
Institut für Meereskunde der
Universität Hamburg
Troplowitzstrasse 7
2000 Hamburg 54
Federal Republic of Germany

*Modeling the lower trophic levels of pelagic ecosystems*

**ROSENTHAL, H.K.**
Biologische Anstalt Helgoland
Zentrale Hamburg
Notkestrasse 31
2000 Hamburg 52
Federal Republic of Germany

*Aquaculture: water quality management in intensive farming systems; mass production of marine fish fry. Environmental Protection: effects of environmental pollutants on early life-history stages of fish; development of effects monitoring strategies; problems associated with introductions of exotic species*

**ROTHSCHILD, B.J.**
Chesapeake Biological Laboratory
University of Maryland
Box 38
Solomons, MD 20688
USA

*Variability in fish populations*

**ROUGHGARDEN, J.**
Dept. of Biological Sciences
Stanford University
Stanford, CA 94305
USA

*Ecology, especially theoretical ecology including the population dynamics of intertidal marine populations*

**SHEPHERD, J.G.**
MAFF Fisheries Laboratory
Pakefield Road, Lowestoft
Suffolk NR33 0HT
England

*Marine fish stock assessment*

**SISSENWINE, M.P.**
National Marine Fisheries Service
Northeast Fisheries Center
Woods Hole Laboratory
Woods Hole, MA 02453
USA

*Fisheries ecology: the factors that control fish productivity and are relevant to multispecies exploitation strategies*

**SMETACEK, V.S.**
Institut für Meereskunde an der
Universität Kiel
Düsternbrooker Weg 20
2300 Kiel 1
Federal Republic of Germany

*Biological oceanography: production and sedimentation of plankton; pelagic food web structure and function*

**SUGIHARA, G.**
Environmental Sciences Division
Oak Ridge National Laboratory
P.O. Box X
Oak Ridge, TN 37831
USA

*Mathematical ecology: food web structure and dynamic population modeling*

**THUROW, F.R.M.**
Bundesforschungsanstalt für Fischerei
Institut für Küsten- und Binnenfischerei
Wischhofstrasse 1-3
2300 Kiel 14
Federal Republic of Germany

*Development of the fish stocks in the Baltic ecosystem, fisheries management*

**TROADEC, J.-P.**
Institut Scientifique et Technique
des Pêches Maritimes
POB 1049
Rue de l'Ile d'Yeu
44037 Nantes Cedex
France

*Fisheries development and management, fishery resource evaluation*

**URSIN, E.A.**
Danish Institute for Fisheries and
Marine Research
Charlottenlund Castle
2920 Charlottenlund
Denmark

*Biological species interaction in exploited fish stocks*

**WALTERS, C.J.**
Institute of Animal Resource Ecology
University of British Columbia
Vancouver, BC V6T 1W5
Canada

*Adaptive management of renewable resources*

**WIEBE, P.H.**
Redfield 226
Woods Hole Oceanographic Institution
Woods Hole, MA 02543
USA

*Biological oceanography*

**ZEITZSCHEL, B.F.K.**
Abt. Marine Planktologie
Institut für Meereskunde an der
Universität Kiel
Düsternbrooker Weg 20
2300 Kiel 1
Federal Republic of Germany

*Biological oceanography; plankton ecology; sedimentation processes*

# Subject Index


Abundance, 2, 17, 20, 26, 27, 43-47, 60-66, 77-83, 101-104, 114-117, 120-123, 132, 133, 143, 145, 160-167, 173, 176, 177, 180, 183, 197, 198, 200-203, 210-221, 228-233, 238-240, 246, 250, 254, 256, 264, 277, 283, 288, 291, 292, 308, 314, 315, 325, 328
Adaptation, passive, 264
Adriatic Sea, 176-179
Adult phase, 20, 25, 26, 49
Advisory Committee on Marine Resources Research (ACMRR), 156, 275, 279, 282, 327
Africa, Northwest, 160, 165, 167, 173, 175, 179
-, West, 5, 156-172, 229, 238
Agreements, international, 1, 39, 280, 300, 310, 316, 317, 330
Aggregated Schaefer model, 139-142
Aggregation, 2, 5, 8, 37, 65, 73, 82, 131, 132, 136-146, 191, 193, 195, 198, 267, 297, 298
- structure, 136
Allometry, 139, 148
Alternate stable states, 144, 210, 211, 217, 220-222
Analyses, single-species, 8, 14, 40, 41, 46, 132, 133, 146, 176, 178, 181-183
Analysis, error, 137, 139, 149
-, multispecies, 44, 132, 133
Analysis, policy, 277-284, 306
-, Virtual Population, 42-44, 49, 101-106
Anchoveta, 5, 24, 37, 47, 70, 71, 78, 142, 173, 174, 182, 202, 231, 247, 249, 252, 277, 335, 336
Anchovies, 5, 18, 22, 24, 37, 47, 62, 67-78, 142, 173, 174, 182, 183, 202, 203, 217, 221, 231, 246-252, 304, 321, 335, 336
Anchovy, Northern, 18, 22, 62, 77
-, Peruvian (see Anchoveta)
Antarctic, 46, 175, 178, 294, 327-330
Aquaculture, 269, 294
Arctic cod (Gadus morhua), 25, 30, 64
Artificial reefs, 293
Artisanal fishermen, 287, 292
- fishery, 156, 158, 171, 176
Assemblages, 5, 146, 157, 165, 192, 198-204, 216, 298
-, species, 5, 157, 165, 298
Assessment, 16, 30, 31, 35-49, 80, 83, 95-101, 107, 170, 176-180, 203, 220, 247, 264-266, 306, 326, 328, 336, 337
Atlantic Ocean, North, 1, 24, 37, 101, 203, 251
- salmon (Salmo solar), 30, 32
Atlanto-Scandinavian herring, 17, 36, 65, 72, 220, 231, 232
Availability of data, 95-108
Aversion, risk, 265, 272

Balanus glandula, 114-121
Baleen whales, 8, 31, 329
Balistes carolinensis, 159, 163-165, 175, 229
Barnacles, 112-127
Bay, Monterey, 111, 116-120
Bering Sea, 82, 174, 181, 203
Beverton and Holt function, 63, 69
Biases in estimates, 43, 84, 135, 265
Biological interactions, 38, 45, 46, 65, 99, 100, 107, 155, 173, 180, 184, 228, 247, 314, 323, 325, 331
Biomass, 19-21, 25, 41, 43, 59-64, 68-72, 79-83, 98, 104, 107, 113, 136-142, 148, 157, 161-170, 175, 177, 181, 212, 231, 250, 265, 298, 304, 320, 322, 327, 328
Blue whale, 175
Boundaries, 68, 112, 179, 291, 309
Buffering plans, wealth, 270
By-catch, 170-172, 277, 296, 303

California, 17, 26, 37, 67, 73, 78, 111, 116-121, 142, 173, 182, 202, 232, 246, 251, 252
- sardine (Sardinops sagan caerulea), 17, 26, 37, 78, 142
Cannibalism, 18-24, 47, 63, 70, 77-82
Capacity, carrying, 19, 61, 162, 163, 192, 196, 238
Capital investment strategies, 270
Capture, 29, 30, 42, 75, 79, 80, 105, 170, 240, 315
Carangids, 157, 159
Carrying capacity, 19, 61, 162, 163, 192, 196, 238
Catch data, 95-108, 141
Catch per unit effort (CPUE), 17, 35, 41-45, 50, 104, 160-167, 177, 181, 220, 234, 248, 249, 298
Catchability, 26, 34-38, 43, 104, 105, 161, 162, 248, 258, 267
Catches, 5-8, 15, 22, 30, 35, 38, 42, 48, 65, 66, 70, 81, 95-108, 132, 141, 155-169, 173-184, 209, 219-221, 229, 239, 240, 248, 253, 269-
Catches (cont.), 272, 281, 288-296, 303-306, 310, 314-323, 327, 336
Catfish, channel, 32
Cautious management, 272, 273
Cephalopods, 3, 46, 65, 81, 157, 160, 165-167, 175, 177, 211, 257
Change and persistence in marine communities, 131, 133, 142-144
Changes, environmental, 4, 24, 44, 61-66, 70, 107, 161, 173, 203, 210, 216-222, 234, 237, 246-259, 263, 264, 323, 328
Channel catfish (Ictalurus punctatus), 32
Chile, 17, 173, 182
Clupeids, 26, 27, 157, 159, 254
Coast, Ivory, 160-164, 181
-, Senegal-Mauritanian, 5, 163
Cod, 5, 8, 24-30, 47, 64-66, 75-79, 175, 178, 182-184, 200, 217, 221, 257, 290, 303
-, Arctic, 25, 30, 64
-, North Sea, 5, 47, 66
Coelenterate, 48
Coexistence, 111, 124, 199
Collapse, 7, 17, 25-27, 34, 40, 42, 133, 142, 161-164, 173, 183, 220, 229, 239, 240, 269, 272, 277, 283, 336
- of Sardinella aurita, 161-164
Commission, International Whaling, 1, 328, 329
Communities, change and persistence in marine, 131, 133, 142-144
-, multispecies marine, 209, 219-222
Community, sparid, 163, 165, 172
- structure, 111, 112, 117, 121, 126, 131, 142, 155, 197, 204, 218, 249, 250, 292
Compartmentation of marine systems, 145, 199-201
Compensation, 18, 20, 25, 41, 60-63, 69, 82-85, 178
Competition, 5, 19, 29, 46, 61, 63, 69, 83, 111, 121-124, 132, 141-145, 155, 170-172, 196, 198, 202, 203, 288-295, 329
Complex systems, 2, 46, 80, 131, 136,

Complex systems (cont.), 143, 148, 211, 212, 222, 245, 246, 256, 279-282, 307-310
Complexity, management of, 134, 145-147, 276
-, model, 134, 135, 143, 147, 213, 214, 246, 329, 331, 336
Composition of landings, 65, 95, 158-160
-, species, 7, 35, 100, 156-169, 173, 177, 181, 191, 197-201, 210-213, 218-221, 288-292, 296, 298, 314, 330-322, 336
Condition, Lange-Hicks, 140, 141
Conflicts, 282, 335
Conservation, 28, 178, 179, 295, 305, 319, 327, 328
Constraints, 7, 61, 147, 148, 303-311, 313-320, 325-327
Content, information, 4, 38, 42, 45, 95-108
Continuous equilibrium, 266
Contrast in stock sizes and policies, 264, 267
-, informative, 264, 267
Control, Dual, 241, 263, 271
- of predators, 212, 218, 219, 293, 294
-, stochastic, 271, 272
-, system, 97, 201
Coral reefs, 27, 127, 199, 229, 291
CPUE, 17, 35, 41-45, 50, 104, 160-167, 177, 181, 220, 234, 248, 249, 298
Critical or keystone species, 124, 146, 194, 256, 330

Data, 3, 8, 14, 19-22, 30-35, 39-43, 46-50, 59, 63-70, 75, 78, 81, 95-108, 134, 136, 139-142, 147, 161-165, 179-181, 185, 194, 197, 203, 219-221, 230-234, 247, 251, 254, 259, 264-273, 281, 295, 299, 306, 322, 330, 336
-, availability of, 95-108
-, catch, 95-108, 141
-, fisheries, 42, 95-108, 164
Decision-making, 14, 33, 34, 42, 132, 182, 236, 237, 269, 275-284, 298, 314, 318, 322, 331
Decision theory, 234, 310
Decomposition, hierarchical, 149
Demography, 4, 111-116, 125, 250
Densities, ultra-low, 26-28, 49
Density-dependence, 4, 6, 14, 18, 19, 22-31, 49, 59-63, 70-72, 77, 78, 82, 127, 214-217, 228, 234, 241, 248
-/independence, 18, 59-63
Depensation, 60, 61, 82
Discard, 38, 99, 160, 161, 165, 180, 277, 291, 292, 296, 303
Distribution, spatial, 65, 74, 100, 120, 121, 127, 306
Disturbance, management, 268
Diversity, species, 28, 121, 124, 156-160, 197, 230, 305
Dover sole (microstomus pacificus), 20, 67
Dual Control, 241, 263, 271
- effects, 271
Dynamics, 2-8, 13-50, 60, 80, 96, 111-127, 131-149, 193, 245, 248, 297-299, 307, 308, 315, 318, 322-325, 331, 337
-, ecosystems, 13, 131-149, 330
-, population, 3-8, 14-18, 25, 28, 48, 60, 80, 114, 120, 125, 127, 245, 248, 308, 323, 325

Ecological interactions, 46, 126
Economic zone, exclusive, 280, 281, 309
Economics, 2, 6, 9, 41, 95, 97, 107, 108, 136, 139, 155, 176-185, 195, 201, 203, 236-240, 248, 249, 258, 259, 268, 270, 277, 282, 289, 292, 295, 299, 303-305, 313-332, 337
Ecosystem model, 192-195, 308
- structure (Topology), 131, 144, 305
- web, 145
Ecosystems, 2, 5, 8, 13, 46, 80-82, 112, 113, 126, 131-149, 174, 178, 192-195, 203, 209-216, 228, 236,

Ecosystems (cont.), 249-256, 294, 303-310, 327-331
- dynamics, 13, 131-149, 330
-, marine, 13, 46, 112, 113, 126, 136, 142-147, 178, 236, 256, 303-310, 328
Eel, sand, 45, 48, 175, 184, 203
EEZ, 280, 281, 309
Effects, dual, 271
Efficiency, growth, 61
Effort (see Fishing effort)
El Niño, 24, 133, 173, 202, 229, 246, 249, 277
Energy flow, 144, 181, 194
Enforcement, 170, 184, 185, 240, 280, 296, 309, 310, 318, 326, 336
Entry, limited, 184, 270, 294-296, 315-317, 323-326, 336
Environmental changes, 4, 24, 44, 61-66, 70, 107, 161, 173, 203, 210, 216-222, 234, 237, 246-259, 263, 264, 323, 328
- noise, 6, 14, 255-258, 264, 265
- stability, 35, 38
- variability, 2-6, 32, 59, 61, 99, 245-260
- variables, 18-20, 25, 49, 66-70, 85, 143, 161, 162, 217, 228-232, 266-269, 308
Equilibria, multiple, 143, 144, 210, 214
Equilibrium, continuous, 266
Error analysis, 137, 139, 149
- structure, 104, 233, 263
Errors in variables problem, 265
-, measurement, 264-268
-, process, 266, 267
Escapement, fixed, 264, 269, 305, 306
ESS, 44, 49, 125
Estimates, biases in, 43, 84, 135, 265
Estimation, parameter, 5, 35, 42-44, 49, 69, 105, 134, 139, 263-267, 271, 299, 307
Evolutionarily Stable Strategy (ESS), 44, 49, 125
Exclusive economic zone (EEZ), 280, 281, 309
Exotic species, 294
Experimental management, 8, 263, 264, 271-273, 294, 298
- - policies, 271-273, 298
- manipulation, 132-136, 146, 147, 196, 201, 204, 239
Experiments, Monte Carlo simulation, 267

Factors, political, 2-8, 107, 195, 238, 259, 275-283, 288-297, 304, 313, 318, 319, 331, 332, 335
Feedback policies, 263-271
Fisheries data, 42, 95-108, 164
- management, 2-6, 14, 32, 38, 63, 66, 95-108, 112, 126, 131, 132, 141-143, 155, 170, 200, 218, 220, 236-239, 254, 258, 263-273, 275-284, 287-300, 303-311, 313-332, 335-337
-, multispecies, 4-7, 44, 50, 131, 136, 139, 155-185, 211, 218, 238, 277, 287-300, 303-311, 313-332
-, pelagic, 7, 36, 37, 65, 146, 163, 170, 174, 177, 182, 314, 320-323
- regulation, 97, 264, 270, 292-295, 309, 310
-, single-species, 5, 13-50, 173, 179, 180, 192, 292, 304, 313, 323, 331
-, temperate zone, 8, 134
-, tropical, 5, 8, 132, 134, 143, 146, 156, 177, 309
Fishermen, artisanal, 287, 292
Fishery, artisanal, 156, 158, 171, 176
- managers, 4, 8, 131, 134, 147, 191, 204, 239-241, 265, 276, 277, 313, 320, 325, 331
Fishing effort, 2, 6, 7, 17, 33, 39, 42, 66, 70, 95-104, 133, 141, 143, 162-170, 174-180, 184, 239, 240, 259, 267, 271, 281, 288, 292-296, 305, 306, 316-318, 322-327, 336
- power, 101, 294, 296, 322-326
- strategy, 158-161
Fit, model, 63, 69, 70, 134, 266
Fixed escapement, 264, 269, 305, 306
Flexibility, 270, 272, 303, 306

Flow, energy, 144, 181, 194
Food web, 2-8, 34, 48, 80, 131, 144, 145, 191, 195-202, 212, 228, 231, 246, 252-257
Function, Beverton and Holt, 63, 69
-, production, 264-266
-, Ricker, 63, 69, 70, 263

Gears, nonselective, 29, 176
Genetics, 2-4, 15, 26-33, 49, 111, 125, 127, 134, 142, 267, 304, 305, 336
Georges Bank, 20, 45, 61-67, 75-84, 220, 280
- - haddock, 20, 61-67, 75, 77, 84
Ghana, 161-165
Growth, 3, 4, 17, 25-32, 47, 49, 59-65, 73-84, 99-107, 113-121, 132, 136-139, 172, 184, 192, 214, 216, 231, 239, 250, 265
- efficiency, 61
- overfishing, 172
- rate, 4, 25, 29, 30, 61, 73-79, 85, 107, 117, 136-138, 214, 250, 265
Guild, 136, 139, 145, 191-194, 198, 199
Guinea Bissau, 159, 163, 165
Gulf of Thailand, 46, 177-183, 222

Habitat, 5, 7, 19, 23, 27, 28, 34, 36, 46, 49, 61, 78, 111, 112, 124-127, 144, 147, 203, 255, 290-293, 315
Haddock (Melanogrammus aeglefinus), 20, 24, 27, 36, 47, 61-67, 75, 77, 84, 175, 184, 200, 253, 257
-, Georges Bank, 20, 61-67, 75, 77, 84
-, North Sea, 24, 36, 47, 61
Harvest policies, 264, 270, 272
Hatcheries, 270, 271
Herring, 5, 7, 16, 17, 24-27, 36, 45, 48, 62-66, 78-84, 106, 133, 142, 175, 182-184, 197-203, 220, 221, 232, 246-259, 290, 304, 321
-, Atlanto-Scandinavian, 17, 36, 65, 72, 220, 231, 232
-, Hokkaido, 17, 26, 27
Herring, Icelandic, 26, 232
-, North Sea, 5, 16, 26, 27, 36, 45, 66, 220, 232
-, Norwegian, 26, 78, 203, 253
Hierarchical decomposition, 149
Hierarchy, 132, 147-149, 291
- theory, 147-149
History, life, 3-7, 22-26, 34-37, 47-50, 62, 111, 112, 125, 126, 142, 222, 228, 229
Hokkaido herring (Clupea harengus), 17, 26, 27
Homing, 15, 16, 27

Icelandic herring (Clupea harengus), 26, 232
ICNAF, 295
Indicator species, 5, 134
Information content, 4, 38, 42, 45, 95-108
Informative contrast, 264, 267
Infrastructures, 296, 319
Instability, 336, 337
Insurance programs, 237, 239, 270
Interactions, biological, 38, 45, 46, 65, 99, 100, 107, 155, 173, 180, 184, 228, 247, 314, 323, 325, 331
-, ecological, 46, 126
-, operational, 44, 155, 184
-, species, 8, 14, 15, 44-50, 107, 132-134, 140-146, 177-184, 191, 192, 199, 201, 210-212, 228, 251, 256, 287-290, 297, 315, 322, 323, 327-331, 337
International agreements, 1, 39, 280, 300, 310, 316, 317, 330
- Whaling Commission, 1, 328, 329
Intertidal zone, rocky, 27, 34, 111-121, 126, 133
Interspecific linkages, 132, 133, 146
Investment strategies, capital, 270
Investments, 34, 97, 266-271, 280, 292
Ivory Coast, 160-164, 181

Japanese sardine (Sardinops melanosticta), 26, 27, 232
Juvenile phase, 22, 24, 36, 48, 49, 65, 145, 246-250, 299

Keystone or critical species, 124, 146, 194, 256, 330
Krill, 8, 46, 156, 172, 173, 211, 303, 304, 315, 327-331

Landings, composition of, 65, 95, 158-160
Lange-Hicks condition, 140-141
Larvae, 3, 22, 24, 47, 48, 59, 60, 67-85, 106, 111-127, 161, 167, 173, 233, 245-249, 299, 322
Larval phase, 22, 24, 125, 246, 249, 299
Levels, trophic, 14, 79, 80, 139, 140, 146, 181, 191-193, 197-202, 219, 222, 250, 291, 306, 329
Life history, 3-7, 22-26, 34-37, 47-50, 62, 111, 112, 125, 126, 142, 222, 228, 229
Limited entry, 184, 270, 294-296, 315-317, 323-326, 336
Linkages, interspecific, 132, 133, 146

Mackerel, 26, 45, 62-69, 80, 98, 106, 161, 163, 168, 170, 174, 183, 221, 231, 251-253, 257, 321
-, Pacific, 26, 62, 69
Management, 2-9, 13-19, 24-44, 63, 66, 95-108, 112, 126, 131-134, 141-147, 172, 176-179, 193-202, 209, 210, 212, 218, 220, 227-241, 245-249, 254-260, 263-273, 275-284, 287-300, 303-311, 313-332, 335-337
-, cautious, 272, 273
- disturbance, 268
-, experimental, 5, 263, 264, 271-273, 294, 298
-, fisheries, 2-6, 14, 32, 38, 63, 66, 95-108, 112, 126, 131, 132, 141-
Management, fisheries (cont.), 143, 155, 170, 200, 218, 220, 236-239, 254, 258, 263-273, 275-284, 287-300, 303-311, 313-332, 335-337
-, multispecies, 7, 8, 13, 131, 132, 141-143, 287-300, 303-311, 313-332
- objectives, 7, 28, 33, 38, 40, 95, 97, 107, 178, 179, 231, 239, 269, 287-293, 303-311, 313, 314, 318-325, 329-332, 335
- of complexity, 134, 145-147, 276
- policies, experimental, 271-273, 298
Managers, fishery, 4, 8, 131, 134, 147, 191, 204, 239-241, 265, 276, 277, 313, 320, 325, 331
Manifold theory, 148
Manipulation, experimental, 132-136, 146, 147, 196, 201, 204, 239
Marine communities, change and persistence in, 131, 133, 142-144
- -, multispecies, 209, 219-222
- ecosystems, 13, 46, 112, 113, 126, 136, 142-147, 178, 236, 256, 303-310, 328
- systems, compartmentation of, 145, 199-201
Marketing, 2, 7, 8, 44, 234, 282
Markets, 2, 7, 8, 44, 155, 159, 160, 170, 175-179, 184, 227, 234, 237, 240, 277, 282, 296, 304, 319
Maturation, 4, 19, 25-32, 46, 49
Maximum Sustainable Yield (MSY), 178, 179, 279, 304-306
Measurement errors, 264-268
Measures, technical, 97, 100, 147
Mediterranean, 160, 174-179
Mesh size, 97-100, 170, 184, 185, 293, 296, 309
Method, Paloheimo, 266
-, Schnute, 266
MEY, 178, 179
Minimum Sustainable Whinge (MSW), 178
Minke whales, 173, 175
Model, aggregated Schaefer, 139, 141, 142
- complexity, 134, 135, 143, 147, 213,

Model complexity (cont.), 214, 246, 329, 331, 336
-, ecosystem, 192-195, 308
- fit, 63, 69, 70, 134, 266
Modeling, 60, 70, 80, 114, 134, 149, 183, 191-204, 237, 251, 265, 272, 297, 298, 308, 310, 320, 322, 329-331
Models, 3, 5, 17, 18, 39, 45, 47, 59, 63, 69-78, 96, 99, 111-117, 121-126, 131-146, 162, 174-185, 191-204, 210, 216, 221, 233, 245, 251, 256, 263-269, 276, 279, 288, 297-299, 307, 315, 320, 322, 327-330, 336
-, multispecies, 5, 140
-, scale specificity of, 139-142
Monitoring systems, 267-269, 272, 305-309, 326
Monte Carlo simulation experiments, 267
Monterey Bay, 111, 116-120
Morocco, 168, 174, 232
Mortality, 14, 22-24, 30, 38, 42-49, 59-65, 73, 77, 78, 81-84, 96-99, 103-107, 113-117, 120-123, 136, 139, 140, 161, 167-172, 182, 184, 209, 212, 214, 220, 228, 229, 231, 239, 240, 249, 250, 254, 264-266, 293, 298, 299, 306, 313, 317, 336
-, natural, 23, 38, 42-47, 59-65, 96, 99, 103-107, 229, 231, 249
- rate, 22-24, 30, 42-44, 60-63, 73, 78, 83, 84, 96, 121, 122, 136, 184, 229, 251, 266
MSW, 178
MSY, 178, 179, 279, 304-306
Multiple equilibria, 143, 144, 210, 214
Multiple stable states, 143, 144, 209-219
Multipurpose vessels, 176, 290, 315, 323-327
Multi-single-target, 175, 176, 290
Multispecies analysis, 44, 132, 133
- fisheries, 4-7, 44, 50, 131, 136, 139, 155-185, 211, 218, 238, 277, 287-300, 303-311, 313-332
- management, 7, 8, 13, 131, 132,
Multispecies management (cont.), 141-143, 287-300, 303-311, 313-332
- marine communities, 209, 219-222
- models, 5, 140
- systems, 4, 5, 14, 124, 156, 179, 191-204, 209-222
- -, taxonomy of, 191-204, 298
Multispecific resource, 156, 314, 315, 320, 325-330

Natural mortality, 23, 38, 42-47, 59-65, 96, 99, 103-107, 229, 231, 249
- regulation, 13, 18-28, 40, 44, 49
Noise, environmental, 6, 14, 255-258, 264, 265
Nonselective gears, 29, 176
North Atlantic Ocean, 1, 24, 37, 101, 203, 251
- Sea, 1, 5, 16, 20-30, 36, 45-48, 61, 66, 73, 82, 83, 132, 142-146, 175-184, 200, 220, 221, 231, 232, 250, 252, 259
- - cod (Gadus morhua), 5, 47, 66
- - haddock, 24, 36, 47, 61
- - herring (Clupea harengus), 5, 16, 26, 27, 36, 45, 66, 220, 232
- - plaice (Pleuronectes platessa), 16, 20-24, 30, 36, 73, 83, 142
- - sole (Solea solea), 20, 36
Northern anchovy (Engraulis mordax), 18, 22, 62, 77
Northwest Africa, 160, 165, 167, 173, 175, 179
Norwegian herring (Clupea harengus), 26, 78, 203, 253

Objectives, management, 7, 28, 33, 38, 40, 95, 97, 107, 178, 179, 231, 239, 269, 287-293, 303-311, 313, 314, 318-325, 329-332, 335
Ocean, North Atlantic, 1, 24, 37, 101, 203, 251
-, Southern, 1, 8, 172, 179, 181, 211, 229, 327, 329
Octopus, 165-167

Operational interactions, 44, 155, 184
Otters, 198-202
Outflow, river, 161-164
Overfishing, growth, 172
Ownership, 97, 277, 300, 316, 327
-, sole, 316

Pacific mackerel (Scomber japonicus), 26, 62, 69
- salmon, 20, 29, 293, 294, 309, 310
- sardine (Sardinops sagax sagax), 17, 47, 67, 68
Paloheimo method, 266
Paneaus japonicus, 293
Parameter estimation, 5, 35, 42-44, 49, 69, 105, 134, 139, 263-267, 271, 299, 307
Parameters, 5, 17, 18, 35, 42-44, 49, 69, 96, 100, 105, 114, 115, 120, 122, 126, 134, 139, 170, 263-272, 297, 299, 307, 330, 331
Passive adaptation, 264
Pelagic fisheries, 7, 36, 37, 65, 146, 163, 170, 174, 177, 182, 314, 320-323
Perch (Perca fluriatilis), 26, 174, 231
Persistence, 2, 39, 41, 63, 265
- and change in marine communities, 131, 133, 142-144
Peru, 5, 17, 24, 37, 47, 70, 71, 78, 133, 142, 173, 181, 182, 202, 231, 247, 249, 335, 336
Peruvian anchovy (Engraulis ringens) (see Anchoveta)
Phase, adult, 20, 25, 26, 49
-, larval, 22, 24, 125, 246, 249, 299
-, juvenile, 22, 24, 36, 48, 49, 65, 145, 246-250, 299
Physics, 245
Plaice, 7, 16, 20-24, 30, 31, 36, 48, 73, 78, 83, 142, 175, 183
-, North Sea, 16, 20-24, 30, 36, 73, 83, 142
Plankton, 126, 127, 161-164, 201, 202, 246, 250-254
Plans, wealth buffering, 270
Platyfish (Xiphophorus maculatus), 32
Policies and stock sizes, contrast in, 264, 267
-, experimental management, 271-273, 298
-, feedback, 263-271
-, harvest, 264, 270, 272
-, taxation, 7, 270, 319
Policy analysis, 277-284, 306
Policymaking, 9, 276-278, 300, 337
Political factors, 2-8, 107, 195, 238, 259, 275-283, 288-297, 304, 313, 318, 319, 331, 332, 335
Population Analysis, Virtual, 42-44, 49, 101-106
- dynamics, 3-8, 14-18, 25, 28, 48, 60, 80, 114, 120, 125, 127, 245, 248, 308, 323, 325
Post-recruit, 62, 102-104
Power, fishing, 101, 294, 296, 322-326
Prerecruit, 60, 65, 79, 83, 84, 101-103, 228
Predator-prey system, 28, 46-50, 59-62, 74, 78-85, 116, 117, 132, 155, 156, 172, 173, 179, 196, 212, 217, 219, 249, 250, 258, 290-294, 329
Predators, control of, 212, 218, 219, 293, 294
Probing, 264, 265, 271, 272
Problem, errors in variables, 265
Process errors, 266, 267
Production function, 264-266
-, surplus, 61, 266
Programs, insurance, 237, 239, 270

Quotas, 2, 7, 8, 132, 172, 182, 184, 239, 240, 247, 270, 282, 295, 307, 309, 315, 316, 320-329, 336, 337

Rate, growth, 4, 25, 29, 30, 61, 73-79, 85, 107, 117, 136-138, 214, 250, 265
-, mortality, 22-24, 30, 42-44, 60-63, 73, 78, 83, 84, 96, 121, 122, 136, 184, 229, 251, 266

Rate, survival, 75, 266
Ratio, RNA/DNA, 77
-, yield/biomass, 98, 107
Recruit, yield per, 17, 31, 41, 99, 245, 247, 318
Recruitment, 6, 14, 18-25, 32-42, 48, 49, 59-72, 77-85, 98-107, 111, 115, 120, 121, 143, 161-168, 178, 179, 183, 220, 229-234, 239, 246-251, 258, 259, 264-268, 296, 299, 305, 328
- variation, 60, 63, 69, 82, 85, 143, 161, 229-231, 247, 258, 266, 268
Red salmon (Onchorynchus nerka), 31
Reefs, artificial, 293
-, coral, 27, 127, 199, 229, 291
Regulation, natural, 13, 18-28, 40, 44, 49
-, fisheries, 97, 264, 270, 292-295, 309, 310
Relationship, stock-recruitment, 18-20, 25, 35, 38-41, 47, 49, 62-64, 69-72, 77, 78, 99, 126, 127, 234, 241, 247, 299, 322
Relationships, 18-20, 25, 35, 38-41, 47, 49, 62-64, 69-72, 77, 78, 99, 104, 126, 127, 141, 161-165, 169, 178, 179, 191-197, 202, 213, 229, 233, 234, 241, 247, 265, 299, 322, 328
Replacement, 63, 133, 167, 168, 183, 194, 198, 259, 295, 315, 319, 320, 326
Research vessel survey, 43, 79, 80, 100-106, 177
Resource, multispecific, 156, 314, 315, 320, 325-330
Responses, system, 264
Reversibility, 209, 216-220, 228
Ricker function, 63, 69, 70, 263
Risk aversion, 265, 272
River outflow, 161-164
RNA/DNA ratio, 77
Rocky intertidal zone, 27, 34, 111-121, 126, 133
Salmon, 20, 29-32, 67, 174, 269, 270, 293, 294, 304, 305, 309, 310
-, Atlantic, 30, 32
-, Pacific, 20, 29, 293, 294, 309, 310
-, red, 31
Sand eel (Ammodytes marinus), 45, 48, 175, 184, 203
Sardina, 168, 170, 254
Sardine, 17, 26, 27, 37, 47, 67, 68, 78, 142, 161, 168-183, 221, 232, 246, 251, 252, 254
-, California, 17, 26, 37, 78, 142
-, Japanese, 26, 27, 232
-, Pacific, 17, 47, 67, 68
Sardinella aurita, collapse of, 161-164
Scale specificity of models, 139-142
Scales, spatial, 16-19, 40, 49, 124, 131, 140, 142, 147, 197, 201, 203, 217, 221, 229, 239, 245-249, 256, 258, 267, 306
-, temporal, 14, 33-42, 49, 65, 107, 131, 132, 140, 142, 147, 148, 197, 217, 221, 229, 234, 239, 241, 245-249, 255-259, 265, 306, 307
Scaling, 139, 141, 142
Schaefer model, aggregated, 139-142
Schnute method, 266
Sea, Adriatic, 176-179
-, Bering, 82, 174, 181, 203
-, North, 1, 5, 16, 20-30, 36, 45-48, 61, 66, 73, 82, 83, 132, 142-146, 175-184, 200, 220, 221, 231, 232, 250, 252, 259
Seabirds, 5, 6, 48, 62, 81, 116, 202, 203, 211, 294
Seals, 46, 172-174, 198, 211, 220, 294
Sebastes (S. marinus), 30
Selectivity, 3, 28-33, 43, 47, 49, 125, 267, 310
Senegal, 5, 157-163, 170-176
Senegal-Mauritanian coast, 5, 163
Senegal-Mauritanian shelf, 156, 157
Series, time, 14, 69, 100-106, 126, 162, 163
Sex-ratio, 30, 31, 192
Shelf, Senegal-Mauritanian, 156, 157

Simulation experiments, Monte Carlo, 267
Single-purpose vessels, 289-291
-/species analyses, 8, 14, 40, 41, 46, 132, 133, 146, 176, 178, 181-183
-/- fisheries, 5, 13-50, 173, 179, 180, 192, 292, 304, 313, 323, 331
Size, mesh, 97-100, 170, 184, 185, 293, 296, 309
-, stock, 38-48, 68, 83, 95-107, 147, 220, 231, 232, 240, 241, 248, 249, 258, 264-272, 316
Sole, 7, 20, 31, 36, 48, 67, 157, 159, 172, 175
-, Dover, 20, 67
-, North Sea, 20, 36
- ownership, 316
Southern Ocean, 1, 8, 172, 179, 181, 211, 229, 327, 329
Sparid, 159-168, 172, 175
- community, 163, 165, 172
Spatial distribution, 65, 74, 100, 120, 121, 127, 306
- scales, 16-19, 40, 49, 124, 131, 140, 142, 147, 197, 201, 203, 217, 221, 229, 239, 245-249, 256, 258, 267, 306
Species assemblages, 5, 157, 165, 298
- composition, 7, 35, 100, 156-169, 173, 177, 181, 191, 197-201, 210-213, 218-221, 288-292, 296, 298, 314, 320-322, 336
- diversity, 28, 121, 124, 156-160, 197, 230, 305
-, exotic, 294
-, indicator, 5, 134
- interactions, 8, 14, 15, 44-50, 107, 132-134, 140-146, 177-184, 191, 192, 199, 201, 210-212, 228, 251, 256, 287-290, 297, 315, 322, 323, 327-331, 337
-, keystone or critical, 124, 146, 194, 256, 330
-, target, 43, 45, 132, 133, 142, 155, 170-175, 277, 289-291, 314, 315, 320, 327, 330
Specificity of models, scale, 139-142
Sperm whales, 31
Sprat (Sprattus sprattus), 45, 48, 61
Squid, 3, 46, 65, 81, 157, 165, 177, 257
Stability, 2, 34-38, 49, 66, 73-75, 83, 131, 142-147, 155, 161, 196-199, 211-217, 229, 270, 336
-, environmental, 35, 38
Stable states, alternate, 144, 210, 211, 217, 220-222
- -, multiple, 143, 144, 209-219
Status quo, 39, 95, 98, 107, 292, 297
Stochastic control, 271, 272
Stock size, 38-48, 68, 83, 95-107, 147, 220, 231, 232, 240, 241, 248, 249, 258, 264-272, 316
- sizes and policies, contrast in, 264, 267
Stock-recruitment (S-R) relationship, 18-20, 25, 35, 38-41, 47, 49, 62-64, 69-72, 77, 78, 99, 126, 127, 234, 241, 247, 299, 322
Stocks, 1-9, 15-20, 25-27, 30-35, 38-49, 62-64, 68-72, 77-85, 95-107, 126, 132-134, 142-147, 155, 163, 168, 172-174, 180, 183, 193, 198, 200-202, 217-220, 228-240, 246-259, 265-272, 275, 279, 283, 290, 291, 304-310, 315-317, 323-327, 337
Strategies, capital investment, 270
Strategy, Evolutionarily Stable, 44, 49, 125
-, fishing, 158-161
Structure, aggregation, 136
-, community, 111, 112, 117, 121, 126, 131, 142, 155, 197, 204, 218, 249, 250, 292
-, ecosystem, 131, 144, 305
-, error, 104, 233, 263
Surplus production, 61, 266
Survey, research vessel, 43, 79, 80, 100-106, 177
Surveys, 40, 43, 79, 80, 95, 100-106, 166, 177, 234, 265, 272, 273, 322, 328
Survival rate, 75, 266
Switching, 84, 155, 173-178, 249,

Switching (cont.), 277, 303
System control, 97, 201
-, predator-prey, 28, 46-50, 59-62, 74, 78-85, 116, 117, 132, 155, 156, 172, 173, 179, 196, 212, 217, 219, 249, 250, 258, 290-294, 329
- responses, 264
Systems, compartmentation of marine, 145, 199-201
-, complex, 2, 46, 80, 131, 136, 143, 148, 211, 212, 222, 245, 246, 256, 279-282, 307-310
-, monitoring, 267-269, 272, 305-309, 326
-, multispecies, 4, 5, 14, 124, 156, 179, 191-204, 209-222, 298
-, taxonomy of multispecies, 191-204, 298

TAC, 210, 234, 295, 298, 316, 317
Tagging (Mark-recapture), 43, 95, 100-106
Target species, 43, 45, 132, 133, 142, 155, 170-175, 277, 289-291, 314, 315, 320, 327, 330
Targets, 43, 45, 132, 133, 142, 155, 159, 170-179, 277, 289-303, 314, 315, 320, 327, 330
Taxation policies, 7, 270, 319
Taxonomy of multispecies systems, 191-204, 298
Technical measures, 97, 100, 147
Temperate zone fisheries, 8, 134
Temporal scales, 14, 33-42, 49, 65, 107, 131, 132, 140, 142, 147, 148, 197, 217, 221, 229, 234, 239, 241, 245-249, 255-259, 265, 306, 307
Thailand, Gulf of, 46, 177-183, 222
Theory, decision, 234, 310
-, hierarchy, 147-149
-, manifold, 148
Tilapia mossambica, 32
Time series, 14, 69, 100-106, 126, 162, 163
Togo, 161, 162
Topology, 131, 136, 144-146
Total allowable catch (TAC), 210, 234, 295, 298, 316, 317
Tragedy of the commons, 300
Trophic levels, 14, 79, 80, 139, 140, 146, 181, 191-193, 197-202, 219, 222, 250, 291, 306, 329
Tropical fisheries, 5, 8, 132, 134, 143, 146, 156, 177, 309
Tychonoff, 148

Ultra-low densities, 26-28, 49
Uncertainty, 3, 6-9, 14, 34, 63, 176, 227-241, 245-260, 263-273, 275-284, 306-310, 320, 322, 329-331
UNCLOS, 281
Upwelling, 34, 37, 67-69, 73, 158-163, 167-169, 173, 181, 196, 201, 222, 229-233, 246, 268, 314

Variability, 13, 18, 20-24, 34, 40-44, 59-85
-, environmental, 2-6, 32, 59, 61, 99, 245-260
Variables, environmental, 18-20, 25, 49, 66-70, 85, 143, 161, 162, 217, 228-232, 266-269, 308
- problem, errors in, 265
Variation, recruitment, 60, 63, 69, 82, 85, 143, 161, 229-231, 247, 258, 266, 268
Vessel survey, research, 43, 79, 80, 100-106, 177
Vessels, multipurpose, 176, 290, 315, 323-327
-, single-purpose, 289-291
Virtual Population Analysis (VPA), 42-44, 49, 101-106

Wealth buffering plans, 270
Web, ecosystem, 145
-, food, 2-8, 34, 48, 80, 131, 144, 145, 191-202, 212, 228, 231, 246, 252-257
West Africa, 5, 156-172, 229, 238

Whales, 3, 8, 31, 35, 37, 46, 147, 156, 172-175, 181, 198, 204, 211, 217, 220, 250, 259, 303, 315, 327-330
-, baleen, 8, 31, 329
-, blue, 175
-, Minke, 173, 175
-, sperm, 31
Whaling Commission, International, 1, 328, 329

Yield/biomass ratio, 98, 107
-, Maximum Sustainable, 178, 179, 279, 304-306
Yield per recruit, 17, 31, 41, 99, 245, 247, 318
Yields, 4, 6, 17, 31, 33, 39, 41, 44, 63, 97-99, 107, 133, 141, 142, 178, 179, 199, 209, 239-241, 245, 247, 256-258, 271, 279, 304-310, 318, 322, 323, 331

Zone, exclusive economic, 280, 281, 309
-, rocky intertidal, 27, 34, 111-121, 126, 133

# Author Index


Arntz, W.E.; 227-244
Bailey, R.S.; 227-244
Beddington, J.R.; 209-225, 227-244
Beverton, R.J.H.; 13-58
Brewer, G.D.; 227-244, 275-285
Clark, C.W.; 287-301, 303-312
Cooke, J.G.; 13-58, 341-348
Csirke, J.B.; 13-58
Daan, N.; 287-301
Doyle, R.W.; 13-58
Dutt, S.; 287-301
Gaines, S.; 111-128
Garcia, S.; 131-153, 155-190
Glantz, M.H.; 227-244
Gulland, J.A.; 131-153, 155-190, 335-337
Hempel, G.; 13-58
Holt, S.J.; 13-58
Hongskul, V.; 287-301
Iwasa, Y.; 111-128
Larkin, P.A.; 287-301
Laurec, A.J.Y.; 227-244
Lawton, J.H.; 131-153
Levin, S.A.; 287-301
MacCall, A.D.; 13-58
Maske, H.; 131-153
May, R.M.; 1-10, 227-244
Nellen, W.P.; 227-244
Newman, G.G.; 287-301, 313-333
Paine, R.T.; 131-153, 191-207
Pauly, D.M.; 287-301
Platt, T.; 131-153
Policansky, D.J.; 13-58
Rachor, E.; 131-153
Radach, G.; 287-301
Rosenthal, H.K.; 287-301
Rothschild, B.J.; 131-153
Roughgarden, J.; 13-58, 111-128
Shepherd, J.G.; 13-58, 95-109
Sissenwine, M.P.; 13-58, 59-94
Smetacek, V.S.; 227-244
Steele, J.H.; 245-262
Sugihara, G.; 131-153
Thurow, F.R.M.; 227-244
Troadec, J.-P.; 227-244
Ursin, E.A.; 131-153
Walters, C.J.; 227-244, 263-274
Wiebe, P.H.; 13-58
Zeitzschel, B.F.K.; 131-153

# Dahlem Workshop Reports

## Life Sciences Research Reports (LS)

LS 32 Exploitation of Marine Communities
Editor: R. M. May (1984, in press)

LS 31 Microbial Adhesion and Aggregation.
Editor: K. G. Marshall (1984)

LS 30 Leukemia.
Editor: I. L. Weissman (1984, in press)

LS 29 The Biology of Learning.
Editors: P. Marler, H. S. Terrace (1984)

LS 28 Changing Metal Cycles and Human Health.
Editor: J. O. Nriagu (1984)

LS 27 Minorities: Community and Identity.
Editor: C. Fried (1983)

LS 26 The Origins of Depression: Current Concepts and Approaches.
Editor: J. Angst (1983)

LS 25 Population Biology of Infectious Diseases.
Editors: R. M. Anderson, R. M. May (1982)

LS 24 Repair and Regeneration of the Nervous System.
Editor: J. G. Nicholls (1982)

LS 23 Biological Mineralization and Demineralization.
Editor: G. H. Nancollas (1982)

LS 22 Evolution and Development.
Editor: J. T. Bonner (1982)

LS 21 Animal Mind - Human Mind.
Editor: D. R. Griffin (1982)

LS 20 Neuronal-Glial Cell Interrelationships.
Editor: T. A. Sears (1982)

## Physical, Chemical and Earth Sciences Research Reports (PC)

PC 5 Patterns of Change in Earth Evolution.
Editors: H. D. Holland, A. F. Trendall (1984, in press)

PC 4 Atmospheric Chemistry.
Editor: E. D. Goldberg (1982)

PC 3 Mineral Deposits and the Evolution of the Biosphere.
Editors: H. D. Holland, M. Schidlowski (1982)

Springer-Verlag Berlin Heidelberg New York Tokyo

# Dahlem Workshop Reports

## Life Sciences Research Reports (LS)


LS 1 The Molecular Basis of Circadian Rhythms.
Editors: J. W. Hastings, H.-G. Schweiger

LS 2 Appetite and Food Intake.
Editor: T. Silverstone

LS 3 Hormone and Antihormone Action at the Target Cell.
Editors: J. H. Clark et al.

LS 4 Organization and Expression of Chromosomes.
Editors: V. G. Allfrey et al.

LS 5 Recognition of Complex Acoustic Signals.
Editor: T. H. Bullock

LS 6 Function and Formation of Neural Systems.
Editor: G. S. Stent

LS 7 Neoplastic Transformation: Mechanisms and Consequences.
Editor: H. Koprowski

LS 8 The Bases of Addiction.
Editor: J. Fishman

LS 9 Morality as a Biological Phenomenon.
Editor: G. S. Stent

LS 10 Abnormal Fetal Growth: Biological Bases and Consequences.
Editor: F. Naftolin

LS 11 Transport of Macromolecules in Cellular Systems.
Editor: S. C. Silverstein

LS 12 Light-Induced Charge Separation in Biology and Chemistry.
Editors: H. Gerischer, J. J. Katz

LS 13 Strategies of Microbial Life in Extreme Environments.
Editor: M. Shilo

LS 14 The Role of Intercellular Signals: Navigation, Encounter, Outcome.
Editor: J. G. Nicholls

LS 15 Biomedical Pattern Recognition and Image Processing.
Editors: K. S. Fu, T. Pavlidis

LS 16 The Molecular Basis of Microbial Pathogenicity.
Editors: H. Smith et al.

LS 17 Pain and Society.
Editors: H. W. Kosterlitz, L. Y. Terenius

LS 18 Evolution of Social Behavior: Hypotheses and Empirical Tests.
Editor: H. Markl

LS 19 Signed and Spoken Language: Biological Constraints on Linguistic Form.
Editors: U. Bellugi, M. Studdert-Kennedy

## Physical and Chemical Sciences Research Reports (PC)


PC 1 The Nature of Seawater (out of print)

PC 2 Global Chemical Cycles and Their Alteration by Man.
Editor: W. Stumm



Distributor for LS 1–19 and PC 1 + 2:
**Verlag Chemie,** Pappelallee 3, 6940 Weinheim,
Federal Republic of Germany